Lecture Notes in Computer Science

Edited by G. Goos, J. Hartmanis and J.

Advisory Board: W. Brauer   D. Gries   J. Stoer

Stig I. Andersson (Ed.)

Analysis of Dynamical and Cognitive Systems

Advanced Course
Stockholm, Sweden, August 9-14, 1993
Proceedings

Springer-Verlag
Berlin Heidelberg New York
London Paris Tokyo
Hong Kong Barcelona
Budapest

Series Editors

Gerhard Goos
Universität Karlsruhe
Vincenz-Priessnitz-Straße 3, D-76128 Karlsruhe, Germany

Juris Hartmanis
Department of Computer Science, Cornell University
4130 Upson Hall, Ithaka, NY 14853, USA

Jan van Leeuwen
Department of Computer Science, Utrecht University
Padualaan 14, 3584 CH Utrecht, The Netherlands

Volume Editor

Stig I. Andersson
CECIL
Blå Hallen, Eriksberg, S-417 64 Göteborg, Sweden

CR Subject Classification (1991): F.1, C.2, F.4.1, I.2, F.2, C.1.m, H.1.1

ISBN 3-540-58843-4 Springer-Verlag Berlin Heidelberg New York

CIP data applied for

© Springer-Verlag Berlin Heidelberg 1995
Printed in Germany

Typesetting: Camera-ready by author
SPIN: 10479188   45/3140-543210 - Printed on acid-free paper

Preface

For the fifth consecutive year, the Summer University of Southern Stockholm included a programme on dynamical systems. This time the emphasis was on various types of cognitive systems in a broad sense. Being a powerful tool in modelling a variety of phenomena and in elucidating structural aspects of complex systems, cognitive systems enjoy steadily increasing attention. The natural mathematical framework is that of dynamical systems or cellular neural networks (CNN). Especially CNN have undergone a remarkable development since the concept first made its appearance in the early 1940s.

The contributions in this volume fall into three (partly overlapping!) classes; limits of computability and undecidability questions, universal CNN – structural aspects and applications, and dynamical systems and complexity. Within the first group Gregory Chaitin covered essential aspects of algorithmic information theory and the limits of computability. In this framework Chaitin extends the Hilbert way of formulating decidability and he also generates far-reaching generalisations of Gödel-type theorems. A brief historical background to this field is provided by Jan Tarski's contribution, in which he also presents the original way in which these questions and results were phrased. In the same spirit, although working within the framework of symbolic dynamics, Antonio Perrone treats general theorems for avoiding undecidability. These results have consequences for, e.g., parallel computations.

The CNN universal machine and its ramifications is the topic of the joint contribution of Leon Chua, Tamás Roska, and Tibor Kozek. Other aspects of neural networks are dealt with in the papers of Leo van Hemmen/Raphael Ritz and Stefan Wimbauer/Leo van Hemmen. The first one focuses on the processing of complex information using spiking neurones whereas the second one concentrates on algorithmic properties of unlearning and its structural consequences. The paper by Alexander Murgu is also concerned with (Hopfield) neural networks although the aim is here to map successive approximation methods for Markov decision problems onto such networks. The section ends with a paper by Sören Molander on image sequencing using finite state automata, where segmentation and delineation of the system is achieved via unsupervised learning.

Complexity and dynamical systems are dealt with in the remaining two papers. Petr Kurka's contribution regards finite automata from the dynamical systems point of view and formulates, using the concept of a regular language, simplicity criteria for dynamical systems. In the paper by M. Sintzsoff/Frédéric Geurts the focus is also on discrete dynamical systems but the analysis now proceeds by using predicate transformers defining certain set-valued functions.

The 1993 meeting was, like the other ones in this series, the result of the combined efforts of a number of organisations and individuals. Thanks are due to IBM Svenska AB for generous support via Mr. Carl Tengwall and Mr. Fredrik Holmberg. Also, of course, I thank the Board of the Summer University of Southern Stockholm and project manager Mr. Staffan Ström of the Stockholm City Council. The success of the technical and administrative arrangement – including much of the general logistics of the meeting – is due uniquely to the capacity and ability of Ms. Annika Hofling, Chalmers University of Technology.

Göteborg, November 1994   Stig I. Andersson

Table of Contents

The Limits of Mathematics
Course Outline and Software

G J Chaitin

IBM Research, P O Box 704, Yorktown Heights, NY 10598, USA

Abstract. A remarkable new definition of a self-delimiting universal Turing machine is presented that is easy to program and runs very quickly. This provides a new foundation for algorithmic information theory. This new universal Turing machine is implemented via software written in *Mathematica* and *C*. Using this new software, it is now possible give a self-contained "hands on" mini-course presenting very concretely the latest proofs of the fundamental information-theoretic incompleteness theorems. (To obtain this 127–page report in LaTeX via e-mail, please contact chaitin at watson.ibm.com.)

Historical Background of Gödel's Theorem

by

Jan Tarski

Abstract

The developments of mathematics which led to Gödel's discovery of the incompleteness theorem are reviewed. A short discussion of this theorem and other related results is included. (This note is intended as background for other talks at the conference.)

Gödels incompleteness theorem, like several other related results which are often considered together, belongs to the domain of *metamathematics*. This is a branch of mathematics in which mathematical disciplines themselves are investigated. One deals there with problems which relate to provability. For instance, the questions of consistency and completeness of a given theory fall within the scope of metamathematics.

One can well say that methamathematics displays a higher level of abstraction than the usual mathematical theories. Of course, nowadays we are accustomed to all kinds of abstractions, and it is a mathematician's second nature to organize some correlated discoveries into a deductive theory. In its early stages, however, metamathematics evolved gradually, in response to concrete situations.

Three developments were especially important in this connection. [1] The first is that of non-Euclidean geometry (in the first half of the 19th century). Here we come upon the point of view, that axioms can be arbitrary assertions, whose consequences are to be investigated. The traditional dictum, that axioms are self-evident truths, started to lose its force. The second of these developments concerns formalization (and started in the second half of the 19th century). In the older idea of proof, one had to show in a

---

[1] These three developments, together with some later ones, are described in greater detail, and in a much more dramatic manner, in the following manuscript: A. Tarski, "Some current problems in metamathematics", text of a Harvard lecture (1939/40-?). This manuscript is being edited and prepared for publications.

convincing way that the conclusion has to hold whenever the hypothesis is fulfilled. The evolution of logic, however, led to the precise notion of a proof as a sequence of statements, where each is an axiom or a statement proved previously, or else follows from previous statements in the sequence according to strict rules. Mathematical theories whose proofs are (or can be) constructed in this way could then themselves be objects of systematic study.

The third development is that of set theory, in which various paradoxes and antinomies were formulated (toward the end of the 19th century). One of the best known is the antinomy of Russell, where we introduce:

$$Y = \{x : x \notin x\} \tag{1}$$

Then: $Y \in Y$ *if and only if* $Y \notin Y$, and we see an inconsistency. This particular difficulty can be easily removed, but there remained the possibility that another such anomaly might occur in a given theory. The problem that the theory in question is consistent, was thus brought to a sharp focus.

This problem of consistency was formulated and emphasized by David Hilbert in his famous Paris address of 1900. Afterwards, Hilbert maintained his interest in the foundations of mathematics over long years. In particular, beginning with 1918 he emphasized the need for a new science (now called metamathematics), which would be devoted to such problems and which would be as rigorous as the rest of mathematics.

The above problem of proving consistency was the first major methodological problem to be considered. There are two other general properties that a given deductive theory may possess or not: those of completeness and decidability. (These two latter properties, in contrast to consistency, are not essential.) We recall that completeness, in its simplest form, means that of any two sentences, A and *not-A*, at least one can be proved. (We assume that the sentences are formulated with the help of only the symbols of the theory. If completeness is combined with consistency, then we must have that exactly one of those two sentences is provable.) Decidability refers to having an algorithm for determining whether a given sentence follows from a given set of axioms or not; we say that a theory is decidable if such an algorithm exists.

Early mathematical investigations yielded various simple results regarding consistency, completeness, and decidability. In particular, in 1927 two articles appeared, which established the completeness and the decidability of the theory of the relation $<$ (for real numbers and for natural numbers).[2]

---

[2] C. Langford, Annals of Mathematics **28** pp. 16, 459 (1927),

Let us look again at the characteristics of metamathematics. We examine there (formal) proofs, which are appropriately constructed sequences of expressions. But during the 1920's, expressions themselves had become familiar objects of study. Let us recall that Hilbert worked in Göttingen. Not far away there were the cities of Lwow (now: Lviv) and Warsaw, where the Polish school (of logic, etc.) was centered, and Vienna, where the Vienna Circle was founded in 1925. Problems of language, and in particular linguistic expressions, were extensively discussed by members of both the Polish school and the Vienna Circle.

Another aspect of metamathematics relates to counting and to natural numbers. We count how many symbols there are in a given expression, we count how many expressions there are in a sequence which determines a given proof, etc. However, the essential properties of natural numbers were familiar by 1920's; some results in fact go back to antiquity (like that of the existence of infinitely many primes), and a more systematic study became possible after the publication of Peano's axioms in 1889.

In 1928, a decade after his return to the subject of foundations, Hilbert proposed four problems involving consistency and completeness. [3] By then the time was ripe for a direct attack. Kurt Gödel of the Vienna Circle proceeded to tackle this group of problems, with remarkable success. (These four problems, incidentally, are seldom mentioned nowadays, and correspond only in part to the four theorems which are discussed below).

The four incompleteness-like theorems. - We now abandon our little tour of history, and will describe briefly the results in question. Let us return to expressions and natural numbers, and consider their interplay. We recall the notion of (formal) rules of proof, which alllow us to derive an expression from one or more others. Such rules can be regarded as relations among expressions, to be called *relations of consequence.* We assume, moreover, that a numbering system for expressions has been adopted, so that we can transfer these relations to the adopted numbering. We then obtain relations among numbers. Now, it turns out possible to construct such a numbering system for expressions, that the relations of consequence and the class of provable expressions can be defined with the help of arithmetical and logical terms only (like ".", "0", "$\exists_n$"). We denote this class as *Pr*, and we will refer to it later.

We now have a scheme, where we go from arithmetic to expressions and back to arithmetic, and we can find *self-referential statements:* statements that can be interpreted as assertions about their own provability. Some of these statements are necessarily nonprovable, together with their negations. Descriptions of such statements can be found in

---

[3]See H. Wang, Reflections on Kurt Gödel, The MIT Press, Cambridge, MA, especially pp. 54ff.

mathematical expositions, and even in semi-popular texts.[4] We emphasize the constructions of such statements applies to all sufficiently extensive systems of axioms and rules of proof (for the arithmetic of natural numbers). This is the content of Gödels's first theorem, of 1931.

Let us elaborate on the self-referential statements. One can find a pair of statements, say B and not-B, which are not provable but are such that we can conclude (from metamathematical arguments, without relying on formal proofs) which of the pair is true. We can therefore adjoin this statement as a new axiom. But then we would have a new notion of provability, and we would be led to a new pair B', not-B'. The incompleteness therefore remains.

Two further comments could be made. First, we see that if we want to establish various facts about natural numbers, we may have to allow metamathematical arguments. One hears sometimes, that perhaps such and such property of natural numbers has not been proved because it might be unprovable, but here the speaker would miss the point of Gödel's theorem. Second, the above interplay between expressions and numbers is possible only if multiplications is admitted into our theory. Otherwise, say in case of natural numbers with addition only, we could not construct selfreferential statements. In fact, this restricted theory was shown (in 1930) to be complete. In the subsequent discussion we therefore always assume that the arithmetical theory admits multiplications.

The second theorem refers to the proofs of consistency. If a given theory is inconsistent, then every sentence is provable. Hence, if we could show e.g. that: $'0 \neq 0'$ *is not provable*, we would have a proof of consistency of the theory. But it turns out that such statements themselves are unprovable, in the sense that their numbers are not in Pr. This is the content of Gödel's second theorem which was presented in the same paper as the first theorem, in 1931). We may add that the proofs of both of these theorems presuppose the consistency of arithmetic.

Hilbert, in emphasizing the importance of the problems of consistency in earlier times, called for *finitistic* solutions; in these infinities (in general) should be avoided. Gödel's second theorem, however, appears to rule out the possibility of such a proof for the arithmetic of natural numbers. (We say "appears", since Hilbert was not entirely explicit about what he meant by "finitistic".)

The next theorem that we consider involves the notion of definability. (Here the question of credits and priorities has been debated.) We recall that the class Pr of numbers

---

[4]E. Nagel and J.R. Newman, in *the World of Mathematics*, edited by J.R. Newman (Microsoft Press, Redmond, WE 1988), vol. 3, pp. 1641ff.; R. Penrose, *The Emperor's New Mind* (Oxford University Press, New York - Oxford 1989), expecially pp. 106 ff.

corresponding to provable expressions can be defined (one can also say, can be expressed) with the help of arithmetical and logical terms only. The theorem in question now states that the class Tr of numbers corresponding to true sentences of arithmetic cannot be so definied. A proof can be made to depend on adaption of a self-referential statement, like that of the paradox of the liar: That what I am now saying, is false.

Since all provable sentences have to be true, it follows that: $Pr \subseteq Tr$. But we do not have equality, hence: $Pr \subset Tr$. We see here a different argument for the existence of true but unprovable sentences.[5]

The last theorem of this group asserts the undecidability of the arithmetic of natural numbers. This result was obtained by Church in 1936 (aside from later refinements, especially by Rosser), by extending Gödel's original techniques and arguments. However, a different approach to the decision problem was originated by Turing and by Post (independently) at about the same time. This brings us to the idea of a *Turing machine* and to connections with computers, and to much more extensive later developments, about which we will hear in other lectures.

---

[5] Additional discussion of these ideas can be found in: A. Tarski, *Truth and Proof*, in *L'Age de la Science 1, p. 279 (1969)*.

A Formal Scheme for Avoiding Undecidable Problems. Applications to Chaotic Behavior Characterization and Parallel Computation [*]

Antonio L. Perrone[1,2,3][**]

[1] Dept. of Physics, Univ. of Rome "Tor Vergata", Via della Ricerca Scientifica 1, I-00133 Rome ,Italy
[2] INFN-National Institute for Nuclear Physics, Section of Rome II, Via della Ricerca Scientifica 1, I-00133 Rome, Italy
[3] Pontifical Gregorian University, P.zza della Pilotta 4, I-00187 Rome, Italy

Abstract. In this paper we intend to analyze a chaotic system from the standpoint of its computation capability. To pursue this aim, we refer to a complex chaotic dynamics that we characterize via its symbolic dynamics. We show that these dynamic systems are subjected to some typical undecidable and polinomially uncomputable problems. Our hypothesis is that these limitations depend essentially on the supposition of the existence of a fixed universal alphabet. The suggestion is thus of justifying a contextual redefinition of the alphabet symbols as a function of the same evolution of the system. We propose on this basis a general theorem for avoiding undecidable problems in computability theory by introducing a new class of recursive functions defined on different axiomatizations of numbers obtained by a modification of classical Peano's axiom of successor.
In such a way, from an experimental viewpoint, we are able to obtain a very fast extraction procedure of unstable periodic orbits from a generic chaotic dynamics. The computational efficiency of this algorithm allows us to characterize a chaotic system by the complete statistics of its unstable cycles. Some examples of this technique applied to classical chaotic attractors are discussed.
Finally we discuss some specific topic about pattern recognition and parallel computation. In this context we show how the same class of new recursive functions allows us to avoid some classical limitations demonstrated by Minsky and Papert in *Perceptrons* concerning the possibility of true parallel computations. These results are applied to the problem of *real time* automatic recognition *via* software of particle tracks in high energy physics experiments.

---

[*] This work is partially supported by Italian INFN (National Institute for Nuclear Physics), project "SKYNNET".
[**] E-mail:PERRONE@ROMA2.INFN.IT

1 Introduction

In this work we introduce a new approach in computability theory we are developing by the Italian National Institute of Nuclear Physics (INFN). This approach is devoted to define a new concept of computability to overcome the classical Turing scheme and its limitations in complex computational problems.

The starting point of our approach is the observation that in the classical computability theory the existence of a *universal alphabet* is always supposed. All computations of a Turing Machine (TM) have thus to be described in this universal alphabet. This property is very appealing from the formal standpoint because it allows to define a very large class of computations through the notion of Universal Turing Machine (UTM). Nevertheless, this universality position constrains the classical approach either into the diagonalization problems and into the related undecidability problems of the predicate calculus, or into the $\mathcal{NP}$ problems of the propositional calculus. These two classes of problems are indeed the very same *satisfiability* problem considered within these two different types of calculus [37].

However, if we examine more deeply the problem of the *different contexts* of calculation, we discover that Turing's approach defines as universal a computation performed by two different TM's (i.e., in two different contexts) *iff* these can be posed in a biunivocal correspondence with a UTM. Such a correspondence on its turn supposes the existence of a *unique universal alphabet* allowing the transposition of the symbols of the two former alphabets into the universal one. In other words, all the different contexts have to be nothing but immediate derivations of such a unique and omnipresent alphabet. This supposition of universal derivations has its ultimate roots in the logical scheme underlying the theory of real numbers [38]. In fact, their existence is granted by an *ad hoc* axiom (the so-called "continuum hypothesis") that cannot be derived from simpler axioms such as those governing the existence of natural numbers, i.e., Peano's axioms. Thus, it is in this set theoretical context that we want to discuss the limitations of Turing's approach to computability to search for a new way to avoid them.

In short, instead of using recursive functions defined on a universal set of natural numbers, we introduce recursive functions defined on different axiomatizations of natural numbers, where the axiomatization on which the computation is performed at the step n and that can change at any step, is defined by the recursive function itself. This is obtained through an axiom that allows to change the number axiomatization as a function of the computation at the previous steps. On its turn, the different axiomatizations of natural numbers on which the computation can be performed, are obtained through an extension of Peano's axioms. In this way, also the formal problems of undecidability and diagonalization can be considered in a new light. Indeed, in our axiomatic approach we can solve diagonalization problems because our axiomatization is able to *generate* progressively the set of axioms defining the proper language to complete the computation. It seems thus that our approach can open a new way toward a class of computable functions wider than those computable by a TM, against Church's classical hypothesis. This new perspective in computability theory is discussed in the next Section.

Of course, the possible applications of this new perspective are as wide as the problems of modern computability theory. In this paper we show only two on which we are actually working. They belong however to two fields sufficiently distant each other, to offer a concrete example of the generality of our approach. We present these two applications in the Sections 4 and 5 of this work, even though at the actual state of our work they seem not yet completely satisfying as particular "implementations" of the general axioms of the computability theory we discuss in the next Section.

The first application we present in Section 4 concerns a new technique of extraction of unstable cycles from a chaotic dynamics. The specialists of computability theory can be more acquainted with this type of problem if we present our approach as a general strategy of solution of *non-linear equations*. In fact, given a chaotic system of equations $\frac{dx}{dt} = f(x)$ where $x \in \mathbb{R}^n$ and f is *non-linear*, we integrated numerically the equation so to obtain the solution $x(t) = F(x,t)$. Then we searched for the cycles of the system by solving the equation $\bar{x} = F(\bar{x})$, so to represent $x(t)$ as a series expansion of these periodicities. Our result consists thus in being able to solve the equation $\bar{x} = F(\bar{x})$ by a computationally effective algorithm because it is capable to extract a cycle of any period T, starting from an arbitrary point $x(t_0) \in x(t)$. The essential property of this algorithm is that it needs only information relative to a time t with $t \ll T$. The exposition of this new method is introduced by a formal discussion on the origins of a generic chaotic dynamics from undecidability problems, so to link the problem of a chaotic dynamics with the previous axiomatic discussion on computability theory.

The second application discussed in Section 5 concerns the problem of the parallel computation. To discuss this problem we start from the deep analysis of this problem developed by M.Minsky and S. Papert [32] who demonstrated that even very simple predicates, such as to decide whether a figure on the Euclidean plane is connected or not, cannot be calculated by a parallel architecture while they are easily calculated by a Turing-Von Neumann sequential architecture. Our axiomatic approach implemented into this context allows a parallelization of such predicates.

2 A Non-Gödelian Approach to Undecidable Problems

2.1 The Notion of Recursive Function and its Relationship with Gödel's Theorems

Let us define as *effectively calculable* a function for which there exists an algorithm, i.e., a *finite* calculation procedure, able to calculate its values. It is generally accepted that this intuitive definition of class of effectively calculable functions is equivalent to the most rigorous definition of class of the functions *computable* by a Turing Machine (TM). Whereas,

- Turing's notion of computable function is equivalent to Gödel's-Herbrand's-Kleene's notion of *general recursive function* [9]; and

– the notion of general recursiveness is, on its turn, equivalent to Church's notion of λ-*definable function*.

The simplest way to introduce the problem of *the effective computability* is to reflect upon a famous class of equations, generally known as *Diophantine equations* [5].
Let us consider the simplest ones, that is the linear Diophantine equations. We want to know whether there exist some integers x and y such that the equation:

$$a \cdot x + b \cdot y = c \quad a, b, c, x, y \in \mathcal{N} \tag{1}$$

is verified for integer values of (a, b, c). It is immediately evident that if a and b are even numbers of the form $2n$, while c is an odd number of the form $2n + 1$ no solution exists. Indeed, an even number (and the first term of the Eq.(1) is even, because it is the sum of two even numbers) can never be equal to an odd number (i.e., the second term of the Eq.(1)). A further problem is thus to know whether there exists an effective computable procedure (i.e., an *algorithm*) that allows us to decide in a finite time whether, for any x or y, Eq.(1) has a solution. A result in elementary number theory [4] ensures us that Eq.(1) has an integer solution *iff* the greatest positive integer that is divisor both of a and b is a divisor also of c. Then the searched algorithm is the algorithm of the greatest common divisor that is certainly effective. An immediate generalization is to search for these integer solutions not for linear equations but for generic polynomials of n-th order with k variables, of the form:

$$\sum_{i_1, i_2, \ldots, i_k = 0}^{n} a_{i_1, i_2, \ldots, i_k} x_1^{i_1} \ldots x_k^{i_k} \tag{2}$$

where $a_{i_1, \ldots, i_k}$ has an integer value. Generally no effective algorithm is known that is able to solve the problem. Moreover, under appropriate hypotheses, it is indeed possible to demonstrate not only that an effective algorithm for such a problem cannot be known, but also that this algorithm does not exist [5].
Obviously, each algorithm must be expressed by a set of instructions in a given language (e.g., English language). K.Gödel proposed thus a very efficient method to make each function f defined on the positive integers correspondent to its own instruction set codified by an integer number (i.e., the "Gödel number").
To illustrate this method, let us order according to their lengths, the set of instructions describing in English language the algorithms of each computable function f. If there exist algorithms of the same length, they will be ordered according to the alphabetical sequence, from the beginning till to the first letter that is different in the two considered cases. Let us associate now with each natural number the i-th set of instructions of such a list E_i. We have generated in such a way a biunivocal correspondence between numbers and utterances in English [9, 10] [4]. A very powerful means to handle such a correspondence is the

---

[4] For instance, let us consider the following correspondence g between the integer numbers and the symbols u of a language K belonging to a first order predicate

use of *recursive functions*. Intuitively, a function is defined as *recursive* wherein values of the function for given arguments are directly related to values of the same function for "simpler" arguments or for "simpler" functions. Generally, the constant functions are considered as the simplest ones. E.g., the function f defined by:

$$f(0) = 1$$
$$f(1) = 1 \tag{4}$$
$$f(x + 2) = f(x + 1) + f(x)$$

gives the Fibonacci sequence: $1, 1, 2, 3, 5, 8, 13 \ldots$ Moreover, a function is defined as *total recursive* if, roughly speaking, it is a n-ary function of which domain is the set of all the n-ples. E.g., the procedure for squaring numbers originates a unary function indicated by x^2. The domain of x^2 is the set of all integers; its range is the set of *all* perfect squares and hence x^2 is total. If a recursive function is not total or it is not known whether it is total (e.g., it is not defined for all the n-ples), this function is defined as *partial recursive* [11, 9].

We can thus relate the problem of the computability of a generic function f with the problem of recursiveness. Indeed, if we consider from this standpoint the illustrated enumeration strategy of the computable functions, it is to be noticed that for Gödel the only computable functions are the total functions. In this way, if we want to use a *recursive strategy* to enumerate completely the set of the computable functions, we can obtain an essential characterization of the core of famous Gödel's *incompleteness theorems* [12, 13]. Let us reconsider the previous list of all the computable functions, each with its own index and each described in a given language (e.g., English), by the set of instructions for computing it. Let us imagine to define the following total function:

$$\phi(x) = \Psi_x(x) + 1 \tag{5}$$

where by ϕ we want to define the function that, given the index x, runs through all the list. When it finds the index x, executes the set of instructions Ψ_x on the

---

theory, where the symbols of negation, implication as well as the punctuation marks ["(", ")", ","] are defined [10]:

$$g(() = 3; g()) = 5; g(,) = 7; g(\sim) = 9; g(\supset) = 11$$
$$g(x_k) = 5 + 8k \quad k = 1, 2, \ldots$$
$$g(a_k) = 7 + 8k \quad k = 1, 2, \ldots$$
$$g(t_k^n) = 9 + 8k \quad k, n \geq 1$$
$$g(A_k^n) = 11 + 8k \quad k, n \geq 1$$

then, for whichever utterance $u_1 u_2 u_3 u_4 \ldots u_r$ we have the correspondent Gödel number given according to the form:

$$g(u_1 u_2 \ldots u_r) = 2^{g(u_1)} 3^{g(u_2)} \ldots P_{r-1}^{g(u_r)} \tag{3}$$

where P_i is the i-th prime number and $P_0 = 2$.

input x; i.e. computes the function $\Psi_x(x)$. Finally, if and when the computation terminates, it adds 1 to the value obtained $\Psi_x(x)$.

Now, because the function ϕ is describable in the same language, this function must belong to the list of computable functions. That is, it is necessary that there exists an index x_0 in the list coding the set of instructions for computing ϕ. That is:

$$\Psi_{x_0}(x_0) = \phi(x_0) \tag{6}$$

This implies evidently a *contradiction* [11] because from the definition in Eq.(5) of the function ϕ we obtain the following:

$$\Psi_{x_0}(x_0) = \phi(x_0) = \Psi_{x_0}(x_0) + 1 \tag{7}$$

At this point, the classical solution generally proposed is related with S.C. Kleene's definition of *partial recursive function* [9]. Essentially, it is allowed that the function $\phi(x)$ be defined *not for all the values of x* but only for some of them. In this restricted range the function is total recursive, while in the other values we can assign it an arbitrary value because it is even impossible to know whether it has values in this points.

In the next Subsection we propose a different approach to undecidable problems.

2.2 A Theorem to Avoid Undecidability by the Mutual Recursive Definition of Numbers and Functions

An Enlarged Axiomatic Theory of Natural Numbers. Our demonstration [6, 7, 8] will pursue the aim of making evident how the diagonalization problem (7) admits a different solution from the classical one of the partial recursive functions [9].

Let us consider a simple diagonalization scheme of the form :

$$\Psi_{x_0}(x_0) = \phi(x_0) = \Psi_{x_0}(x_0) + 1 \tag{8}$$

where Ψ is the partial recursive function defined from the $(x + 1)$- th set of instructions Q_x. Let x_0 be chosen so that Ψ_{x_0} is the partial recursive function ϕ defined for the following set of instructions: "to calculate $\phi(x)$, find Q_x, calculate $\Psi_x(x)$, and *if and when* a value for $\Psi_x(x)$ has been found, let $\Psi_x(x) + 1$ the value for $\phi(x)$".

Our analysis of the problem is different from the Gödelian one, because we do not share his supposition that the set of natural numbers is given a priori. On the contrary, we suppose that it is possible to construct a generative theory of different and reciprocally irreducible sets of natural numbers. We individuated the core of such a generative theory of natural numbers in a modification of Peano's axiom of successor and, consequently, in a modification of his axiom of identity[5]. Essentially, we think that it is possible to maintain the axiomatic structure of classical arithmetic defined by the Italian mathematician, by integrating it with a modification of the axiom of successor in view of allowing the number theory

---

[5] See Peano's Axiom 2 defined afterwards.

the availability of several different axiomatizations of *natural* numbers.[6] Each axiomatization, on its turn, is ruled by the classical axioms of natural numbers in $\mathbb{N}$.

Moreover, we formulate an axiom of identity relative to the successor allowing, at least partially, number theory to define some operations among numbers defined on different axiomatizations. Of course, because Peano's axioms remain unmodified within each particular axiomatization, the usual algebraic notions of "grouppoid", "group", "ring", etc. equally hold within each axiomatization. On the contrary, the properties of these algebraic notions, if applied on operands belonging to different axiomatizations, are still to be studied more deeply.

According to these general principles, let us illustrate now our strategy to deal with the diagonalization problem according to the Gödelian formulation given in (8).

Though arithmetic and natural numbers are governed by Peano's axioms:

1. $\forall x \ Sx \neq 0$
2. $\forall x \ Sx = Sy \Rightarrow x = y$
3. $\forall x \ x + 0 = x$
4. $\forall x \ x + Sy = S(x + y)$
5. $\forall x \ x \cdot 0 = 0$
6. $\forall x \ x \cdot Sy = x \cdot y + x$

let us define a *"collection"* [14] $G = \{g_1, g_2, \ldots, g_i, \ldots\}$ of axiomatizations of natural numbers [7] obeying to Peano's axioms with the following modification that allows us to link such different axiomatizations in a coherent framework. Instead of considering Peano's axiom of successor:

$$S(x + y) = x + S(y) \tag{9}$$

where S is the *succession operator*, i.e. $S(4+5) = (4+5)+1 = 4+S(5) = 4+6$, we pose:

$$\exists i, j; \ i \neq j \mid \forall x_{g_i}, y_{g_i} \in \mathcal{N}_{g_i} \ \forall x_{g_j}, y_{g_j} \in \mathcal{N}_{g_j}$$
$$S_{g_i}(x_{g_i} + y_{g_i}) = S_{g_j}(x_{g_j} + y_{g_j})$$
$$(x_{g_i}, y_{g_i}) \neq (x_{g_j}, y_{g_j}) \tag{10}$$

---

[6] This term of axiomatizations of natural numbers could seem inappropriate, because western mathematics, from Plato's age on, is acquainted to consider the natural numbers as univocally and universally defined. But, as we will see, this is not the only way to grant universality in calculations.

[7] Notice that our approach from the set theory standpoint is in some sense a midway between G.Cantor's *constructive* definition of sets and P.J.Cohen's *generic* definition of sets [13]. In fact, our number sets are constructive because each of the different axiomatizations of natural numbers we introduce, follows Peano's axioms. On the contrary, in our approach it is "generic" the definition of the different axiomatizations we introduce by the Axiom1.

That is, we define the succession operation on *different axiomatizations*, where each axiomatization g_i defines a set of natural numbers $\mathcal{N}_{g_i}$. Of course, within each axiomatization Peano's axiom of succession holds according to the following:

$$S_{g_i}(x_{g_i} + y_{g_i}) = x_{g_i} + S_{g_i}(y_{g_i}) \tag{11}$$

At this point, the diagonalization problems of the form (7) can be solved in a straightforward way. When a undecidable situation could occur on an axiomatization $\mathcal{N}_{g_i}$ of natural numbers, *the axiomatic itself*, owing to the axiom defined by (10), is allowed to *skip* on another axiomatization $\mathcal{N}_{g_j}$ with $i \neq j$ to terminate the computation.

The Theorem.

Definition 1

$$G = \{g_1, g_2, \ldots, g_i, \ldots\} \tag{12}$$

Definition 2 $\mathcal{N}_{g_i} = \{0_{g_i}, 1_{g_i}, 2_{g_i}, 3_{g_i}, \ldots\}$ *is the set of naturals obeying to Peano's axioms in the axiomatization g_i.*

Axiom 1

$$\exists i, j; \; i \neq j \mid \forall x_{g_i}, y_{g_i} \in \mathcal{N}_{g_i} \; \forall x_{g_j}, y_{g_j} \in \mathcal{N}_{g_j}$$
$$S_{g_i}(x_{g_i} + y_{g_i}) = S_{g_j}(x_{g_j} + y_{g_j})$$
$$(x_{g_i}, y_{g_i}) \neq (x_{g_j}, y_{g_j}) \tag{13}$$

where S_{g_i} is the succession operator as it is defined in the precedent Subsection. In the case that the two operands are defined on the same axiomatization g_i, the following holds[8]:

Lemma 1

$$S_{g_i}(x_{g_i} + y_{g_i}) = x_{g_i} + S_{g_i}(y_{g_i}) \tag{14}$$

Proof. It is sufficient to consider the Axiom 1 by posing $i = j$. This makes applicable the Peano axiom of successor. □

---

[8] The next formula is commonly known as the Peano *axiom of successor*: precisely, it is the fourth axiom in the list given in the last Subsection. It is remarkable that by our axiomatic approach the Peano axiom of successor becomes a lemma descending from the Axiom 1.

Notes and Comments. Notice that although we are using the equality sign between the first and the second member of an arithmetic operation, this does not implies any "coincidence" of the two members because the Peano *axiom of identity* $S(x) = S(y) \Rightarrow x = y$ falls owing to the new axiom of successor, i.e., Axiom 1. The equality sign is thus indicating the possibility of obtaining the equality after the application of the succession operator S_{g_i}. This means that the sum operation and more generally any operation between different axiomatizations is defined as *relative* to the succession operation. We would have to write thus $\underset{S_{g_i}}{=}$ instead of $=$. Nevertheless, for sake of simplicity, we continue to use the symbol $=$, but with the explained different meaning of *relative* and no longer *absolute* equality.

Because the following property does no longer hold:

$$S(x) = S(y) \Rightarrow x = y \qquad (15)$$

it is possible now to define the sum operation with respect to two numbers a_{g_i}, b_{g_j} that are defined on two different axiomatizations g_i and g_j. That is, we are able to solve the following equation: $a_{g_i} + b_{g_j} = c_{g_k}$. Indeed, we are allowed to choose *arbitrarily*[9] the axiomatization k of the result: $k = i$, $k = j$ or $k \neq i, j$.

Case $k = i$ or $k = j$. Let us choose that the result is expressed in the same axiomatization of a: $k = i$. We have thus that $a_{g_i} + b_{g_j} = c_{g_i}$, but the Axiom 1 allows us to obtain the searched result because of the following:

$$S_{g_i}(c_{g_i} - a_{g_i}) = S_{g_j}(b_{g_j}) \qquad (16)$$
$$\Rightarrow c_{g_i} - a_{g_i} = b_{g_j}$$

that determines c_{g_i} univocally. The first member $S_{g_i}(c_{g_i} - a_{g_i})$ is computable because of the Lemma 1 since the two operands are defined on the same axiomatization g_i.

Case $k \neq i, j$. In this case the introduction of a new temporary variable d defined on one of the two axiomatizations of the operands a, b. Synthetically, we have:

$$a_{g_i} + b_{g_j} = c_{g_k} \quad k \neq i, j$$
$$S_{g_i}(d_{g_i} - a_{g_i}) = S_{g_j}(b_{g_j}) \Rightarrow a_{g_i} + b_{g_j} = d_{g_i} \qquad (17)$$
$$S_{g_i}(d_{g_i}) = S_{g_k}(c_{g_k}) \Rightarrow d_{g_i} = c_{g_k} \Rightarrow a_{g_i} + b_{g_j} = c_{g_k}$$

At this point we are allowed to introduce the definition of a new class of recursive functions. Let us redefine the function Ψ of Eq.(8) so that it can profit by the properties of the Axiom 1 because the function Ψ is defined now on different axiomatizations of natural numbers.

---

[9] Notice that such an arbitrariness is allowed only at the axiomatic level. At the physical level the nature of the same physical phenomenon will suggest us the more appropriate axiomatization to obtain the result.

Definition 3

$$\Psi^{(n)} : \mathcal{N}_{g(n)} \to \mathcal{N}_{g(n)}$$
$$g^{(n)} = g_{i(n)} = g_{i(n-1)+\Delta i(n)} \in G = \{g_1, \ldots, g_i, \ldots\} \tag{18}$$
$$\Delta i(n) = i(n) - i(n-1)$$

where $n \in \mathbb{N}$ is an integer number indicating the axiomatization $\mathcal{N}_{g_{i(n)}}$ of the recursive function [10].

In the precedent definition the procedure by which the recursive function $\Psi^{(n)}$ passes through the different axiomatizations $\mathcal{N}_{g(n)}$ is not defined. This task is devoted to the term $\Delta i(n)$. Indeed, the Axiom 1 grants only the *existence* of the different axiomatizations, while the following axiom rules their use during the computation.

Axiom 2 $\Delta i(n) \propto \Delta \Psi^{(n-1)} = \Psi_{g_{i(n-1)}} - \Psi_{g_{i(n-2)}}$

The logic underlying this axiom is simple. If the computation of the recursive function $\Psi^{(n)}$ does not change notwithstanding it is calculated on different axiomatizations so that $\Delta \Psi^{(n-1)} = 0_{g_{i(n-1)}}$, then we reached the fixed point in the computation of $\Psi^{(n)}$. This strategy is similar but not identical to the well-known computational techniques of the variable step size in which the size of the computation step ε is progressively halved as a function of the fluctuations in the results obtained with previous different step sizes. The difference between the two methods is discussed in §4.6.

At this point we have all the definitions necessary to state the following:

Theorem 1 *The function* $\phi(x_{g_i}) = \Psi_{x_{g_j}}^{(j)}(x_{g_j}) + 1_{g_j}$, *where the initial point of the computation is* $x_{g_{i(0)}} = x_{g^{(0)}}$ *and* $\Psi^{(j)}$ *is defined according to Eq. (18), is general recursive.*

Proof. To demonstrate that the function ϕ is *general recursive* it is sufficient to demonstrate that there exists an index $x_{g_k}^0$ such that the following holds:

$$\Psi_{x_{g_k}^0}^{(k)}(x_{g_k}^0) = \Psi_{x_{g_j}^0}^{(j)}(x_{g_j}^0) + 1_{g_j} \tag{19}$$

that solves the undecidability expressed by the Eq. (8). But this condition can be always satisfied in the collection G because of the following application of the Axioms 1 and 2

$$\exists j, k \; j \neq k \mid S_{g_k}\left(\Psi_{x_{g_k}^0}^{(k)}(x_{g_k}^0) - 1_{g_k}\right) = S_{g_j}\left(\Psi_{x_{g_j}^0}^{(j)}(x_{g_j}^0) + 0_{g_j}\right) \tag{20}$$

By applying Peano's axiom of successor, from this last expression, we obtain:

$$\Psi_{x_{g_k}^0}^{(k)}(x_{g_k}^0) + S_{g_k}(-1_{g_k}) = \Psi_{x_{g_j}^0}^{(j)}(x_{g_j}^0) + S_{g_j}(0_{g_j}) \Rightarrow \Psi_{x_{g_k}^0}^{(k)}(x_{g_k}^0) + 0_{g_k} = \Psi_{x_{g_j}^0}^{(j)}(x_{g_j}^0) + 1_{g_j} \Rightarrow$$

$$\Psi_{x_{g_k}^0}^{(k)}(x_{g_k}^0) = \Psi_{x_{g_j}^0}^{(j)}(x_{g_j}^0) + 1_{g_j}$$

$\square$

---

[10] In the present formulation such an index is not yet defined on its turn on different axiomatizations. For this reason we *still* need the Definition 1.

2.3 A Dynamic Approach to Coding Theory and its Relationship with $\mathcal{NP}$ Problems

We can apply what we said till now to revise a classical method of *coding theory*. This revision could explain the informational capabilities of unstable dynamics such as a chaotic dynamics. The method we want to reconsider is the *Gödelian method of language coding*. This coding makes correspondent to each statement a *numerical integer code* that marks *univocally* the statement.

It is sufficient to associate to each symbol $a_i \in \mathcal{A}$ the odd number $2i + 1$. Let $W = a_{i_1} a_{i_2} \ldots a_{i_k}$ be a generic word in the considered alphabet $\mathcal{A}$. Let us indicate by $gn(W)$ the Gödel number of W:

$$gn(W) = \prod_{j=1}^{k} \Pr(j)^{2j+1} \tag{21}$$

where $\Pr(j)$ denotes the j-th prime number. Moreover, let us pose $gn(W \equiv 0) = 1$. Essentially the univocal character of coding is founded on the *uniqueness of the number decomposition into prime factors*, even though other coding strategies that do not need such a decomposition were demonstrated [15].

We sketch now the core of the coding strategy founded on the decomposition into prime factors. Let $d = (a, b)$ the greatest common divisor of a and b. We know that it can be always written as $d = ka + lb$ with $k, l \in \mathbf{Z}$. We have thus the following:

Lemma 2 *If a prime number p is the divisor of the product ab, it must be the divisor either of a or of b.*

From this lemma the *fundamental theorem of arithmetic* follows immediately:

Theorem 2 *Any integer number $n > 1$ can be decomposed into a product of prime numbers in only one way.*

In this way the correspondence between the code $gn(W)$ and the string W results to be univocal. Essentially, the novelty of the pioneering work of K.Gödel about undecidability consists in linking the coding strategy to this property of arithmetic. The diagonalization technique, indeed, was already well-known since a long time[11]. Nevertheless, if we analyze the demonstration of the precedent Theorem 3 [4], we note that it can be demonstrated from the precedent Lemma 2 by *applying recursively the associative property of arithmetic*. On the contrary, in the formulation of our Theorem 1 in §2.2, *the associative rule is dynamic* because it is defined on numbers belonging to different axiomatixations. The consequence is that the decomposition is no more unique because one only product does not represent univocally one only number. We introduced then a further

---

[11] The diagonalization procedure was made known in the mathematician realm by the works of R.Dedekind and G.Cantor, but this notion has been previously introduced in the work about the theory of functions of P. du Boys-Reymond.

degree of freedom in the coding theory. It consists in the *modality of association of the parts*. Roughly speaking, according to our approach,

$$(ab)c \neq a(bc) \tag{22}$$

In this way, we can make dependent on this modality the extraordinary variety of behaviors from a reduced set of initial conditions that is characteristic of non-linear and/or chaotic dynamics, even though they are very simple like a quadratic map: $x_{n+1} = px_n \cdot (1-x_n)$. From the coding theory standpoint, instead of an *apriori* tessellation of the space leading us to undecidable problems, the suggestion is of using a *variable topology* allowing us to obtain from the dynamics in each context the maximum of information capacity. This is granted by the application of Theorem 1. It can give us a dynamic (recursive) tool, that is a *deterministic* strategy, to escape from undecidability, both in coding theory and in chaotic dynamics characterization. All this will become more evident by the application of this theorem to define a deterministic strategy for the extraction of *unstable* cycles from a generic chaotic dynamics, as we show below (see §4.5). For the moment, however, we want to sketch the possible perspectives that this new axiomatic theory opens in the field of the so-called $\mathcal{NP}$ problems. The passage is almost immediate if we consider that *some* undecidable problems of the predicate calculus have their counterpart in the $\mathcal{NP}$ problem of the propositional calculus. One of these cases is the following ([37], pag. 439):

Theorem 3 *The satisfiability problem for the propositional calculus is $\mathcal{NP}$-complete.*

This problem can be examined within our approach by considering the Axioms 2 and 1 in the context of the propositional calculus. In this framework these axioms allow us a deterministic strategy to determine the right language in which the generic propositional formula can be satisfied. If we use the symbolism defined afterwards in §3.3, we can describe our strategy according to the following formula:

$$\Delta g_i \propto \Delta \mathcal{R}(A, B, g_i) = \mathcal{R}(A, B, g_i(n+1)) - \mathcal{R}(A, B, g_i(n)) = \tag{23}$$

$$\bigvee_P \bigvee_Q \left[(A = gPhQk) \wedge (B = \bar{g}P\bar{h}Q\bar{k})\right]_{n+1} - \tag{24}$$

$$\bigvee_P \bigvee_Q \left[(A = gPhQk) \wedge (B = \bar{g}P\bar{h}Q\bar{k})\right]_n$$

where the subscript index (n) indicates that all symbols in the expression are to be considered according to the *alphabet at the time step* n. To sum up, we are rewriting the meaning of the symbols on the dynamics given by the behavior of the words A and B in different contexts. In the precedent formula the symbols $\wedge, \vee$ have respectively the usual meaning of the functions *and* and *or*.

An application of this strategy to avoid the exponential explosion of computational time is showed in the Subsection 4.5 where we discuss a new method of extraction of unstable cycles of whatever length from a chaotic dynamics. In fact, the classical approach to this problem suffers unavoidable problems of an exponential growing of the requested computational time.

3 The Relationship between Chaotic Behavior and Undecidable Problems

3.1 A Topological Definition of a Generic Chaotic Dynamics

In order to demonstrate the practical significance of the enlarged axiomatic theory of numbers discussed in the precedent Section, in this Section we want to demonstrate firstly how a chaotic dynamics is related with *undecidability* problems. More precisely, it can be demonstrated [1] the equivalence between the complete specification of the range of a chaotic dynamics f, characterized by its symbolic dynamics and the undecidable character of the classic word problem studied at the end of 40's by two famous works of A.A.Markov [2] and E.L.Post [3]. Synthetically, in the demonstration [1] we pass through the following steps:

1. from a generic chaotic dynamics, to
2. a Bernoulli shift map; hence to
3. a symbolic dynamics in a group G; hence to
4. the combinatorial group theory where the undecidability of the "word problem" occurs.

Let us define the main characteristics of a generic chaotic dynamics $f : J \rightarrow J$ to map them in a topological definition in view of the passage to a symbolic dynamics. We can summarize these characteristics as follows:

Definition 4 *f has* sensitive dependence *on* initial conditions:

$$\exists \delta > 0 \mid \forall x \in J \; \forall N(x) = \{y \mid d(x, y) \le \epsilon\}$$
$$\exists m > 0 \; \exists y \in N(x) \mid d\left(f^{(m)}(x), f^{(m)}(y)\right) > \delta \tag{25}$$
$$f^{(m)}(x) = \underbrace{f(f(f(\cdots f(x)\cdots)))}_{m \text{ times}}$$

where $d(x, y)$ is a metric in J.

Definition 5 *f is* topologically transitive *if $\forall U, V \subset J \; \exists m \in \mathbb{N}$ such that*

$$f^{(m)}(U) \bigcap V \ne 0 \tag{26}$$

Definition 6 *f has* periodic points *which are* dense in J.

Topological transitivity implies that the points of J can move under iterated action of f from any arbitrary small neighborhood to any other. In this respect topological transitivity is essentially equivalent to *ergodicity*.
The consequent definition of chaotic behavior is the following:

Definition 7 *We relate chaotic behavior to the existence of recurrent non periodic sequences.*

- $\forall \mu \in \mathbb{N} \, \exists \lambda > \mu$ such that the segments $\{x_{\nu(\alpha)}, x_{\nu(\alpha+1)}, \ldots, x_{\nu(\alpha+\mu)}\}$ are contained in *every* segment of length λ, $\{x_{\nu(\beta)}, x_{\nu(\beta+1)}, \ldots, x_{\nu(\beta+\lambda)}\}$

In this way, we have defined chaos with respect to the possibility of finding a symbol sequence of length μ in *each* other sequence of length λ. This is the well known property of fractal objects of being *self-similar*, i.e., of repeating on a small scale the same structure of the large scale.

3.2 The Passage to a Bernoulli Shift Map

To obtain the passage to the computability theory, let us consider a simple case having the just remembered properties: the Bernoulli shift. Let $\mathcal{B} : S \to S$ be a Bernoulli shift map from a generic ensemble S isomorphic with the unitary interval into itself. This map is defined by: $x_{n+1} = 2x_n \bmod 1$ with $x_n \in S$. The properties of such a map are immediately evident if we write the generic x in its *dyadic* form:

$$x = \sum_{l=0}^{\infty} a_l 2^l = [a_0, a_1, a_2, \ldots] \tag{27}$$

where $a_l \in \{0,1\}, a_0 = 0$. We have thus:

1. *Sensitivity to initial conditions*. If two numbers x and x' are different only at the n-th digit, after n iterations they are different at the first digit.
2. *Randomness*. $\mathcal{B}$ has the very same random feature as the coin tosses. Indeed, it is sufficient to define:

$$x_k = \begin{cases} L \text{ if } x_k \leq \frac{1}{2} \\ R \text{ if } x_k > \frac{1}{2} \end{cases} \tag{28}$$

   to establish easily the very same correspondence with random sequences of coin tosses.
3. *Ergodicity*. The starting point is the number theory theorem according to which, in any base, *almost all irrational numbers* (i.e., all those with probability one) *contain any given finite sequence of digits infinitely many times* [4]. So, we can be sure that, starting from any irrational number x_0, we can approximate any $x \in S$ with any precision $\epsilon > 0$, for a unlimited number of times by the successive application of the dynamics $\mathcal{B}$ [5].

3.3 The Passage to a Symbolic Dynamics in a Group G

Let us search now for a coding strategy allowing us to transport this problem in computability theory. Let us consider for the sake of simplicity the Euclidean space and a *square tessellation* of the plane, that is a simple square lattice Λ. In this Euclidean plane the *geodesics* are raw straight lines l. Let us label the successive intersections of the straight lines of slope $s, s > 1$, with a, if they occur at vertical bonds of Λ, with b, if they occur at horizontal bond. In this way, we associated to the geodesic l a string composed of a and b symbols with the following properties: i) the a symbols will be always isolated; ii) between two a symbols always $[s]$ or $[s+1]$ b symbols occur (where the symbol $[s]$ denotes the largest integer $\leq s$). From this sequence let us construct the word w, obtained by substituting any sequence ab^n with the symbol $\bar{b}$ and by substituting any symbol b with the symbol $\bar{a}$. At this point some definitions are necessary to introduce ourselves in classical Markov's and Post's word problem [5].

Definition 8 *A* semigroup *is an arbitrary set of elements together with a binary function f that has this set as the domain of its variables and that has a subset of this set as its codomain, so that the associative rule*

$$f(x, f(y, z)) = f(f(x, y), z) \tag{29}$$

holds for all the x, y, z of the set.

Definition 9 Generators *of a semigroup are the letters of a generic alphabet on which the semigroup is defined with respect to the operation of juxtaposition.*

Definition 10 *Let $\{g, \bar{g}\}$ be a pair of words (not necessarily non empty) on a given alphabet. Then the pair $\{g, \bar{g}\}$ is defined as a* relation *on the generators constituting the alphabet.*

Definition 11 *Let us consider the semigroup free on n generators with the alphabet $a_1, a_2, \ldots, a_n$. Let $\{g_i, \bar{g}_i\}$, $i = 1, 2, \ldots, m$, a finite set of relations on these generators. Then, if A and B are two words on this alphabet, we write $A \sim B$ if there exist the words P, Q such that for some $i, 1 \leq i \leq m$: or $A = Pg_iQ$ and $B = P\bar{g}_iQ$; or $A = P\bar{g}_iQ$ and $B = Pg_iQ$.*

Definition 12 *We write $A \approx B$ if there exists a word sequence :*
$A = A_1, A_2, \ldots, A_p = B$
such that $A_j \sim A_{j+1}, j = 1, 2, \ldots, p - 1$.

We can now introduce the problem of the word on a generic semi-group Σ. The problem can be formulated as follows:

− Determine for two words A and B on a given alphabet $\mathcal{A}$ if $A \approx B$ or not.

With respect to this problem, the following theorem holds:

Theorem 4 (Markov-Post, 1947) *There exists a semigroup Σ, defined by a finite set of relations and of generators, of which word problem is recursively unsolvable.*

If the considered space is no longer the classic Euclidean space, the alphabet - and hence the set of generators of the corresponding semigroup - becomes more complex. In the general case, we consider the limit set S of G induced by the dynamic system of geodesic flows on $M := \mathrm{H}/G$ or D/G for some discontinuous group G. M is by definition a Riemann surface with constant negative curvature. In this general case we consider thus as the non-Euclidean space the Lobachevskii plane. Namely, this plane is the upper half of the complex plane $\mathbf{H} = \{x + iy/y > 0\}$ with metrics $ds^2 = (dx^2 + dy^2)/y^2$. The geodesics are arcs of semi-circle with center on the x-axis, which is the boundary of $\mathbf{H}$. $\mathbf{D}$ is the Poincaré *unit disk* $\mathbf{D} = \{z \in C, |z| < 1\}$ with the hyperbolic metrics $ds^2 = 4dr^2/(1 - r^2)^2$. The geodesics are arcs of semi-circle orthogonal to the circle $|z| = 1$, which is the boundary of $\mathbf{D}$. The group G we introduced generates the manifold M as a quotient with one of its normal subgroups R, $M = G/R$. M is thus a topological space on which G acts as a one-to-one topological self-mapping group.

Let us consider now on the limit set S a group Φ of homeomorphisms of M acting on S with dense orbits. In this way, the map f and the group Φ are equivalent orbit on S, in the sense that for any given pair of points $x, y \in S$ there exists an element $\phi \in \Phi$ such that: $x = \phi \circ y$ if there exist the integers $p, q \geq 0$ such that $f^{(p)}(x) = f^{(q)}(y)$. The corresponding sequences of the map f can thus be replaced by sequences of symbols *indexed by the integers* — more properly, by a Gödelian numbering — ranging over the subalphabet of $\mathcal{A}$ constituted by the set of generators that provide the complete presentation defining Φ.

In this way, the correspondence between the sequences of a map f and the sequences of a generic alphabet $\mathcal{A}$, constructed on the geometry of the map and computable by a generic Turing Machine (TM), is complete.

3.4 The Undecidable "Word Problem" and a Chaotic Dynamics

If we reconsider now the definition of chaotic behavior given an §3.1 we can notice that this problem is perfectly correspondent to the "word problem". Indeed, let us consider a sequence A of length μ and let us test whether it is contained in *each* sequence B of length λ greater than μ. To make this test, we would have to be able to establish the correspondence $A \approx B$. That is, we would have to establish whether there exist generators of the group Φ linking each other by f the two considered sequences. But this is exactly the "word-problem" that we know to be recursively non-computable. In other terms, we are able to see that Φ is really generated by a finite group of relations. Nevertheless, *not all* the relations in the range of a generic function f on Gödel numbering of f can be derived from this group. To see this, let us consider the simple characteristic function h of f defined as follows:

$$h(n) = \begin{cases} 0 \text{ if } n \in \text{ range } f \\ 1 \text{ if } n \notin \text{ range } f \end{cases} \tag{30}$$

It is evident that it is impossible to specify *globally* the properties of the map f, namely, its (detailed) whole range. In this way, the chaotic behavior of the map f depends massively on the *intrinsic limits* of the theory and not from an extrinsic ignorance of the dynamics as it is generally supposed.

However, what is essential to point out with respect to this approach is that if the extrinsic *unpredictable* character of a chaotic dynamics depends on its intrinsic *undecidability*, this undecidability, on its turn, depends on the supposition of the *actual* existence of an *universal alphabet* in which *all* the possibilities of the chaotic dynamics generated by f are given.

According to these general principles, let us consider a particular case of "word problem". More precisely, we consider the case in which we want to know the *eventual* periodicity of an arbitrary point of a generic chaotic dynamics. To sum up, we are considering the case in which $B \equiv A$.

We are thus searching for a sequence that starts and terminates into A: $A = A_1, A_2, ..., A_p = A_1 = A$ con $A_j \sim A_{j+1}$, $j = 1, 2, ..., p - 1$. In the next Section we show that, owing to the limitation imposed by the "word problem" to such a search, the time necessary to individuate a periodicity property for a generic chaotic dynamics grows "exponentially" because it implies combinatorial techniques of searching. These techniques limit effectively the search to very short periodicities. On the other hand, we show in the Section §2.2 that this limit can be *automatically* avoided if we use our general principle of a mutual redefinition between numbers and functions (see §2.2). Following this approach we demonstrate in the same Section that it is possible to test in a very short time the periodicity of any point in a generic chaotic dynamics with evident computational advantages. Finally, we introduce the relationship between such computational techniques with the problem of compression and coding of the information through a chaotic dynamics.

4 A new Approach to the Problems of Information Transmission and of Behavior Characterization for a Chaotic Dynamics

4.1 Information Theory and Markov-Post Word Problem

In this Section, we sketch the main problems deriving from the application of the Shannon-Khinchine information theory to chaotic systems. We show that such a theory is in principle insufficient to give us a suitable axiomatic system to deal with the problem of information transmission on chaotic channels or, more precisely, to deal with the problem of coding/decoding a chaotic signal.

Underlying the definition of every *information source* is the set $\mathcal{A}$ of symbols used by it, that we call its *alphabet*, and that we always assume to be *finite* [33]. The separate symbols x_i of this alphabet, are called its *letters*. Let us consider a sequence of letters, infinite on both sides

$$x = (\ldots, x_{-1}, x_0, x_1, \ldots) \tag{31}$$

which represents a possible "life history" of the given source. We shall regard it as an elementary event in a certain (infinite) probability space of which specification characterizes the sequence (31) as a random process. The set of all sequences (31) (i.e., the set of all elementary events of this space) will be denoted by $\mathcal{A}^I$. Any subset of the set $\mathcal{A}^I$ represents an event of this space, and conversely. Generally, if $t_1, t_2, \ldots, t_n$ are integers and $\alpha_1, \alpha_2, \ldots, \alpha_n$ are letters of the alphabet $\mathcal{A}$, then the event: "The source emits the letter α_i at time t_i $1 \leq i \leq n$)" corresponds to the set of all sequences x for which:

$$x_{t_i} = \alpha_i \ (1 \leq i \leq n) \tag{32}$$

Let us define this set of elementary events x as a *cylinder*. At this point, it is sufficient to know the probabilities $\mu(Z)$ of all the cylinders Z to define the sequence (31) as a random process. If we know the probability $\mu(Z)$ of each cylinder Z, we can determine univocally the probability $\mu(S)$ of each set $S \in F_A$ composed by elementary events x; where F_A is *the Borel intersection* of the alphabet A (see [33]). I.e., it is the intersection of all Borel fields containing all the cylinders of the alphabet A. Obviously, as a basic property, $\mu(A^I) = 1$ holds. Moreover, since the alphabet A and the probability measure μ characterize completely the statistical nature of the source, we denote it by the symbol $[A, \mu]$.

4.2 The Coding Theory

Let us discuss now the coding problem. Generally, we have to deal with situations in which source alphabet and channel alphabet for the transmission are different. Let us define as $[A_0, \mu]$ the source of which output has to be transmitted by the channel $[A, \nu_x, B]$, where generally the alphabets A_0 and A are different. Before transmitting a signal on the channel, it is necessary to translate ("*encode*") the letter sequence of A_0 alphabet from the source into a letter sequence of A alphabet of the channel. We denote the source sequence as:

$$\theta = \ldots, \theta_{-1}, \theta_0, \theta_1, \ldots \tag{33}$$

with $\theta_i \in A_0$. In the classical theory each θ is *univocally* transformed into a sequence:

$$x = \ldots, x_{-1}, x_0, x_1, \ldots \tag{34}$$

where $x_i \in A$. The rules according to which such a transformation occurs constitutes the used *code*. So, from the mathematical standpoint, a code is simply any function such that :

$$x = x(\theta) \tag{35}$$

where $\theta \in A_0^I, x \in A^I$. Notice that each coding operation can be considered as a transmission on a channel with input alphabet A_0 and output alphabet A. Such a channel is characterized by the property of being *noiseless*, that is at each input sequence θ has to correspond one output sequence $x = x(\theta)$. We can denote such a channel by the usual symbols: $[A_0, \rho_0, A]$, where the measure ρ_0 takes in this case the interesting form of a *characteristic function*:

$$\rho_0(M) = \begin{cases} 1 & x(\theta) \in M \\ 0 & x(\theta) \notin M \end{cases} \tag{36}$$

with $M \in A^I$.

4.3 The Informational Problem in Chaotic Systems

The classical information theory [33, 34] is faced with the problem of measuring the transmitted information or, more generally, with the problem of giving a measure of information. The solution of both problems, however, supposes as already given a *priori* the coding function $x(\theta)$.

The following and more interesting problem to be solved would be to develop an information theory in which the same coding function *is generated* by the information system. In fact, the solution of such a problem is essential to understand how to use the chaotic systems to transmit information or more generally to understand how the information is encoded in systems displaying a chaotic behavior. The problem becomes even more interesting and urgent to solve the information problem in biological systems where the chaotic behavior seems to be the rule and not the exception.

Let us consider a generic chaotic dynamics and let us imagine to have to study how it associates a given message $\theta = \ldots, \theta_{-1}, \theta_0, \theta_1, \ldots$ of the alphabet $\mathcal{A}_0$ to another string $x = x(\theta) = \ldots, x_{-1}, x_0, x_1, \ldots$ of another alphabet $\mathcal{A}$. In other terms we want to understand which is the coding strategy that a chaotic system uses to transmit information. As a first step, let us try to solve this problem according to the classical Khinchine-Shannon approach to information theory. We send thus a test message θ to understand how the chaotic system encodes it either by following it or by trying of finding out it from the attractor so as to be able to decode it. The possible advantages of using a chaotic channel would be at least two:

- The signal robustness to noise
- The possibility of transmitting several messages *simultaneously* on the same channel by associating each message with a different unstable periodicity of the chaotic signal.

Notice that we are supposing here a univocal coding according to which Shannon coding theorem still holds. Following this theorem, the source entropy must be smaller than the length of the coded message. If we abandon such an univocal coding the use of a chaotic channel could be advantageous also for information compression. However, we do not want to deal with this further possibility in the context of the present work.

When we want to use a chaotic channel we are faced with the limitation depending on Markov-Post theorem. In fact, it becomes impossible to determine whether a given string $x(\theta) = \{x_{\nu(\beta)}, x_{\nu(\beta+1)}, \ldots, x_{\nu(\beta+\lambda)}\}$ is or is not the coding of the message $\theta = \{x_{\nu(\alpha)}, x_{\nu(\alpha+1)}, \ldots, x_{\nu(\alpha+\mu)}\}$, where $\lambda \geq \mu$ and the symbology we are here using is the same of the precedent Section. From the numerical standpoint, we can observe that if we pose our input string θ as the initial condition of the chaotic dynamics after a short time it is spontaneously encoded, owing to the unstability of the dynamics, into a new string $x(\theta)$ that is surely much longer than the initial one but also with the disagreeable property that is impossible to find out from it the whole initial message. On the contrary, the essential requisite of a good coding is the possibility of deducing from a sub-sequence of the coded message $x(\theta)$ the whole message of the source θ without waiting for a time indefinitely long. This is in principle impossible with a chaotic dynamics, since if we want to come back to the origin from a subset of the coded message $\{x_{\nu(\beta)}, x_{\nu(\beta+1)}, \ldots, x_{\nu(\beta+\lambda)}\}$ the sensibility of a chaotic dynamics to the initial conditions would carry us on another input message θ different from the initial one.

To profit by the richness of a chaotic behavior for using it as a transmission channel we need to "domesticate" it, that is we need to stabilize the system so to obtain a code that can be decoded. The characteristics of such a procedure would be essentially two:

- We have to be able to extract *any* unstable cycle of the dynamics;
- We have to be able to perform such an operation in a time that is linear with cycle length.

4.4 The Problem of the Characterization of Chaotic Signals: the Stabilization of Unstable Periodic Orbits

The general problem of the characterization of a chaotic dynamic and its attractor is related with the impossibility of a series expansion of simple (non fractal) elements.

The idea suggested by Auerbach et al [18] is to make possible a series expansion by approximating the dynamics through cycles ever more complex (i.e., with ever higher period). In this way, the characteristic quantities of the dynamics are calculated by expanding them on the values estimated on these long cycles. What is critical in this approach is the cycle extraction to characterize the dynamics. Auerbach and his collaborators suggested to search combinatorially for the recurrences within a given string of the dynamics values.

More exactly, let it be[12]

$$\dot{x} = f(x, \mu) \tag{37}$$

the equation of the dynamic system $f \in \mathbb{R}^N$ we are considering with μ a set of control parameters. If we integrate such a dynamics by a temporal step $dt \ll 1$, we obtain a time series $x(t)$. At this point, we can represent the dynamic system in an "embedding" space; so, by determining an embedding dimension m and a delay time τ, we are now able to represent the precedent string in a new form according to the method of time-delay [20]:

$$\mathbf{x}(t) = (x(t), x(t+\tau), \ldots, x(t + [m-1] \cdot \tau)) \tag{38}$$

Notice that there exists no universal method to determine the embedding dimension and the delay time. Especially the latter is almost arbitrarily chosen. Let us define, for sake of simplicity, $x_i = x(t_i) = x(t + i \cdot \tau)$. After its reconstruction, the series becomes $\mathbf{x}(t) = \mathbf{x_i} = \{x_k\}_{k=0,\ldots,m-1}$, $\mathbf{x_i} \in \mathbb{R}^m$. The basic

---

[12] The following part is related with continuous systems and it has not been developed directly by Auerbach et al but by Pawelzik et al. [19].

principle of Auerbach extraction method is strictly related with the evidence that the chaotic orbits are closures of the set of unstable periodic orbits [21]. Before all, the fixed points are individuated by searching for successive pairs of points $x_j x_{j+1}$ distant less than a fixed distance r. Once this search is completed on the whole temporal series, all the points that are close after an iteration are available. All these pairs are grouped into clusters of diameter R. Within these clusters the fixed points are estimated as mass centers of all the selected positions. Of course, R must be larger than r and, simultaneously, R must be sufficiently small for distinguishing among different periodic orbits, while r has to be sufficiently large for including different pairs of the series. At this point, however, the extraction of the *unstable* periodic orbit is completed. Other interesting information can be pull out from the multiplicity of the cycles, that is from the number representing how many times a given cycle appears in the chaotic series. This operation can be repeated for cycles of any order whenever we define a cycle of period n, that is, whenever we find a pair of points x_i and x_{i+n} such that:

$$\| x_i - x_{i+n} \| < r \tag{39}$$

where $\| x \|$ is the usual Euclidean metric. For an informational use of chaos, the main problem of such an approach is the requested time to identify a cycle. Time grows *exponentially* with the number of cycle points since we are constrained to identify the cycle by combinatorial techniques. We show in the next Section that this problem is solved by applying the theorem of Section (2.2) to the problem of the extraction of unstable orbits from a given chaotic dynamics.

4.5 A New Method to Extract Unstable Periodic Orbits from a Chaotic Dynamics

The Main Characteristics of Our Approach to a Chaotic Dynamics. In this Subsection we apply the principle of mutual redefinition between numbers and processes (see §2.2) to the case of the extraction of a unstable cycle from a chaotic dynamics. The more relevant novelty is that such a stabilization is obtained by a mutual redefinition of a delay time τ with respect to the fluctuations of the dynamics x. This allows us an amazing simplification of classical techniques of a chaotic dynamics characterization. That is:

- It allows the extraction of a cycle directly from a continuous chaotic temporal series *without using embedding coordinates*. In other terms, our method allows to avoid the necessity of estimating a priori both an embedding dimension and a delay time that are both arbitrary and time-consuming.
- This selection is obtained in a very short time, effectively it is obtained in real-time. Moreover, the CPU-time necessary to extract the cycles *grows linearly* with the length of the extracted cycle.
- The number of calculated points of the dynamics requested to extract a cycle of period n is *much shorter* than the global number of points of the same cycle (it is of the order of some per cent unit of them) [13] .

---

[13] More precisely, we have to distinguish two possibilities:

– The method distinguishes *automatically* among different cycles of the same period.

– Owing to the previous properties, it is possible to obtain by our method a *detailed and almost complete statistics of the unstable periodicities of whichever length* of a generic chaotic dynamics.

An Application to the Extraction of a Unstable Chaotic Orbit: Some Numerical Results. Let $f \in \mathbb{R}^n$ be the chaotic dynamic system of which we want to extract or to stabilize or to synchronize a cyclic point of a generic period T. Instead of "sampling" the system at fixed intervals dt, we sample it according to a time series defined as follows:

$$t_n = t_{n-1} + \tau_n$$
$$t_0 = 0 \tag{40}$$
$$\tau_0 = \varepsilon \ll 1$$

The new variable time step τ_n is defined as

$$\tau_{n+1} = \tau_n \cdot (1 + \Lambda)\,\theta(v) + \tau_{\min}\,[1 - \theta(v)]$$
$$\tau_{\min} = \text{const} \tag{41}$$

where

$$\lambda_i = \frac{1}{\tau_n} \frac{|\,\delta x_i(t_n)\,|}{|\,\delta x_i(t_n - \tau_n)\,|} \quad i = 1, \ldots, n \tag{42}$$

$$\Lambda = \max_i |\,\lambda_i\,| \tag{43}$$

$$v = \sum_i^N \theta(\lambda_i) \tag{44}$$

$$\theta(x > 0) = 1, \theta(x \le 0) = 0 \tag{45}$$

We easily verify that the new separation

$$\delta x_i = x_i(t_n) - x_i(t_{n-1}) \tag{46}$$

reduces itself below a pre-assigned value in a reliable way. That is, by this procedure the observer is synchronizing himself to the trajectory in the sense illustrated by L.M. Pecora and T.L. Carrol in a recent paper [23] without their

---

● The case in which we know the analytic form of the dynamic system f. In this case, the method grants the extraction of the cycle by generating *only* the points of the examined cycle (this demonstrates the linearity of the method: the requested time grows linearly with the length of the cycle, better, it is equal to the length itself!).

● If we do not know the analytic form of the dynamic system but we have only a data vector, our method grants the computational benefit that it is not necessary to compare the point of which we are searching the periodicity with *all* the other points but *only with a very restricted subset of them*.

constraint of negative sub-Lyapunov exponents. In fact, in our method, the ordered sequence t_n is no longer a regular sequence so that, if we plot the quantity τ_n versus t, we have an erratic signal (see Fig. 1).

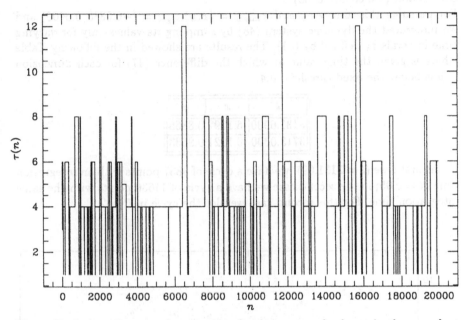

Fig. 1. Evolution of τ_n as a function of n for the same cycle shown in the precedent figures.

In view of an informational use of our general approach to chaotic systems, if we use instead of Eq. (46) the original form [24] of the redefinition function between space and dynamics for chaotic systems, we can develop our method as a technique of *extraction* of the *unstable* chaotic orbits:

1. We changed the form of the redefinition function between space and dynamics as follows:

$$\delta x_i = x_i(t_n) - x_{\text{target}} \tag{47}$$

   with $x_{\text{target}} = \text{const}$.
2. We chose casually a point on the attractor and we searched for which values of time t the difference (47) becomes less than a given value. In this way we found *if and where* the point x_{target} of the attractor is periodic.

We used for the simulations the classic Rössler 3D chaotic attractor:

$$\dot{x} = -y - z$$
$$\dot{y} = x + 0.2y \tag{48}$$
$$\dot{z} = 0.2 + z(x - 5.7)$$

We generated previously 10^5 points of the dynamics so that we can have a skeleton of points of the attractor to be used as x_{target} to search for the periodic points. Finally, we chose the $X - Y$ projection of the dynamics and then we chose to study the possible periodicities of the point defined for instance by the coordinates $(-6.0885, 0.5400)$.

For pursuing this aim, we posed $\mathbf{x}_{target} = (x, y)_{target} = (-6.0885, 0.5400)$ and we integrated the dynamic system (48) by sampling its values only for varying time intervals τ_n defined by (41). The results are showed in the following Table where is given the time value at which the difference (47) for each dimension x, y is below the fixed threshold 0.4.

t	x	y
3131.6900	-6.1292	0.8865
3713.0600	-6.1921	0.8633

In the first case ($t = 3131.69$), we have a cycle of 3837 points with an integration time $dt = 0.05$; in the second case we have a cycle of 11630 points with the same integration time. For this second case we plot the cycle in Fig. 2.

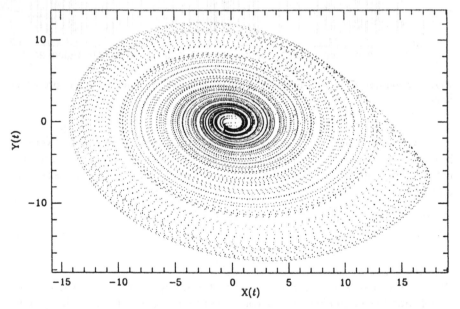

Fig. 2. Plotting of the 11630 points cycle for the Röessler 3D attractor.

The relevant feature of our method is that this cycle has been found by the algorithm *in only 555 time steps* plotted in Fig. 3. So a compression factor of $\approx 95.3\%$ with respect to the number of points of the cycle has been obtained. Notice that in the second case the extracted cycle was of order 80. This means that the dynamics performs 80 orbits before passing newly through the same point.

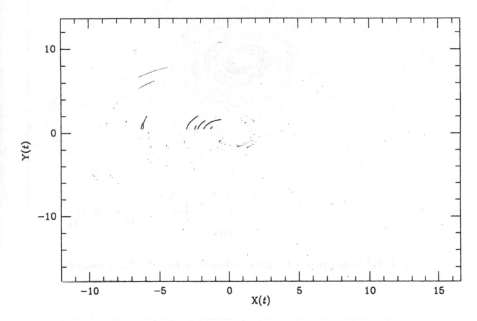

Fig. 3. Plotting of the 555 points necessary to find the cycle of the previous figure.

The Histogram of the Unstable Periodic Orbits within a Chaotic Attractor Owing to the amazing simplicity of our method to extract a so long cycle, we tried to perform an *exhaustive analysis* of the structure of unstable orbits in the Röessler 3D attractor we are studying. So, we searched for *all the possible cycles* passing through 10^4 points of this attractor (see Fig. 4). In this way, we obtained the histogram of Fig. 5 in which the number of cycles as a function of their periods are reported. In fact, we performed a *series expansion of the unstable cycles* of that given chaotic dynamics.

A similar analysis has been developed also for the Lorenz attractor. The results are shown in Figs. 6.

In fact, the method we have just discussed still undergoes the limit that some periodicities are not found. Essentially, this problem depends on the fact that sometimes the Eq. (41) produces some variations of τ_n too strong so that the time "jump" results to be too large. We can observe this phenomenon in the Fig. 7. In this figure we show the temporal evolution of x (up) and of y (down) for the Lorenz attractor:

$$\dot{x} = -\sigma \cdot (x - y)$$

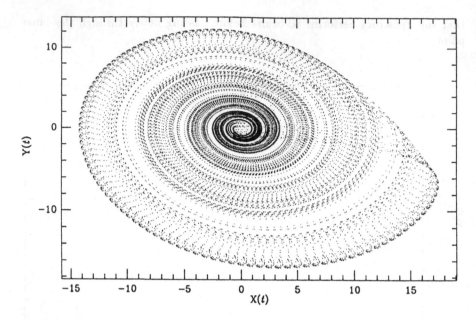

Fig. 4. 10^4 target points for Eq.47 taken from Röessler 3D attractor.

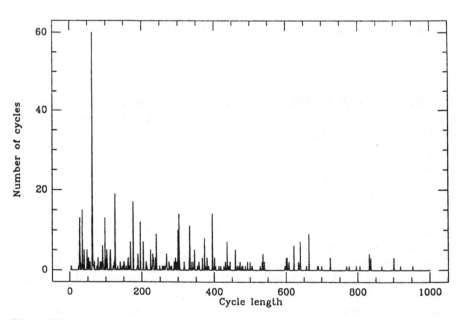

Fig. 5. Histogram of the number of cycles as a function of their length for all the 3500 cycles that have been found.

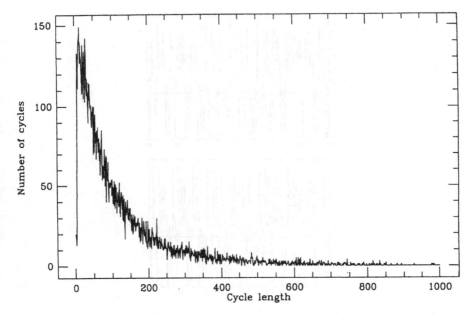

Fig. 6. An histogram of the number of periodicities as a function of the length of their period for the Lorenz attractor.

$$\dot{y} = -xz + Rx - y$$
$$\dot{z} = xy - bz$$

with $\sigma = 10, b = 8/3, R = 28$.

We are studying an optimization of our method to avoid this limitation and to obtain the complete histogram.

4.6 A Short Comparison with the Multi-Step Method and the Extrapolation Method of Integration

In view of applying the general principle of mutual redefinition between functions and their arguments to the definition of a computational scheme for the *integration* of ordinary differential equations also of the stiff-type, let us sketch a short comparison between our method of redefinition of the calculation step as a function of the quantity calculated by the step itself and the standard methods used till now for the integration of ordinary differential equations. This comparison could help also to enhance ever better the peculiar characteristic of our idea with respect to more classical approaches. It is well-known in literature [35, 39] the standard *multi-step* method for the integration of ordinary differential equations. This method, to calculate an integral I, calculates successive approximations I_n. Each of them corresponds to an integration step ε_n halved with respect to the precedent one: $\varepsilon_{n+1} = \varepsilon_n/2$. The procedure terminates iff the

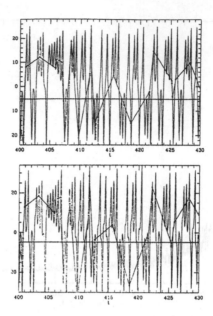

Fig. 7. A typical situation in which a too large variation of τ implies the loss of a searched periodic point.

difference $|I_{n+1} - I_n|$ becomes less than a pre-assigned limit. There exist several improvements of this method among which the more advanced [36] combines it with extrapolation methods through rational functions in order to find the value of ε_n for which the difference $|I_{n+1} - I_n|$ is null.

The main difference between this method and our method concerns evidently the evolution of the ε_n step. In fact, both the analytical demonstrations and the practical implementations of these methods combining multi-step and extrapolation techniques, imply that the ε_n series so generated be *monotonically decreasing*: $\varepsilon_n < \varepsilon_{n-1} < \ldots < \varepsilon_1$. On the contrary, in our method, *a nonmonotone evolution* of ε_n is not only possible, but necessary because we linked the definition of ε and of f (see Eq. 41, where instead of ε we have τ_n and instead of f we have Λ). In this connection, see Fig. 1 where we show the evolution in time of the variable step τ_n. Its irregular behavior is evident.

Of course the computational advantage of our method depends precisely on this irregular behavior: instead of exploring function intervals absolutely useless in the calculus, we allow the computation step to jump *only* through zones of the functions that are significant for the calculus we are developing. In this connection, the example showed in the Fig. 8 is eloquent to illustrate the effectiveness of our method. The continuous line connects the points singled out by the variable step τ_n. It is patent that by our method the computation can avoid zones of the function in which ranges containing the searched values (denoted in the figure by full dots) are absent.

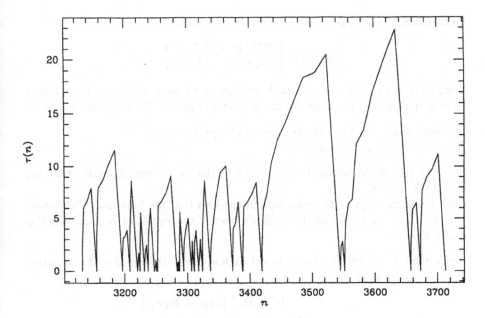

Fig. 8. Non-monotone evolution of τ_n to avoid unuseful computations.

5 An Application to Parallel Computation: Dynamic Definition of Net Topology for Parallel Architectures

5.1 The Problem: Parallel Computation in the "Geometrical" Perceptron.

A further application of the same general idea of mutual redefinition between functions and their arguments concerns a class of topological problems that M.Minsky and S.Papert demonstrated at the end of 60's to be not computable for parallel architectures, in spite of their simple formulation and solution for Turing-Von Neumann sequential architectures [32]. Let us consider a classical perceptron computing on a lattice R a linear function $\Psi(X)$ of a generic *input* $X \subset R$. The problem we want to discuss is whether there exists a predicate Ψ able to assume the input characteristic function, given a fixed topology according to which the perceptron "reads" the input. In other terms, given some partition $\{A_i\}$ of the retina R, our aim is to ascertain whether there exists a predicate Ψ able to decide whether a generic input X has at least one point in each cell A_i. Formally,

Definition 13 *Let the retina R be an arbitrary set of points (it can be also finite).*

Definition 14 *We define as input X a generic subset of R: $X \subset R$.*

We indicate by bold letters the *collections* of subsets of R

Definition 15 *Let $\varphi_{\mathbf{F}}(X)$ be a predicate or a mask:*

$$\varphi_{\mathbf{F}}(X) = \begin{cases} TRUE & iff \ X \subset \mathbf{F} \\ FALSE & iff \ X \not\subset \mathbf{F} \end{cases}$$

Effectively, a predicate is a variable statement whose truth ($\varphi = 1$) or falsity ($\varphi = 0$) depends on the choice of the geometric figure X drawn on the retina R.

Definition 16 *Let $S(\varphi)$ be the support of a predicate φ:*
$S(\varphi) \mid \forall X \subset R \ \ \varphi(X) = \varphi(X \cap S)$

Where $X \cap S$ is the intersection of X and S, that is, the set of points belonging both to X and S.
This definition helps us to make explicit the idea that the function φ "depends" only on a certain subset of R. So, $S(\varphi)$ denotes the subset of R upon which φ *really depends*.

Definition 17 *Let Ψ be a linear threshold function with respect to the collection Φ of predicates φ in R:*

$$\exists \theta \in \Re, \exists \{\alpha(\varphi) \in \Re\} \forall \varphi \mid$$
$$\Psi(X) = \lceil \sum_{\varphi \in \Phi} \alpha(\varphi) \, \varphi(X) > \theta \rceil = L(\Phi) \tag{49}$$
$$\text{with } \lceil x \geq 0 \rceil = +1, \ \ \lceil x < 0 \rceil = 0$$

Obviously, the predicates of which support is very small are too *local* to be interesting. The primary interest is in fact for those predicates of which supports are the whole retina R, but that can be defined as linear threshold combinations of predicates with much smaller supports. To quantify such an exigency, let us introduce the notion of *order k* of a predicate. This notion indicates that no one of the predicates composing the linear threshold function has a support containing more than k points. Formally:

Definition 18 *We define as order of a predicate the smallest integer number k for which there exists a set Φ of predicates φ satisfying Ψ :*

$$\begin{cases} \mid S(\varphi) \mid \leq k, \ \forall \varphi \in \Phi \\ \Psi(X) = L(\Phi) \end{cases} \tag{50}$$

where $\mid S(\varphi) \mid$ is the number of points in $S(\varphi)$.

Definition 19 *We define two predicates φ and φ' as equivalent with respect to the group G, $\varphi \overset{G}{=} \varphi'$, if there exists a member g of G such that $\varphi(gX)$ and $\varphi'(X)$ are the same for each X.*

Definition 20 *We define as equivalence class of predicates the set Φ of predicates φ equivalent with respect to the group G.*

We can thus formulate without ambiguity the following theorem " $\Psi_{One-in-a-box}$ " [32]

Theorem 5 *Let $A_1, ..., A_m$ be disjoint subsets of R and define the predicate:*

$$\Psi(X) = \left[\left| X \bigcap A_i \right| > 0, \forall A_i \right] \tag{51}$$

that is, there is at least one point of X in each A_i. If for all i, $|A_i| = 4m^2$, then the order of Ψ is $\geq m$.

Corollary 1 *If $R = A_1 \bigcup A_2 \bigcup ... \bigcup A_m$ the order of Ψ is at least $\left(\frac{1}{4}|R|\right)^{1/3}$.*

5.2 A Possible Solution: the "Dynamic" Perceptron

An Intuitive Presentation of the Proposed Solution Our suggestion is that it is possible to maintain the polynomial associated with $N_j(X)$ within the limit m imposed by the topology of the single partition A_i, on condition that the topology given by the set $\mathbf{F}$ of the different masks φ (see Definition 15) is allowed to modify itself with respect to their input A_i. Let us consider the linear expression defining in a general form whichever predicate Ψ:

$$\Psi(X) = \left[\sum_{\varphi \in \Phi} \alpha(\varphi)\varphi(X) \right] \tag{52}$$

the main problem is the *a priori* evaluation of the summation with respect to the input and/or with respect to the predicate we are actually considering. Group theory notions are very appropriate in this context because they make invariant the masks φ under the actions of group elements. Following this idea, we can simplify the precedent summation by grouping the masks according to *equivalence class* Φ_j. I.e., the class composed by all the predicates invariant with respect to a given group G. In this way, the summation is simplified because we can demonstrate [32] that masks belonging to the same equivalence class have the same weights. So, we can rewrite the predicate Ψ in the simpler form:

$$\Psi(X) = \left[\sum_j \alpha_j \sum_{\varphi \in \Phi_j} \varphi(X) \right] \tag{53}$$

At this point let us evaluate the number $N_j(X)$ of the masks (characteristic functions) $\varphi(X)$ satisfied by the generic input X for each equivalence class. This term grows up combinatorially with the number of the input points $|X|$ and with the number of the mask points $|S(\varphi)|$, $\varphi \in \Phi_j$:

$$\binom{|X|}{|S(\varphi)|} = \frac{|X| \cdot (|X| - 1) \, ... \, \cdot (|X| - |S(\varphi)| + 1)}{|S(\varphi)|!} \tag{54}$$

This growth represents an insuperable difficulty because it requires a mask with a support as large as the retina R. Of course, this destroys in principle the perceptron capacity of parallel computation because a calculation unit able "to see" the *whole* input becomes necessary.

This limitation occurs every time we want to work mathematically with sets in which all the elements are completely specified. According to us, a possible solution is to substitute such a definition of a set via the previous complete specification of all its elements (in this case, all the masks $\varphi \in \Phi_j$) with a dynamic procedure of *mutual element-specification/set-definition*. In this way, starting from the initial specification of one element, we can construct progressively the definition of the whole set. Or, better, we can always reach a set definition as general as we *effectively* need to perform the calculation.

In the next paragraphs we offer some formal demonstrations of this procedure in some cases that result to be irresolvable for the classical "geometric" perceptron.

The Dynamic Perceptron and the Solution of "Ψ-one-in-a-box" Problem Let us consider the combinatory term providing us with the number of masks satisfied by the generic input X:

$$N_j(X) = |\{\varphi \mid \varphi \in \Phi_j \text{ AND } S(\varphi) \subset X\}| \qquad (55)$$

Minsky and Papert themselves stress that the evaluation of this term would become significantly easier if an appropriate pre- processing was available. Nevertheless, the authors abandon the study of such a solution because in their cognitive scheme this would imply an a- priori knowledge of the whole input.

On the contrary, we can obtain the same result without any a priori knowledge. Indeed, if we consider only the density and the distribution of the input points, it is sufficient to vary the support S of each mask φ according to the density variation of the input X "read" by the mask itself in two successive instants with two different topologies. Formally [14]:

$$|\Delta S_X(n)| \propto \varphi_{S_X(n)} - \varphi_{S_X(n-1)} \qquad (56)$$

This process stops if we have the *right* topology in S for the input X; in this case the equation for ΔS reads:

$$\varphi_{S_X(n)}(X) = \varphi_{S_X(n-1)}(X) \Leftrightarrow |\Delta S_X(n)| = 0 \qquad (57)$$

We give now a list of formal definitions to make rigorous such a procedure. We premise, however, that from now on we consider only *square geometries*. In fact, it is easy to obtain extensions for every type of geometry by the simple substitution of the *linear* unit of Δl with a *curvilinear* unit.

From now on we consider a *bidimensional* retina $R = \{x, y\}$. More exactly,

Definition 21 *R is a bidimensional lattice $\{x, y\}$ with step p.*

Definition 22 $X \subset R$

Let us consider a generic element i of the input set X corresponding to the pair $\{x_i, y_i\} \in R: i \in X$. For each point i of the set X we construct the sets $S_X^i \subset R$ defined as follows:

---

[14] For the rigorous definition of $\Delta S_X(n)$ see below the Definition 25.

Definition 23

$$S_X^i(0) = \emptyset$$
$$S_X^i(1) = \text{square of side } 2p \text{ centered at } i \qquad (58)$$
$$S_X^i(n+1) = S_X^i(n) + \Delta S_X^i(n), \ n > 1$$

Definition 24 $\varphi_{S_X^i(n)}(X) = \text{number of } X \text{ points in } S_X^i(n)$.

Definition 25

$$\Delta S_X^i(n) = \begin{cases} \emptyset \text{ if } \varphi_{S_X^i(n)}(X) - \varphi_{S_X^i(n-1)}(X) = 0 \\ \text{the set constructed by subtracting to the square centered in } i \text{ and of side } 2np \\ \text{the square centered in } i \text{ of side } 2(n-1)p \\ \text{if } \varphi_{S_X^i(n)}(X) - \varphi_{S_X^i(n-1)}(X) > 0 \end{cases}$$

$$(59)$$

Definition 26

$$n_{max} = n : \Delta S_X^i(n) = 0 \ \forall i \qquad (60)$$

We have thus all the necessary definitions to enunciate the following.

Theorem 6 *Let* $A_1, A_2, ..., A_m$ *be disjoint square subsets of* R *and let us define the predicate:*

$$\Psi^D(X) = \left[\left| X \cap A_i \right| > 0 \text{ for any } A_i \right] \qquad (61)$$

that is, there is at least a point of X *in each* A_i. *We indicated by* $\Psi^D =$
$$\left[\sum_{\varphi_{S_{A_i}^j(n)}} \alpha(\varphi_{S_{A_i}^j(n)}) \varphi_{S_{A_i}^j(n)}(X) \right]$$ *the "dynamic" perceptron where the masks* $\varphi_{S_{A_i}^j(n)}(X)$ *are determined following the definitions (23-25). Hence, if for all* i $| A_i |= 4m^2$, *then the order of* Ψ^D *is* $4m^2$.

Proof. Let $A_1, \ldots, A_m$ a set of disjoint square sets $A_i \cap A_j = \emptyset \ \forall i \neq j \ i = 1, \ldots, m$
The "dynamic" equations (23-25) rewritten by considering as their input set the sets A_i are:

$$S_{A_i}^k(0) = \emptyset$$
$$S_{A_i}^k(1) = \text{square of side } 2p \text{ centered at } i \qquad (62)$$
$$S_{A_i}^k(n+1) = S_{A_i}^k(n) + \Delta S_{A_i}^k(n), \ n > 1$$

$$\varphi_{S_{A_i}^k(n)}(A_i) = \text{number of } A_i \text{ points in } S_{A_i}^k(n) \qquad (63)$$

$$\Delta S_{A_i}^k(n) = \begin{cases} \emptyset \text{ if } \varphi_{S_{A_i}^k(n)}(A_i) - \varphi_{S_{A_i}^k(n-1)}(A_i) = 0 \\ \text{the set constructed by subtracting to the square centered in } i \\ \text{and of side } 2np \text{ the square centered in } i \\ \text{of side } 2(n-1)p \text{ if } \varphi_{S_{A_i}^k(n)}(A_i) - \varphi_{S_{A_i}^k(n-1)}(A_i) > 0 \end{cases}$$

$$(64)$$

From it, the following holds:

$$\min_{k \in A_i} \left| S_{A_i}^k(n_{max}) \right| = |A_i| = (2m)^2 = L \tag{65}$$

because $|A_i| = 4m^2 \ \forall i$.

$$\max_{k \in A_i} \left| S_{A_i}^k(n_{max}) \right| = 4 \cdot |A_i| = (4m)^2 = M \tag{66}$$

The demonstration of precedent statements is immediate.
It is evident that: $\forall A_i \ \exists 1! \ S_{A_i}^k(n_{max}) = S_i(n_{max}) : |S_i(n_{max})| = L$.
Let us define $p_s(X) = \lceil |X \cap S| > 0 \rceil$, posing that $p_i(X) = p_{S_i(n_{max})}(X)$.
Let P be the equivalence class defined as follows:

$$P = \{p_i, i = 1, \ldots, m\} \tag{67}$$

Let us develop the predicate Ψ^D in its linear form (see Definition 17) as follows:

$$\Psi^D = \left[\sum_{p \in P} \alpha(p) \cdot p(X) \right] = \left[\sum_{i=1}^m \alpha(p_i) \cdot p_i(X) \right] \tag{68}$$

The order (see Definition 18) of the predicate Ψ^D is:

$$\text{ord}\left(\Psi^D \right) = \max_i \left(|S_i(n_{max})| \right) = L = (2m)^2 \tag{69}$$

□

A Comparison with Multi-Layer Architectures The results till now obtained lighten also the classical problem of the comparison between single-layer and multi-layer architectures. Indeed it is well-known in literature [42] that the problems of *unbounded order*, e.g., $\Psi_{\text{One-in-a-box}}$, can be easily solved by multi-layer architectures in which the predicate φ is on its turn a linear threshold function:

$$\varphi_i = \left[\sum_j \beta_{ij} x_j > \theta_i \right] \tag{70}$$

so that, the function Ψ becomes:

$$\Psi = \left[\sum_i \alpha_i \left[\sum_j \beta_{ij} x_j > \theta_i \right] > \theta \right] \tag{71}$$

For instance, let us consider the problem of Ψ_{Parity}. It is possible to demonstrate that it can be computed by a simple muti-layer architecture. The trick is simply of determining a mask $\varphi_{(n)}$ that counts the number of points n of its support S and that gives as output 0 or 1, respectively if this number is or is not exceeded by the number of points $|X|$ of its input X. We are thus defining a predicate practically for each possible number n of points of the input. Obviously, if we can define an indefinite number of these predicates, any input X is allowed to have a number $|X|$ of points $x_i \in \{0,1\}$ that is recognized by some of these masks. Formally:

$$\varphi_{(n)}(X) = \lceil |X| > n \rceil = \left[\sum x_i > n \right] \tag{72}$$

Moreover, let us define (Ψ is defined below):

$$\alpha_0 = \Psi(0)$$
$$\alpha_1 = \Psi(1) - \alpha_0$$
$$\vdots$$
$$\alpha_{n+1} = \Psi(n+1) - \sum_{i=0}^{n} \alpha_i$$

Then, Ψ_{Parity} can be re-written in the form:

$$\Psi(X) = \left[\sum_{i=1}^{|R|} \alpha_i \cdot \varphi_{(i)} > 0 \right] \tag{73}$$

where all the masks $\varphi_{(i)}$ have an order $|S(\varphi_{(i)})| \leq n$ bounded by n. It seems then that the multi- layer architectures can solve any problem posed by the single-layer. The problem is that the requested number of predicates $\varphi_{(i)}$ grows up with the dimension of R. Indeed, in the summation of the Equation (73) we count $|R|$ predicates. Practically, by a multi-layer architecture we simply displaced the unbounded growth from the *order* of the single predicate in the single-layer architecture to the *number* of these predicates in the multi-layer architecture.
In the multi-layer architectures it would be possible to recognize any pattern F simply by defining the single predicates φ as follows:

$$\varphi_F = \left[\sum_{x_i \in F} x_i - \sum_{x_i \notin F} x_i \geq |F| \right] \tag{74}$$

In this way a *generic* class $\mathbf{F}$ of figures F could be recognized by:

$$\Psi_{\mathbf{F}} = \left[\sum_{F \in \mathbf{F}} \varphi_F > 0 \right] \tag{75}$$

but this formula is trivial because it means simply that the class $\mathbf{F}$ has a characteristic function $\Psi_{\mathbf{F}}$ that can be expressed if we define *as many predicates φ_F as the figures F of the class $\mathbf{F}$ are.* Obviously, such a number grows too fast for any realistic implementation.
In this context, the philosophy of our *single-layer but dynamic* architecture (see the Definitions 23- 25) begins to reveal itself as a winning strategy. Indeed, also our architecture defines a new set of predicates $\varphi_{S'_X(n_{max})}(X)$ for each input

X. However, these predicates are *dynamic*. That is, because they depend on a dynamics that is a function of the input X, they *do not need* the simultaneous definition of all the predicates as, on the contrary, it is requested in a multi-layer architecture. In other terms, instead of defining all the characteristic functions $\varphi_{(i)}$ as it is the case of the multi-layer architecture, it seems more advantageous to use the result[15] of the dynamics of mutual redefinition between the characteristic functions $\varphi_{S_X^i(n)}(X)$ and their supports $S_X^i(n)$, driven by the individual input X according to the prescriptions of the Definitions 23- 25. In this case, indeed, we do not consider any longer all together the elements of the generic class $\mathbf{F}$, but only those *actually present* as inputs of the net. Notwithstanding this drastic limitation of predicates, the occurrence of a new figure $F \in \mathbf{F}$ is equally recognized by the net within the class $\mathbf{F}$ because the modalities of the dynamic redefinition of the predicates $\varphi_{S_F^i(n)}(F)$ and of their supports $S_F^i(n)$ with respect to the new figure F result to be the same of the other figures considered till now. This holds, as long as all these figures, e.g. F_i and F_j, belonging to the same class $\mathbf{F}$ result to be linked by a generic reversible transformation $T : F_j \underset{T^{-1}}{\overset{T}{\rightleftharpoons}} F_i$ defined on the retina R^{16}. Effectively, in this case we can suppose that the dynamic rule for the mutual redefinition between supports S_X^i and predicates $\varphi_{S_X^i}$ (see Eq. (56) and Definition 25), is of the form:

$$| \Delta S_X^i(n) | \propto f(\Delta \varphi(n)) = f(\varphi_{S_X^i(n)} - \varphi_{S_X^i(n-1)}) \tag{76}$$

where $f : \mathbb{R}^+ \longrightarrow \mathbb{R}^+$ is a generic function defined on the positive real numbers and expressing the correspondence between the figures F_i and F_j at the level of the supports and of the predicates describing them. The limit of the convergence of the dynamics, however, does not change under the action of f since this obvious relation holds:

$$\varphi_{S_X^i(n)} - \varphi_{S_X^i(n-1)} > 0 \iff f\left(\varphi_{S_X^i(n)} - \varphi_{S_X^i(n-1)}\right) > 0 \tag{77}$$

For this reason, the form of the Definition 25 does not change under the action of f. This means that the supports $S_X^i(n_{max})$ of the functions $\varphi_{S_X^i(n_{max})}$ at the end of the dynamics for the two inputs F_i ed F_j of the same class $\mathbf{F}$ are comparable, so that their classification by the weights α becomes possible. This result is made more evident by the application discussed in the next paragraph.

Another complementary strategy to solve the problem of the unbounded order of single-layer architectures and/or of the unbounded number of predicates of multi-layer architectures is pursued by the "new" *connectionist* architectures and particularly by the so-called *back-propagation* algorithm. Effectively, this algorithm can give good results on *real problems* because generally they can be simpler than the ideal case discussed in the perceptron theorems. Nevertheless,

---

[15] For instance, the support $S_X^i(n_{max})$ of the dynamic predicate $\varphi_{S_X^i(n_{max})}(X)$.

[16] For instance, the two figures are coincident if rotated, translated, etc.

it is well-known that the convergence time of learning dynamics of the back-propagation algorithm can be often excessively long. It seems thus that the use of non- linear transfer functions in the learning dynamics, if sometimes can result useful in practice, generally tends to transfer the computational problem from the "space" domain of the unbounded number of the predicates, to the "temporal" domain of the necessary convergence time. This depends on the fact that the use of a non-linear term in the learning dynamics does not solve formally the computational root of the problems here discussed.

However, a formal demonstration of this hypothesis would require a more deep and attentive discussion because it ultimately involves the problem of the formal demonstration of the so-called *NP completeness* theorem in computability theory. Nevertheless we outline in the final Section of this work the main lines that we intend to follow in the next future for dealing with this essential question from the standpoint of our "dynamic" approach to computability theory. For the moment it is sufficient to demonstrate experimentally in the next Section that our algorithm, in dealing with a recognition task characterized by a high complexity of the input patterns (simultaneous presence of many scales in the pattern space), does not suffer the temporal limitations of the non-linear architectures just as it does not suffer the formal limitations of the linear architectures before discussed.

5.3 An Application to High-Energy Particle Recognition

Introduction We applied the precedent theorems to solve a problem common to the great majority of the experiments in high-energy physics: the automatic discrimination/recognition of the produced events. Several works have proposed to use also neural net architectures for this aim [40]. Generally, the best results do not exceed the 70%-85% of successful discrimination/recognition on Monte-Carlo simulations of real data.

A fundamental research field in high-energy physics is the understanding of the nucleon structure. This aim can be achieved through the study of nucleon electromagnetic form factors. In the time like region ($q^2 > 0$) the nucleon (N) form factors can be measured through the "branching ratio" of the annihilation process:

$$e^+ e^- \longrightarrow N \bar{N} \tag{78}$$

The differential cross-section of this reaction in the "one photon exchange" approximation, is given by:

$$\left(\frac{d\sigma}{d\Omega} \right)_{c.m.} = \frac{\alpha^2 \beta}{4s} \left[|G_M|^2 \left(1 + cos^2\theta \right) + \frac{4M^2}{s} |G_B|^2 sin^2\theta \right] \tag{79}$$

where: $s = q^2$ is the square of the energy in the center of mass, θ is the angle between the electron beam and the nucleon direction, β and M are respectively the velocity and the mass of the nucleon and α is the fine structure constant.

The actual experimental knowledge of the nucleon electromagnetic form factors is not very satisfactory. The proton form factors have been extensively investigated by high statistics experiments in the space-like and, recently, time-like region[17]. On the contrary neutron data are still poor in the space-like region and before the measurements recently performed at ADONE e^+e^- storage ring (Frascati, Rome), there was no experimental information about the neutron electromagnetic structure in the time-like region.

Perturbative QCD (PQCD), Extended Vector Meson Dominance Models (EVDM) and other theoretical models give very different predictions on the neutron form factors in the time like region. A suitable quantity to compare different models is the following ratio of the cross-sections σ:

$$r = \frac{\sigma\left(e^+e^- \longrightarrow n\bar{n}\right)}{\sigma\left(e^+e^- \longrightarrow p\bar{p}\right)} \tag{80}$$

According to QCD sum rules, a value of $r = 0.25$ is expected at high q^2. EVDM-inspired theories predict $r \simeq 2 \div 100$ because of the not well established location of the vector meson recurrences.

Clearly, both the theoretical and the experimental state of art claim for data about $e^+e^- \longrightarrow n\bar{n}$ in view of a better understanding of the nucleon structure. The FENICE[18] experiment, at the upgraded ADONE e^+e^- storage ring, has been especially designed for the measurement of the cross-section of annihilation processes (78) and hence for the determination of the neutron form factors in the time-like region. The FENICE experimental apparatus is described in details elsewhere [41]. The experiment is producing every month from 1 to 10 millions of events and globally it will produce 300 millions. At first level triggering, only 30 millions of them will survive. Approximately, a sample of $\sim 10^5$ must be analyzed by humans.

To characterize some of the main categories of events to be distinguished, events such as neutron-antineutron ($e^+e^- \longrightarrow n\bar{n}$), proton-antiproton ($e^+e^- \longrightarrow p\bar{p}$) and multi-hadron productions belong to the non-collinear events category. Events such as Bhabha scattering ($e^+e^- \longrightarrow e^+e^-$), gamma-gamma and muons anti-muons $e^+e^- \longrightarrow \mu^+\mu^-$ productions belong to the collinear events category. The so-called "machine-noise" events constitute a further numerous family. These events are the result of the following interactions: $e^+e^- \longrightarrow p\bar{p}\eta$, $e^+e^- \longrightarrow p\bar{p}\pi^0$, $e^+e^- \longrightarrow \pi\pi$, KK, $\pi\pi\pi^0$.

In our specific application, we began by distinguishing between muons and Bhabha scattering. This task is difficult because of the topology similarities in the two classes of events. The only difference is the presence of a limited swarming at the extremities of Bhabha tracks in opposition with the almost perfect collinear character of muon tracks. On the other hand, the elimination of muon tracks is really essential because, despite of the presence of veto-revelators, they effectively constitute till the 70% of the produced events. Indeed the μ particles, often originating from cosmic rays, are able to go through the revelation chamber practically with any possible bearing. In this way, they switch an absolutely

---

[17] Experiment PS-170 at LEAR and experiment E760 at Fermilab.

[18] Fattori Elettromagnetici del Neutrone In Collisioni Elettrone-positrone

casual number of revelators and/or interact with little swarming effects. We give in Fig. 9 some examples of such an occurrence.

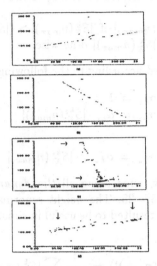

Fig. 9. Typical μ and *bhabha* (e^+e^-) tracks going through FENICE revelation chamber. Bhabha events can be recognized for their characteristic swarming effect at the extremities (see arrows in the figure).

Our task was thus double. Firstly, it was necessary to discriminate the Bhabha events from the muon events, because only the former are relevant for the physics of the process. Secondly, it was necessary to recognize and to filter cosmic events with strange topologies. The discrimination difficulty consists in the impossibility of using as parameters both the point number and the linear regressions on a straight line, because the point number is varying for each track so that it is not class discriminating.

5.4 The Network Structure.

Following the main ideas presented in the last Subsection and discussed elsewhere [31], we designed a discrete state $\{0,1\}$ net of which input is the acquisition chamber bidimensional XY image. This image consists of 300×300 points, each with two possible states: $\{0,1\}$. The only available information is thus the *topological* information. No further physical information is used such as released energy, fly time, ADC, etc. The net after having received its input, evolves its topology according to the "dynamic" perceptron scheme of Definitions 23 - 25. When it reaches a fixed topology, the simple summation of the "geometric" perceptron is calculated according to the following:

$$\sum_k \alpha_k \left(\varphi^k_{S_{X(n_{max})}} \right) \cdot \varphi^k_{S_{X(n_{max})}}(X) > \theta \tag{81}$$

where the $\varphi^k_{S_X(n_{max})}$ are determined by the "dynamic" perceptron and the weights $\alpha_k \left(\varphi^k_{S_X(n_{max})} \right)$ are defined according to the following:

$$\alpha_k = \begin{cases} |S^k_X(n_{max})| & \text{if } |S^k_X(n_{max})| > \langle |S_X(n_{max})| \rangle \\ \alpha_{k_n}|S^k_X(n_{max})| & \text{otherwise} \end{cases} \qquad (82)$$

The term α_{k_n} is determined according to the following succession:

$$\begin{aligned} \alpha_{k_1} &= 1 \\ \alpha_{k_2} &= \alpha_{k_1} \cdot |S^{k_2}_X(n_{max})| \\ &\vdots \\ \alpha_{k_n} &= \alpha_{k_{n-1}} \cdot |S^{k_n}_X(n_{max})| \end{aligned} \qquad (83)$$

where the double index k_i runs along all the *dynamic* masks satisfying the condition of Definitions 23 - 25. The weight renormalization by the factorial term $(\text{MAX}_k|S^k_X(n_{max})|)!$ resulted to be useful in simulations. The average term $\langle |S^k_X(n_{max})| \rangle$ is given by:

$$\langle |S^k_X(n_{max})| \rangle = \frac{1}{N_X} \sum_k |S^k_X(n_{max})| \qquad (84)$$

with N_X as a suitable renormalization constant. This particular strategy of weight definition allows the net to enhance input zones characterized by higher densities (scattering) typical of Bhabha events so that the net is successful in discriminating these tracks from μ collinear tracks.

It is important to notice that two generic events X_1 and X_2 belonging to the same class, e.g., μ, are identified *by two different sets of weights* α_{X_1} e α_{X_2}. This derives directly by the proposed dynamics (83) for the computation of the weights α_k. Because they depend on the individual input X[19], all this suggests to us a procedure of *dynamic* definition also for the weights α that is similar to the procedure described in the Definitions 23 - 25.

This point requires a further inquiry and it is not developed here.

Notwithstanding the algorithm simplicity, we reached the encouraging results shown in Table I on a sample of 3860 *real* particle tracks (2784 Bhabha and 1076 μ). The plotting of the net non renormalized output (i.e., $\sum_k \alpha_k \left(\varphi^k_{S_X(n_{max})} \right) \cdot \varphi^k_{S_X(n_{max})}(X) > \theta$ without the Heaviside function $\lceil x \rceil$) is given in Fig. 10. In Fig. 11 we show the plotting of the number of classified particles as function of the discrimination threshold θ.

Obviously we chose for θ the value of 1230 because it results to be the best compromise for the discrimination of the two particle classes. The net algorithm is written in F77 FORTRAN language. With the weight dynamics included, it requires approximately 5 minutes of CPU time on a 4 MIPS (*Peak* value) machine

---

[19] Notice that such a relationship between the net weights and the predicates is classical. Indeed, it corresponds to the original learning rule of the classical perceptron: see [32] ch.11.

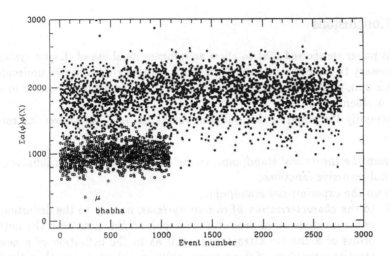

Fig. 10. Plotting of the net non-renormalized output (see text) for μ and bhabha tracks discrimination.

to recognize the above described set of 3860 real (not Monte-Carlo) particle tracks. This means that the algorithm is able to recognize the events *in real time by software*, because the event acquisition rate of the FENICE apparatus is approximately 10 Hz.

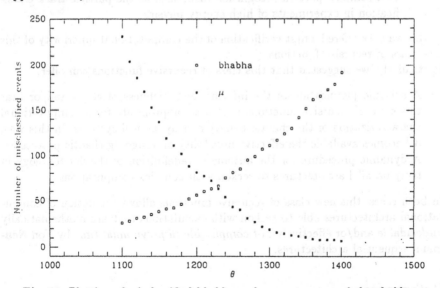

Fig. 11. Plotting of misclassified bhabha and μ events versus θ threshold.

6 Conclusions

In this paper we showed that to allow an informational use of chaotic systems it is necessary to solve mathematical problems generally classified as undecidable. For this aim, we discussed a new class of *mutual recursive* functions able to avoid these problems.

To exemplify the power of this new computational tool, we showed its applicability:

- *From the theoretical standpoint*, to the diagonalization of an infinite list of total recursive functions;
- *From the experimental standpoint*,
  1. to the *characterization of chaotic systems*, namely to the definition of a new synchronization procedure in a polynomial time onto the unstable orbits of a chaotic attractor, as well as to the definition of a new extraction procedure of these same orbits in a shorter time than the same period of the orbit and without arbitrary definitions of any embedding space;
  2. to the *solution of some formal problems of parallel calculus*, namely to the design of a new "dynamic" architecture of perceptron able to solve both some formal problems of the classical linear perceptron and some convergence problems of the multilayer non-linear perceptron (i.e., back-propagation) in pattern recognition. The effectiveness of this new "dynamic" architecture has been demonstrated by its successful application to a complex pattern recognition task, such as the particle track classification in experiments of high energy physics.

In this way, we offered a first verification of the computational superiority of this new class of recursive functions.

Particularly, we suggested that this class of recursive functions can offer:

- A suitable justification of the informational richness of chaos and/or that this class of recursive functions is at least appropriate for a computational characterization of the chaotic behavior in dynamical systems. In this way, it becomes available the effective possibility of designing chaotic processors.
- A dynamic procedure for the automatic redefinition of the net topology in truly parallel architectures to reckon with complex computations.

In both cases, this new class of recursive functions allows the design of computational architectures able to reckon with calculations that are mathematically *undecidable* and/or effectively *not computable in polynomial time* by Von Neumann sequential architectures.

References

1. Agnes, C., Rasetti M.: Undecidability of the word problem and chaos in symbolic dynamics. Il Nuovo Cimento **106 B**, 8 (1991) 879– 907

2. Markov A.A.: On the impossibility of certain algorithms in the theory of associative systems. Doklady Akademii Nauk S.S.S.R. **55** (1947) 587–590
3. Post E.L.: Recursive unsolvability of a problem of Thue. The Journal of Symbolic Logic **12** (1947) 1–11
4. Niven I., Zuckerman H.S., Montgomery H.L.: An introduction to the theory of numbers. Wiley & Sons, New York (1991)
5. Davis M.: Computability and Unsolvability. New York (1958)
6. Basti G., Perrone A.: Teorema di computabilitá dinamica per macchine finite. Internal Report Dept. of Physics, Unv.of Rome "Tor Vergata" Rome **ROM2 F/92/0** (1992)
7. Basti G., Perrone A.: A theorem of computational effectiveness for a mutual redefinition of numbers and processes. Proceedings of Int. Symp. on Information Physics (ISKIT'92), Iizuka, Japan, July 12-15, 1992. Kyushu Institute of Technology Press, Iizuka (1992) 122-133
8. Basti G., Perrone A.: Chaotic Neural Nets, Undecidability and Uncomputability. Int. J. of Intelligent Syst. (to appear)
9. Kleene S.C. : Introduction to Metamathematics. North-Holland Pub. Co., Amsterdam (1980^8)
10. Mendelson E.: Introduction to Mathematical Logic. Princeton U.P., Princeton N.J. (1964)
11. Rogers H. Jr.: Theory of Recursive Functions and Effective Computability. MIT Press, Cambridge Mass. (1988^2)
12. Gödel K.: Über formal unentsheidbare Sätze der Principia Mathematica und verwandter Systeme. Montash. für Math. u. Phys., **38** (1931) 173–198
13. Cohen P.J.: Set Theory and the Continuum Hypothesis. New York (1966)
14. Clavelli M., De Giorgi E., Forti M., Tortorelli V.M.: A self-reference oriented theory for the Foundations of Mathematics. In: Analyse Mathematique et applications-Contrbutions en l'honneur de Jacques-Louis Lions. Gauthier-Villars, Paris (1988) 67–115
15. Smullyan R.: Theory of Formal Systems. Princeton N.J. (1962)
16. Ebbinghaus H.D., Hermes H., Hirrzebruch F., Koecher M., Mainzer K., Neukirch J., Prestel A., Remmert R.: Numbers. Springer, New York-Heidelberg (1990)
17. Artuso R.: Scale e cicli in dinamiche caotiche. Ph.D. Thesis, Univ. of Milan Press, Milan (1992)
18. Auerbach D., Cvitanoviè P., Eckmann J.P., Gunaratne G., Procaccia I.: Phys. Rev. Lett. **58** (1987) 2387
19. Pawelzik K., Schuster H.G.: Phys. Rev. A **43** (1991) 1808
20. Packard N.H., Crutchfield J.P., Farmer J.D., Shaw R.S.: Phys. Rev. Lett. **45** (1980) 712
21. Eckmann J.P., Ruelle D.: Rev. Mod. Phys. **57** (1985) 617
22. Arecchi F.T., Basti G., Boccaletti S., Perrone A.L.: Adaptive recognition of a chaotic dynamics. Int. J. of Bifurcation and Chaos (in press).
23. Pecora L.M., Carrol T.L.: Phys. Rev. Lett. **64** (1990) 821
24. Basti G., Cocciolo P., Perrone A.: Controlling chaotic systems by mutual redefinition of space and dynamics. In: State-of-the-Art Mapping, Orlando FL, April 13-15, 1993, SPIE Proceedings Series, Washington **1943** (1993) 199–208
25. McCulloch W.S., Pitts W.H.: A logical calculus of the ideas immanent in nervous activity. Bull. of Math. Biophys. **5** (1943) 115–133
26. Pinkas G.: Symmetric neural networks and propositional logic satisfiability. Neur. Comput. **3** (1991) 282–291

27. Amari S.I.: Neur. Netw. **6** (1993) 161
28. Baum E.B. , Haussler D.: Neur. Comput., **1** (1989) 151
29. Blum A.L., Rivest R.L.: Neur. Netw. **5** (1992) 117
30. Perrone A.L., Castiglione P., Basti G., Messi R.: Dynamic definition of net topology for parallel architectures. An outlook of computational dynamics. Progr. of Theor. Phys. (to appear)
31. Perrone A., Basti G., Chiavoni A.: A neural module for real time simultaneous discrimination and locking on different temporal series in noisy environment. In: Applications of Artificial Neural Networks III, Orlando FL, April 21-24, 1992, SPIE Proceedings Series, Washington **1709** (1992) 926–936
32. Minsky M.L., Papert S.A.: Perceptrons, expanded edition, MIT Press, Cambridge MA (1988)
33. Khinchin A.I.: Mathematical Foundations of Information Theory, Dover Publ., New York (1957)
34. Shannon,C.E.: Bell Syst. Techn. Journ. **27** (1948) 623
35. Stoer J., Bulirsch R.: Introduction to Numerical Analysis, Springer-Verlag, Berlin (1980)
36. Bulirsch R., Stoer J.: Numerical treatment of ordinary differential equations by extrapolation methods. Numer. Math. **8** (1966) 1-13.
37. Lewis H.R., Papadimitriou C.H.: Elements of the Theory of Computation, Prentice Hall, London (1981)
38. Turing A.M.: On Computable Numbers, with an Application to the Entscheidungsproblem Proceedings London Mathematical Society **42** (1936) 230–265 and **43** (1936) 544–546
39. B.P. Demidovič, I.A. Maron: Osnovy vyčislitelnoj matematiki, Nauka, Moscow (1981)
40. B.Denby, Computer Phys. Comm. **49** (1988), 429.
    C.Peterson, Nucl. Inst. & Meth. **A279** (1989), 537
    M.Gyulassy and M.Harlander, Computer Phys. Comm. **66** (1991), 31.
    B.Denby, Fermi National Accelerator Laboratory preprint, FERMILAB-Conf-92/121-E.
41. A.Antonelli *et al.*, LNF **87-18(R)** (1987).
    A.Antonelli *et al.*, Nucl. Inst. & Meth. (to be published).
42. A. Gamba, L. Gamberini, G. Palmieri, R. Sanna, Nuovo Cimento Suppl. **20** (1961) 221.

Cellular Neural Networks – A Tutorial on Programmable Nonlinear Dynamics in Space

L. O. Chua[1], T. Roska[2], T. Kozek[2]

[1] Department of Electrical Engineering and Computer Sciences and the Electronics Research Laboratory, University of California at Berkeley

[2] Computer and Automation Institute of the Hungarian Academy of Sciences, Budapest

Abstract. Cellular Neural/Nonlinear Networks (CNN) are analog, non-linear, mainly locally connected processor arrays placed on a multidimensional grid. In this tutorial the general framework and some application areas are described, mainly for mathematicians and physicists. The new invention, the CNN Universal Machine is exposed as well; its unique capability of implementing stored–programmable nonlinear spatial dynamics is highlighted. Finally, the first silicon VLSI implementations providing enormous computing power (in the order of 10^{12} operations per second on a single chip) are reviewed.

1 Introduction

This paper is a concise tutorial on a new computing paradigm written mainly for mathematicians and physicists. The cellular neural network (CNN) paradigm and the CNN Universal Machine and Supercomputer is described. First, the CNN paradigm is summarized and some instructive examples are shown. A CNN is a regular, analog, nonlinear processor array placed on a multidimensional geometrical grid, where the interactions between the processors are local within a finite neighborhood. The CNN array, whose function is defined by the processor interaction pattern, called cloning template, performs computation in a fully parallel way on the input signal array. Several useful templates were developed for image processing, partial differential equations, etc. Form, shape, color, motion, depth information can be computed, even in a neuromorphic way (i.e. mimicking the structure and the behavior of living visual systems).

Next, the CNN Universal Machine architecture, the first stored–program analog computing array is described briefly as a workhorse for our new type of algorithm, the *analogic* CNN algorithm. A single *analog* instruction, a cloning template, may execute the integration of ten thousand nonlinear differential equations. Such instructions are combined with *logic*. A local logic block is incorporated in every processor in the array, and the control of the instruction sequence is defined by the logic of the "analogic" CNN algorithm, a new type of algorithm just emerging.

Silicon VLSI implementations are also reviewed briefly. The first tested chips exhibit enormous computing power, for the problems best suitable for our computing paradigm. It is about Tera (10^{12}) analog operations per second or about a million frame per second for processing gray scale images with 100x100 resolution.

2 The CNN Paradigm

Based on the main ideas of [1,2], the CNN paradigm is defined [4] possessing the following qualities:

Definition: The CNN is a

(i) 2-, 3-, or n-dimensional **array of**
(ii) mainly identical **dynamical systems**, called cells, which satisfies two properties:
(iii) most **interactions** between the cells **are local** within a finite radius r, and
(iv) all **state variables are continuous valued functions**.

Remarks

1. The space variable is always discretized: each cell is identified by 2, 3 (or n) integers $(i, j, k..., n)$.
2. The time variable t may be continuous or discrete
3. The interconnection effect represented by the *cloning template* may be a *nonlinear* function of the state x, output y, and input u of each cell, within the neighborhood N_r of radius r, as well as that of the time t (e.g. *time delay* or time- varying coefficients). The cloning template may be space variant or space invariant.
4. The dynamical system is determined uniquely by an *evolution law* (e.g. ODE, discrete maps, differential-difference equations, functional equations, etc.) such that given $x(t)$ and $u(t)$ for all $t \in (-\infty, t_0)$, or $t \in (-\infty, \ldots -\Delta t, 0, \Delta t, \ldots n_0 \Delta t)$, $x(t)$ and $u(t)$ are uniquely determined for all $t \geq t_0$, or $n\Delta t$ with $n \geq n_0$. This includes the boundary conditions, as well.

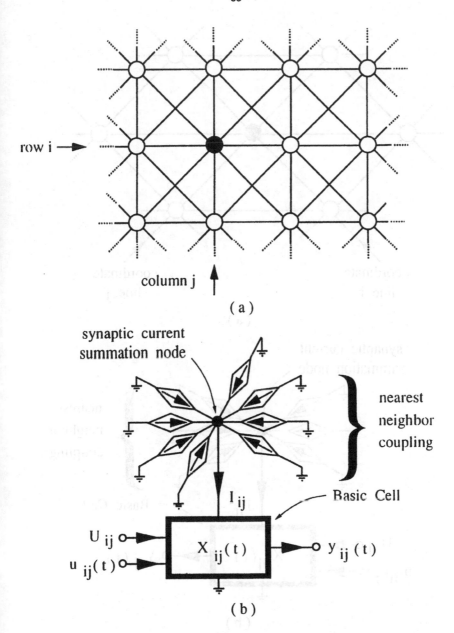

row i →

column j ↑

(a)

synaptic current
summation node

nearest
neighbor
coupling

I_{ij}

Basic Cell

U_{ij}

$u_{ij}(t)$

$X_{ij}(t)$

$y_{ij}(t)$

(b)

Figure 1: A rectangular lattice CNN

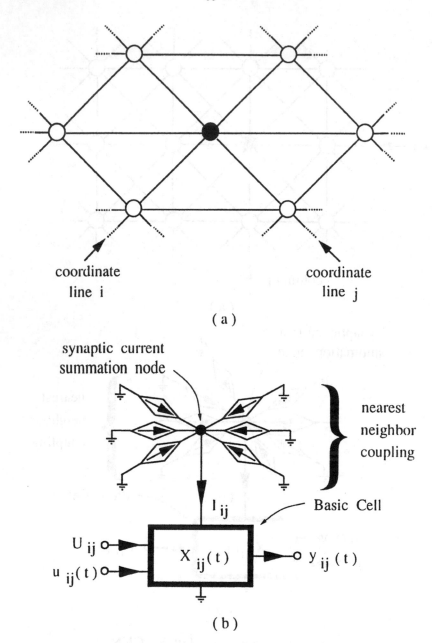

coordinate
line i

coordinate
line j

(a)

synaptic current
summation node

nearest
neighbor
coupling

I_{ij}

Basic Cell

U_{ij}

$u_{ij}(t)$

$X_{ij}(t)$

$y_{ij}(t)$

(b)

Figure 2: A triangular lattice CNN

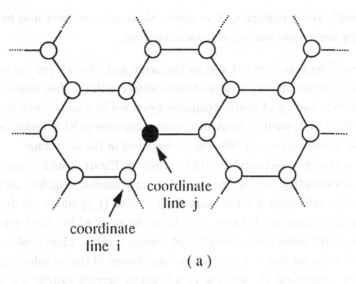

coordinate
line j

coordinate
line i

(a)

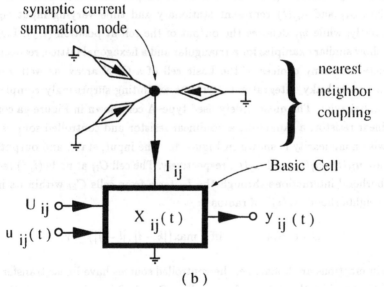

synaptic current
summation node

nearest
neighbor
coupling

I_{ij} — Basic Cell

U_{ij}

$u_{ij}(t)$

$X_{ij}(t)$

$y_{ij}(t)$

(b)

Figure 3: A hexagonal lattice CNN

5. Occasionally, the dynamical system and/or the interconnections may be perturbed by some noise sources of known statistics.

The above definition does not specify the array grid, the cell (or processor) dynamics, the interaction, etc. in an exclusive manner with the clear intention of embracing a wide variety of spatio–temporal dynamics in a single, well defined framework which lends itself for implementation using current VLSI technologies. In this spirit, various types of CNN will be described in the followings.

A CNN array on a rectangular grid is shown in Figure 1 with a symbolic notation of a computing cell. In the case of strictly nearest neighbor coupling only those eight cells influence the state $X_{ij}(t)$ of cell (i, j) which are directly connected to it. Interaction between cells is, in the most widely used types of CNN, implemented using voltage–controlled current sources. This enables us to use Kirchhoff's current law to compute the contribution of the neighboring cells I_{ij} simply by connecting the sources to a synaptic current summation node. Variables U_{ij} and $u_{ij}(t)$ represent stationary and time–varying input signals, respectively, while y_{ij} denotes the output of the cell at position (i, j). Figure 2 and 3 show similar examples for a triangular and a hexagonal lattice, respectively.

There are many choices of the basic cell of a CNN array, as well, ranging from a simple leaky–integrator to systems exhibiting surprisingly complex dynamical bahavior. The most widely used type–A cell shown in Figure 4a consists of a linear resistor, a capacitor, a nonlinear resistor and controlled sources. The piecewise nonlinearity is shown in Figure 4b. The input, state, and output variables are $u_{ij}(t)$, $x_{ij}(t)$, and $y_{ij}(t)$, respectively. The cell C_{ij} at node (i, j) receives neighborhood interactions through the I_{ij} term from cells C_{kl} within its immediate neighborhood $N_r(ij)$ of radius r:

$$C_{kl} \in N_r(ij) \qquad \text{iff} \quad \max\{|k - i|, |l - j|\} \le r \tag{1}$$

If the interactions are linear, i.e. the controlled sources have linear transfer characteristics, and, without loss of generality, $R = 1, C = 1$ is assumed, then the dynamics of a cell array of size $M \times N$ is determined by the state and output equations:

$$\dot{x}_{ij}(t) = -x_{ij}(t) + \sum_{C(kl) \in N_r(ij)} a_{ij,kl} y_{kl}(t) + \sum_{C(kl) \in N_r(ij)} b_{ij;kl} u_{kl}(t) + U_{ij} \tag{2}$$

$$y_{ij}(t) = f(x_{ij}(t)) \tag{3}$$

$$1 \le i \le M, \qquad 1 \le j \le N$$

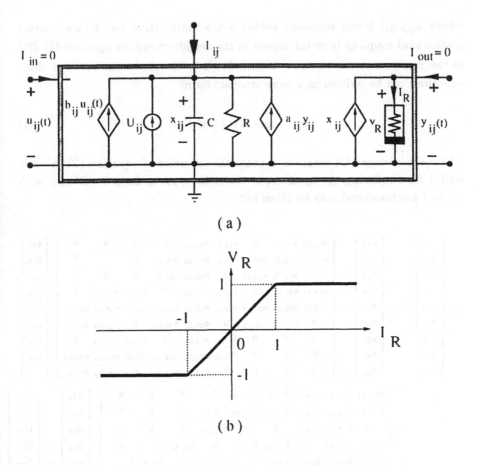

(a)

(b)

Figure 4: A type-A CNN Cell

where $a_{ij,kl}y_{kl}$ terms represent feedback–type interaction and $b_{ij,kl}u_{kl}$ terms feedforward coupling from the inputs in the neighborhood. In equation (2) $f(.)$ is the piecewise nonlinearity of Figure 4b. In a vector–matrix notation these equations can be written in a more compact form:

$$\dot{\mathbf{x}} = -\mathbf{x} + \mathbf{Ay} + \mathbf{Bu} + \mathbf{U} \qquad (4)$$

where the vectors and matrices contain the variables of equation (2) in some well defined ordering. As an example, the state equation for a 3×3 array with a $r = 1$ neighborhood may be given by:

$$\frac{d}{dt}\begin{bmatrix} x_{11} \\ x_{12} \\ x_{13} \\ x_{21} \\ x_{22} \\ x_{23} \\ x_{31} \\ x_{32} \\ x_{33} \end{bmatrix} = - \begin{bmatrix} x_{11} \\ x_{12} \\ x_{13} \\ x_{21} \\ x_{22} \\ x_{23} \\ x_{31} \\ x_{32} \\ x_{33} \end{bmatrix} + \begin{bmatrix} a_{11,11} & a_{11,12} & 0 & a_{11,21} & a_{11,22} & 0 & 0 & 0 & 0 \\ a_{12,11} & a_{12,12} & a_{12,13} & a_{12,21} & a_{12,22} & a_{12,23} & 0 & 0 & 0 \\ 0 & a_{13,12} & a_{13,13} & 0 & a_{13,22} & a_{13,23} & 0 & 0 & 0 \\ a_{21,11} & a_{21,12} & 0 & a_{21,21} & a_{21,22} & 0 & a_{21,31} & a_{21,32} & 0 \\ a_{22,11} & a_{22,12} & a_{22,13} & a_{22,21} & a_{22,22} & a_{22,23} & a_{22,31} & a_{22,32} & a_{22,33} \\ 0 & a_{23,12} & a_{23,13} & 0 & a_{23,22} & a_{23,23} & 0 & a_{23,32} & a_{23,33} \\ 0 & 0 & 0 & a_{31,21} & a_{31,22} & 0 & a_{31,31} & a_{31,32} & 0 \\ 0 & 0 & 0 & a_{32,21} & a_{32,22} & a_{32,23} & a_{32,31} & a_{32,32} & a_{32,33} \\ 0 & 0 & 0 & 0 & a_{33,22} & a_{33,23} & 0 & a_{33,32} & a_{33,33} \end{bmatrix} \begin{bmatrix} y_{11} \\ y_{12} \\ y_{13} \\ y_{21} \\ y_{22} \\ y_{23} \\ y_{31} \\ y_{32} \\ y_{33} \end{bmatrix}$$

$$+ \begin{bmatrix} b_{11,11} & b_{11,12} & 0 & b_{11,21} & b_{11,22} & 0 & 0 & 0 & 0 \\ b_{12,11} & b_{12,12} & b_{12,13} & b_{12,21} & b_{12,22} & b_{12,23} & 0 & 0 & 0 \\ 0 & b_{13,12} & b_{13,13} & 0 & b_{13,22} & b_{13,23} & 0 & 0 & 0 \\ b_{21,11} & b_{21,12} & 0 & b_{21,21} & b_{21,22} & 0 & b_{21,31} & b_{21,32} & 0 \\ b_{22,11} & b_{22,12} & b_{22,13} & b_{22,21} & b_{22,22} & b_{22,23} & b_{22,31} & b_{22,32} & b_{22,33} \\ 0 & b_{23,12} & b_{23,13} & 0 & b_{23,22} & b_{23,23} & 0 & b_{23,32} & b_{23,33} \\ 0 & 0 & 0 & b_{31,21} & b_{31,22} & 0 & b_{31,31} & b_{31,32} & 0 \\ 0 & 0 & 0 & b_{32,21} & b_{32,22} & a_{32,23} & b_{32,31} & b_{32,32} & b_{32,33} \\ 0 & 0 & 0 & 0 & b_{33,22} & a_{33,23} & 0 & b_{33,32} & b_{33,33} \end{bmatrix} \begin{bmatrix} u_{11} \\ u_{12} \\ u_{13} \\ u_{21} \\ u_{22} \\ u_{23} \\ u_{31} \\ u_{32} \\ u_{33} \end{bmatrix} + \begin{bmatrix} U_{11} \\ U_{12} \\ U_{13} \\ U_{21} \\ U_{22} \\ U_{23} \\ U_{31} \\ U_{32} \\ U_{33} \end{bmatrix}$$

As it can be observed in (5), due to the locally connected nature of CNN, the interconnection matrices $\mathbf{A}$ and $\mathbf{B}$ are sparse matrices (it is even more apparent in case of larger arrays) and have a band structure.

In many applications it is desirable to have translation invariance in the interaction pattern along the array. In such a case the value of the interaction coefficients $a_{ij,kl}$ and $b_{ij,kl}$ depends only on the relative position of cells C_{ij} and C_{kl} which they connect. This property clearly implies that any CNN array of arbitrary size is defined by only a few parameters. This set of parameters arranged in matrices is called *cloning template*. In case of the smallest neighborhood $(r = 1)$

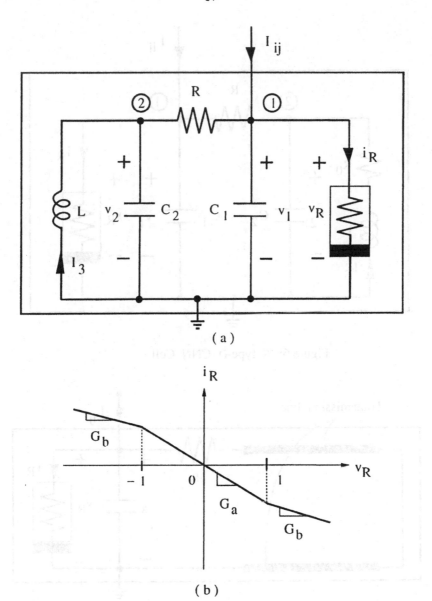

(a)

(b)

Figure 5: A type-C CNN Cell

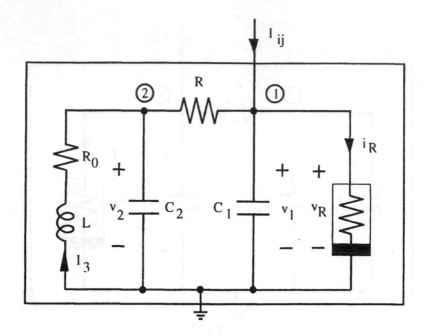

Figure 6: A type-D CNN Cell

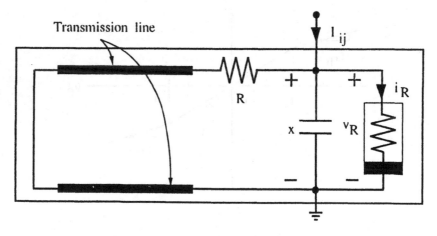

Figure 7: A type- E CNN Cell

$$A = \begin{bmatrix} a_{i-1,j-1} & a_{i,j-1} & a_{i+1,j-1} \\ a_{i-1,j} & a_{i,j} & a_{i+1,j} \\ a_{i+1,j+1} & a_{i,j+1} & a_{i+1,j+1} \end{bmatrix}, \qquad B = \begin{bmatrix} b_{i-1,j-1} & b_{i,j-1} & b_{i+1,j-1} \\ b_{i-1,j} & b_{i,j} & b_{i+1,j} \\ b_{i+1,j+1} & b_{i,j+1} & b_{i+1,j+1} \end{bmatrix}$$

are the feedback and feedforward templates which, together with the $U_{ij} = U$ term (i.e. if U_{ij} is also translation invariant), constitute a cloning template. It can be easily verified that there is a one–to–one correspondence between the large sparse matrices $\mathbf{A}$ and $\mathbf{B}$ and the small matrices A and B.

One way to more complex CNN's is to introduce nonlinear $(\hat{A}_{ij,kl})$ and delay type $(A_{ij,kl}^{\tau})$ interactions [3] which are very useful in a great number of applications. The state equation is modified then as:

$$\dot{x}_{ij}(t) = -x_{ij}(t) + U_{ij} +$$
$$+ \sum_{C(kl) \in N_r(ij)} \hat{a}_{ij,kl}(y_{kl}, y_{ij}) + \sum_{C(kl) \in N_r(ij)} \hat{b}_{ij,kl}(u_{kl}, u_{ij}) +$$
$$+ \sum_{C(kl) \in N_r(ij)} a_{ij,kl}^{\tau} y_{kl}(t) + \sum_{C(kl) \in N_r(ij)} b_{ij,kl}^{\tau} u_{kl}(t) \qquad (6)$$

It is also possible to use them combined (nonlinear operators in delay templates) which are then denoted by $\hat{a}_{ij,kl}^{\tau}(y_{kl}, y_{ij})$ and $\hat{b}_{ij,kl}^{\tau}(u_{kl}, u_{ij})$.

Another road to complexity is through using different basic cells. The most important ones include the second order type–B cell, which is the same as the type–A cell except for an additional capacitor connected across the cell's output, the type–C cell shown in Figure 5 (Chua's circuit), the type–D cell shown in Figure 6 (Canonical Chua's circuit), and the type–E cell shown in Figure 7 (Delayed Chua's circuit). These and similar cells give raise to many interesting nonlinear phenomena and allow for efficient simulation and study of higher order spatio–temporal dynamics.

3 Applications

To demonstrate some application areas of CNN, three examples are given in this section. Although oscillatory and other, more complex behaviors are also investigated, in most current applications fixed–point dynamics is used for solving various problems. This means that given the input and the initial states, the CNN array programmed by the cloning template performs the computation through its transient. After the system reached a fixed–point and the transient decays, the signal array on the output of the cells is what represents some meaningful information. To make its use more evident two examples for image processing are given here.

3.1 Image processing

In Figure 8a an input image is shown where the convex corners are to be extracted. This can be performed efficiently by a type–A CNN with the following cloning template:

$$A = \begin{bmatrix} 0 & 0 & 0 \\ 0 & 2 & 0 \\ 0 & 0 & 0 \end{bmatrix}, \qquad B = \begin{bmatrix} -0.25 & -0.25 & -0.25 \\ -0.25 & 2 & -0.25 \\ -0.25 & -0.25 & -0.25 \end{bmatrix} \qquad U = -3$$

The input image is a signal array whose elements are the grey–scale values of pixels (coded e.g. such that black = +1, white = -1, and grey–levels are in between). This signal array is loaded into the inputs and also the initial states of the cells. After the transient settles on the output signal array the extracted convex corners show up in black while the rest of the image is white (Figure 8b). Note that the time required to extract the corners does not depend on the number of detected locations and to process a picture of arbitrary size needs the same amount of time.

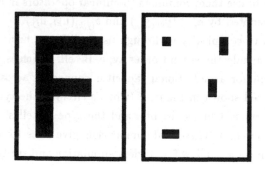

Figure 8: Input (a) and output (b) of a corner detecion CNN

3.2 Solving Partial Differential Equations

A different use of device dynamics allows us to solve partial differential equations by CNN. Here not only the settled output of the cells, but also the evolution of the cell–transients is of interest.

Consider the three basic types of linear second order partial differential equations. A simple spatially discretized approximation of the heat equation

$$\frac{\partial u(\mathbf{x}, t)}{\partial t} = c\nabla^2 u(\mathbf{x}, t)$$

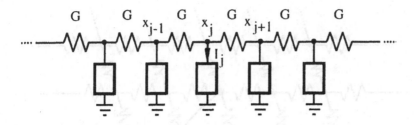

Figure 9: A one-dimensional scalar reaction-diffusion CNN

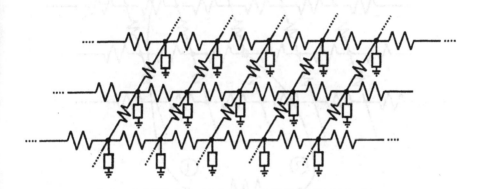

Figure 10: A two-dimensional scalar reaction-diffusion CNN

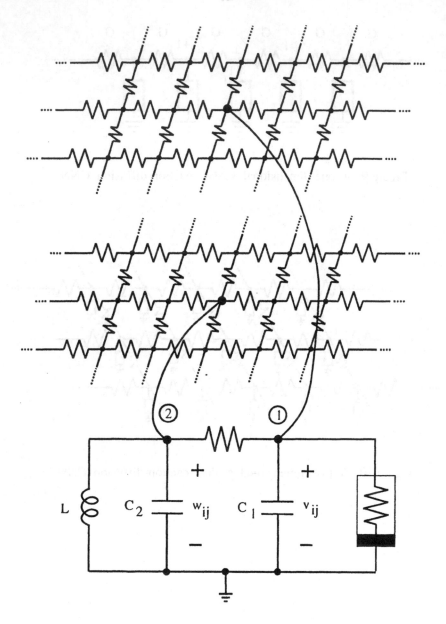

Figure 11: A two-dimensional vector reaction-diffusion CNN

can be solved in 2–D by a type–A CNN programmed by the template:

$$A = \begin{bmatrix} 0 & \frac{1}{h^2} & 0 \\ \frac{1}{h^2} & \frac{-4+1}{h^2} & \frac{1}{h^2} \\ 0 & \frac{1}{h^2} & 0 \end{bmatrix} \qquad B = 0 \quad I = 0$$

where h is the uniform grid size. Provided that the transient remains bounded such that $|u| \leq 1$ (i.e. the cells do not saturate) the time evolution of the cell array's transient gives the solution of the heat equation and at steady state the solution of the Laplace equation

$$0 = c\nabla^2 u(\mathbf{x}, t)$$

is obtained. Similarly, using the same template on a type–B CNN an approximation of the wave equation

$$\frac{\partial^2 u(\mathbf{x}, t)}{\partial t^2} = c\nabla^2 u(\mathbf{x}, t)$$

can be acquired.

Nonlinear PDE's with complex dynamics can also be investigated using cellular neural networks. Reaction–diffusion systems which give rise to many types of nonlinear wave phenomena are a good example. A simple spatially discrete 1–D system of this type, a scalar reaction–diffusion CNN is shown in Figure 9. Its two–dimensional counterpart is shown in Figure 10. When the network parameters are set appropriately, autowaves and spiral waves arise in the system of Figure 11 [19], a two–dimensional vector reaction–diffusion CNN. As it can be observed, type–C CNN cells are connected via two resistive layers.

3.3 Analogic CNN algorithms

Now we return to image processing applications. One can easily think of problems that are solved by not a single image–transformation, but a number of them. A simple example could be the task of finding vertical line segments in an image that are thinner than a given number of pixels. Denote this number by n. The flow diagram of an algorithm that solves this problem is given in Figure 12. It makes use of three image transformations: a peeling, a detection, and a logical OR operator. All operators can be realized by simple CNN templates which are

$$A = \begin{bmatrix} 0 & 0 & 0 \\ 0 & 2 & 0 \\ 0 & 0 & 0 \end{bmatrix}, \qquad B = \begin{bmatrix} 0 & 0 & 0 \\ 3 & 3 & 0 \\ 0 & 0 & 0 \end{bmatrix} \qquad U = -5$$

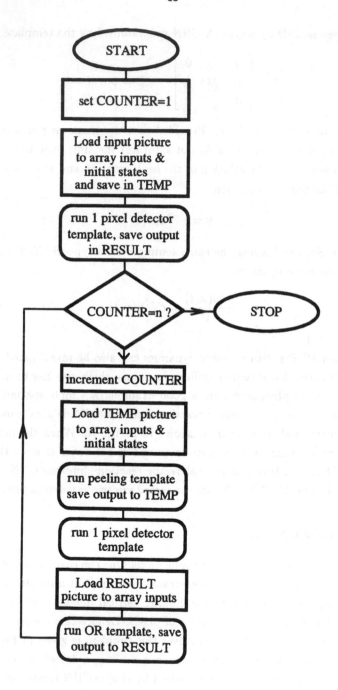

Figure 12: Analogic algorithm for detecting vertical lines thinner than *n* pixels

for peeling,

$$A = \begin{bmatrix} -0.1 & 0.4 & -0.1 \\ 0.4 & 0 & 0.4 \\ -0.1 & 0.4 & -0.1 \end{bmatrix}, \qquad B = \begin{bmatrix} 0.2 & 0 & 0.2 \\ -3 & 2.5 & -3 \\ 0.2 & 0 & 0.2 \end{bmatrix} \qquad U = -0.5$$

for detection, and

$$A = \begin{bmatrix} 0 & 0 & 0 \\ 0 & 3 & 0 \\ 0 & 0 & 0 \end{bmatrix}, \qquad B = \begin{bmatrix} 0 & 0 & 0 \\ 0 & 3 & 0 \\ 0 & 0 & 0 \end{bmatrix} \qquad U = 2$$

is for logical OR. An example for $n = 2$ is shown in Figure 13. Note that such a processing requires moving and storing various images (signal arrays) and "running" of different templates. How this can be efficiently done is described in the following section.

4 The CNN Universal Machine

As we have seen in the last example, a predefined sequence of templates, a kind of CNN algorithm, may be useful in solving complex image processing tasks. There are at least two problems, however, which have to be solved if we try to run these algorithms on silicon chips: (i) we have to store the intermediate results cell by cell (not to loose the speed advantage when moving data from and to the chip) and (ii) we have to store and change the analog templates according to the logic and conditions of the executed algorithm.

The CNN Universal Machine and Supercomputer [5] was invented just to provide a platform for this kind of algorithms which represents a new computing paradigm. It is an analogic (analog and logic) array–computer architecture whose nucleus is the CNN array, but every cell is extended by local analog memory, local logic memory, a local logic output unit and a local analog output unit (these two units combine more than one locally stored values). Each extended cell is controlled by their local communication and control unit. The global programming and control of each cell and the whole array is carried out by the Global Analogic Programming Unit (GAPU). In addition to the machine code of the template sequence, it contains the analog template values in the Analog Program Register (APR). The details of the description can be found in [5].

The key point is that the CNN Universal Machine, and its implementation, the analogic CNN microprocessor, provides for the execution of analogic CNN algorithms. It means, for example, that a single instruction, a template defined

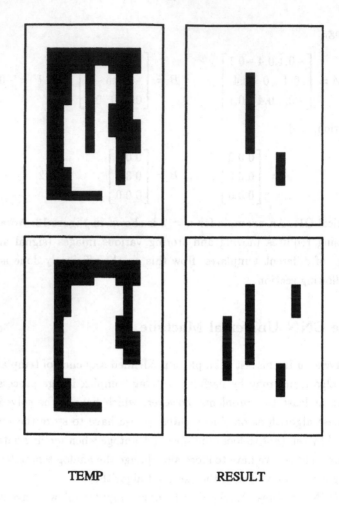

TEMP   RESULT

Figure 13: Stages of the detection algorithm of Figure 12 for $n=2$

by 19 analog values (real numbers), may mean the integration of 10 000 nonlinear differential equations. The running time of such an instruction could be the same as the time required for adding 10 digital numbers. There are many difficult problems which can be solved with reasonable accuracy with the CNN Universal Machine.

In addition to the simple image processing templates, the solution of a complex PDE could be a single instruction as well. Hence, algorithms can be designed with complex nonlinear dynamics in 2 or 2-and-a-half dimensions (or in 3–D when using digital implementation).

The flow diagram of the analogic algorithm, such as the example shown in Figure 12, can be described by an appropriate language, e.g. the Analogic CNN Language (ACL) used in the CNN Workstation hardware/software toolkit [11]. In addition, a compiler, an operating system, a machine code can be defined [11,12].

Most probably the CNN architecture is the only reasonable choice for stored program analog computation. Fully connected arrays seem to be absolutely impractical due to extremely long analog instruction code and reprogramming time.

When implementing the signal sensors cell by cell on the chip (e.g. optical sensors [13]), the fully parallel input needs no scanning overhead. Since the control of the analogic algorithm is digital and the detection–type outputs can even physically be read as a logic memory, the whole system is directly interfaced to digital systems.

We do feel that by using these new analogic algorithms difficult and significant problems will be solved by a completely new approach that involves programming complexities of a few dozen instructions. The theoretical properties of these algorithms are yet to be explored.

5 Silicon VLSI implementations

The fully tested first CNN chip with fixed template showed enormous computing power: using 2 micron CMOS technology, 0.3 Tera XPS (10^{12} analog operations per second) are calculated on 1 cm^2 [14]. Since that time even faster fixed template circuits have been implemented, with optical inputs as well [13]. Several other designs, including discrete-time chips [15] are known [6-10].

The new CNN designs incorporate programmability to different extents. The design of CNN Universal chips are on the way [17,18]. A reasonable specification is as follows (using a 1.2 micron CMOS technology):

– array sizes 50x50 to 100x100 cells

- 50 nsec to 500 nsec settling time for non-propagating templates
- 1-8 W power dissipation
- 1.5-2.5 cm^2 active chip area
- r=1 or 2 neighborhood size
- on chip photosensors
- 5-10 on chip templates in the APR
- 10-100 machine code instructions in the global analogic control unit

Since the architecture and the related circuit–design techniques are brand new, it takes some time to develop the optimal (in terms of area, time, and dissipation) designs.

Acknowledgements

This work has been supported by the Office of Naval Research, under grant N00014-89-J-1402, by the National Science Foundation, under grant MIP-89-12639, and by a joint research grant No. INT 9001336 of the National Science Foundation and the Hungarian Academy of Sciences, as well as by the research grant No. 2578/91 of the National Research Fund of Hungary (OTKA).

References

1. L.O.Chua and L.Yang, "Cellular neural networks: Theory," IEEE Trans. Circuits and Systems, vol. 35, pp. 1257- 72, 1988.
2. L.O.Chua and L.Yang, "Cellular neural networks: Applications," IEEE Trans. Circuits and Systems, vol. 35, pp. 1273-90, 1988.
3. T.Roska and L.O.Chua, "Cellular neurl networks with non–linear and delay–type template elements," Int. J. Circuit Theory and Applications, vol. 20, pp. 469–481, 1992.
4. L.O.Chua and T.Roska, "The CNN paradigm," IEEE Trans. Circuits and Systems–I, vol. 40, pp. 147-156, March 1993.
5. T.Roska and L.O.Chua, "The CNN Universal Machine: An analogic array computer," IEEE Trans. Circuits and Systems–II, vol. 40, pp. 163-173, March 1993.
6. Proc. First IEEE Int. Workshop on Cellular Neural Networks and their Applications (CNN-90), ed. T.Roska, 1990.
7. Proc. 2nd IEEE Int. Workshop on Cellular Neural Networks and their Applications, ed. J.A.Nossek, 1992.
8. Special Issue on Cellular Neural Networks (eds. J.Vandewalle and T.Roska), Int. J. Circuit Theory and Applications, vol. 20, Sept-Oct. 1992.

9. Special Issues on Cellular Neural Networks (eds. J.A.Nossek and T.Roska), IEEE Trans. Circuits and Systems, I: Fundamental theory and applications and II: Analog and digital signal processing, vol. 40, March 1993.

10. T.Roska and J.Vandewalle (eds.), Cellular Neural Networks, J.Wiley and Sons, Chichester, London, New York, 1993.

11. "The CNN Workstation User's Manual, Version 5.1," Computer and Automation Institute of the Hung. Acad. Sci. (MTA-SzTAKI), Budapest, 1993.

12. T.Roska, L.O.Chua, and A.Zarándi, "Language, compiler, and operating system for the CNN supercomputer," UCB/ERL Memo No. M93/93, Berkeley, California, April 1993.

13. A.Rodriguez-Vazquez et. al., "On a 16x16 CNN chip with optical sensors," Internal report, Microelectronics Ceneter, University of Sevilla, Sevilla, 1993.

14. J.M.Cruz and L.O.Cuha, "Design high speed high density CNN's in CMOS technology," CAS 1992

15. H.Harrer, J.A.Nossek, and R.Stelzl, "An analog implementation of discrete-time cellular neural networks," IEEE Trans. Neural Networks, vol. 3, pp. 466-477, 1992.

16. K.Halonen, V.Porra, T.Roska, L.O.Chua, "VLSI implementation of a reconfigurable cellular neural network containing local logic (CNNL)," Int. J. Circuit Theory and Applications, vol. 20, pp. 573-582, 1992.

17. Internal reports of the Nonlinear Electronics Laboratory at ERL, U. C. Berkeley; private communication

18. A.Rodriguez-Vazquez, private communication

19. V.Pèrez-Muñuzurri, V.Pèrz-Villar, and L.O.Chua, "Autovawes for image processing on a two-dimensional array of Chua's circuits: flat and wrinkled labyrinths," IEEE Trans. Circuits and Sytems, vol. 40, pp. 174-181 , March, 1993.

Neural Coding: A Theoretical Vista of Mechanisms, Techniques, and Applications

J. Leo van Hemmen and Raphael Ritz

Physik-Department, Technische Universität München
D-85747 Garching bei München, FR Germany

Abstract. This paper presents an overview of the capabilities of *spiking* neurons in processing complex information. We propose a flexible neuron model (the Spike Response Model), that is amenable to both analytic treatment and straightforward numerical simulation, and analyze the dynamics of a large network consisting of these neurons. We also present tools that, given some homogeneity, enable one to analytically treat the dynamical response of a network, a highly nonlinear system. Finally, we evaluate the underlying mechanisms, such as the dependence upon the axonal delays, the local inhibition, structural feedback, and discuss applications to associative feature linking, pattern segmentation, and context-sensitive binding.

1   Introduction

In processing information and compared with their hardware equivalents, biological neural networks show remarkable properties and capabilities. In addition to their highly parallel organization it is also the neural coding which is responsible for attractive features such as fast feature linking and pattern segmentation. In this paper we analyze the underlying mechanisms, present analytical techniques which allow an efficient description of the dynamical behavior of a network, a highly nonlinear system, and offer applications as diverse as associative feature linking, figure-ground segregation, and context-sensitive binding. A key to the succesful solution is that the data processing is performed by *spiking* neurons.

Before turning to a spike-generating model proper, we review some ideas from the contexts of the Hopfield model and Hebbian learning, a local rule correlating the behavior of the pre- and postsynaptic neuron, and explain the notion of sublattice, that allows an analytic treatment of spatio-temporal behavior of a nonlinear network. We then introduce the Spike Response Model, that is easy to simulate and incorporates both the presynaptic refractory behavior and the post-synaptic response through the dendritic tree (the input of a neuron) present the single-neuron behavior, and build a network out of these neurons. The dynamical response of such a network is analyzed in detail, both stationary incoherent and oscillatory coherent firing are studied, and various scenarios leading to this behavior are presented and related to experiments showing coherent oscillations in the visual cortex. It may be interesting to mention that delays, which are omnipresent in any brain of reasonable size, play an important role. The paper finishes with several applications that take advantage of the new degree of freedom, time, that is inherent to *spiking* neurons.

Numerical simulations are an essential ingredient in understanding the complex behavior of highly nonlinear systems such as neural networks. So a question of prime importance is: How do we generate the dynamics? The neurons are described by two-valued variables $S_i = \pm 1$, whose interpretation will be discussed below. Here $1 \leq i \leq N$ labels the neurons and $\Omega = \{-1, 1\}^N$ is the phase space. Given a dynamical rule for each neuron, one may update the neurons either sequentially or in parallel or as a mixture of these two. In the present paper we will use parallel or synchronous updating only.

Since biological systems are rather noisy we also want to model the noise. This can be done, for instance, by a Glauber dynamics. Given the input $h_i(t)$ of neuron i at time t the *transition probability* to $S_i(t+1) = +1$ is

$$\text{Prob}\{S_i(t+1) = +1 | h_i(t)\} = \frac{1}{2}\{1 + \tanh[\beta(h_i(t) - \vartheta)]\} \tag{1}$$

where ϑ is a threshold and $\beta = 1/T$ so that the formal temperature T measures the strength of the noise. In the limit where both T and ϑ vanish we end up with $S_i(t+1) = \text{sgn}[h_i(t)]$. Throughout what follows updating of (1) is synchronous. In this way we generate a Markov chain. Most of the time the bulk limit $N \to \infty$ is implicitly understood.

2 Hopfield Model and Hebbian Learning

Hopfield [32] introduced a model that aims at storing *stationary* patterns. The neurons have two modes. They fire either at a maximal ($S_i = +1$) or at a minimal ($S_i = -1$) rate. It is plain that in a *rate* description a fine time discretization with $\Delta t = 1$ ms does not make any sense. The phase space Ω is still the same as before in that $\Omega = \{-1, 1\}^N$. A pattern μ is a sequence of random numbers $\{\xi_i^\mu, 1 \leq i \leq N\}$ where the ξ_i^μ are independent, identically distributed random variables which assume the values ± 1 with probability $p = 1/2$. That is, for the time being we use *unbiased* random patterns, which are to be stored in a globally coupled network.

Let us first assume that we want to store only a single pattern, a specific Ising spin configuration. We still have to specify the learning rule, i.e., the couplings J_{ij} as they appear in the local field h_i and the dynamics,

$$h_i(t) = \sum_{j=1}^{N} J_{ij} S_j(t) \quad \Rightarrow \quad S_i(t+1) = \text{sgn}[h_i(t)] \ . \tag{2}$$

Since there is no 'thermal' noise, this is a zero-temperature dynamics. In contrast to Hopfield (see also Little [34]) and throughout what follows we assume a synchronous updating, i.e., we apply '(1)' for all the neurons at the same time (parallel dynamics). In the Hopfield model there are no delays. We try the ansatz $J_{ij} = N^{-1}\xi_i^\mu \xi_j^\mu$ and find, using (2),

$$h_i^\mu = \xi_i^\mu \sum_{j=1}^{N}(N^{-1}\xi_j^\mu)\xi_j^\mu = \xi_i^\mu \qquad (3)$$

so that the pattern is a fixed point under the dynamics. If we now have to store $q > 1$ patterns, we simply try

$$J_{ij} = N^{-1} \sum_{\mu=1}^{q} \xi_i^\mu \xi_j^\mu . \qquad (4)$$

for $i \neq j$ and $J_{ii} = 0$. This a called a "Hebbian" learning rule as will be explained shortly.

We first verify that (4) makes sense. To this end we perform a signal-to-noise ratio analysis. We substitute $S_i = \xi_i^\nu$ into (2) so as to find

$$h_i^\nu = \xi_i^\nu \sum_{j=1}^{N}(N^{-1}\xi_j^\nu)\xi_j^\nu + N^{-1} \sum_{j(\neq i),\mu(\neq \nu)} \xi_i^\nu \xi_j^\nu \xi_j^\mu . \qquad (5)$$

The first term on the right is simply ξ_i^ν, which is what we wanted, plus N^{-1} times a sum of independent, identically distributed random variables. This second term is the noise, which is Gaussian according to the central limit theorem [33] as $q = \alpha N \to \infty$. More precisely, it is a Gaussian with mean zero and standard deviation $\sqrt{\alpha}$. So for α small enough the first term, the signal, dominates the noise with typical amplitude $\sqrt{\alpha}$. As was predicted by Hopfield on the basis of numerical simulations and confirmed analytically by Amit et al. [1], this happens as long as $\alpha \leq \alpha_c = 0.14$; see also van Hemmen and Kühn [27]. Beyond α_c no associative memory exists.

The coupling J_{ij} connects j with i, with direction $j \to i$. It is called a synaptic strength as the synapse at an axon ending of the presynaptic neuron j mediates the spikes to the dendritic tree of the postsynaptic neuron i. So the synapse only sees the pre- and postsynaptic neuron. How, then, can it learn? From a more abstract point of view, the main idea of Hebb [19] was to *correlate* the behavior of the neurons i and j. Following Herz et al. [28, 29, 30] we implement this idea as follows.

When a spike travels from neuron j to neuron i it takes a finite amount of time, say Δ ms. When we then look at our synapse at time t, we correlate the behavior of neuron j at time $t - \Delta$, which is what we see at the synapse at time t, with the response of neuron i at time $t + 1$, where '1' means 1 ms, our time discretization. (We have dropped the rate interpretation and taken spiking neurons with absolute refractory period $= 1$ instead. So $S_i = 1$ tells us that neuron i spikes whereas $S_i = -1$ tells us that it does not.) Our implementation of the Hebbian idea is that during a learning session of duration T_1 the synaptic coupling J_{ij} changes by

$$\Delta J_{ij} = N^{-1} \cdot T_1^{-1} \sum_{0 \leq t \leq T_1} S_i(t+1) S_j(t-\Delta) \tag{6}$$

where T_1 greatly exceeds the longest delay $\Delta^{\max}$ in the system. Through the prefactor T_1^{-1} we take into account saturation effects, i.e., having listened to the theme BACH a whole evening does not imply that all other melodies have gone; in other words, saturation occurs. For the stationary patterns of the Hopfield model we recover (4). We also note that (6) incorporates a correlation in space $(i \leftrightarrow j)$ and time $(t+1 \leftrightarrow t-\Delta)$. In a biological neural network we have a broad *distribution* of delays. This suffices for the learning rule (6) to store (nearly) arbitrary spatio-temporal patterns [14, 29, 30].

Until now we had assumed that the patterns had a vanishing activity $a = \langle \xi \rangle = p - (1-p) = 2p - 1$ with $p = 1/2$. Suppose now that we need a nonvanishing a. This is particularly relevant for biological nets with $p = \text{Prob}\{\xi_i^\mu = +1\} \approx 0$. How should we modify the learning rule (6)? The surprisingly simple answer [21] is that we replace $S_j(t - \Delta)$ in (6) by $[S_j(t - \Delta) - a]$. This is an *asymmetric* learning rule. It is very efficient.

Let us now return to the original Hopfield model with parallel dynamics. We have shown a plausibility argument that a pattern ν is a *fixed* point of the dynamics (2). That is nice but we want more: we want it to be stable. To prove stability we exhibit a Lyapunov function H, i.e., a function that is bounded from below and decreasing under the system's dynamics until it reaches a minimum of H. If a pattern is at a minimum, and for $\alpha \leq \alpha_c$ it is, then stability follows directly.

If $J_{ij} = J_{ji}$ (symmetry), the Lyapunov function is [15, 16],

$$H(\mathbf{S}) = -\sum_{i=1}^{N} \left| \sum_{j=1}^{N} J_{ij} S_j \right| . \tag{7}$$

Keeping in mind the dependence of $\mathbf{S} = (S_i, 1 \leq i \leq N)$ upon t we rewrite (7) in the form

$$H(t) = -\sum_{i=1}^{N} |h_i(t)| = -\sum_{i=1}^{N} S_i(t+1) h_i(t) \tag{8}$$

since the local field at i is $h_i = \sum_j J_{ij} S_j$ and, for parallel dynamics at zero temperature, we have $S_i(t + 1) = \text{sgn}[h_i(t)]$ *for all* i. Due to the symmetry $J_{ij} = J_{ji}$ we can also write

$$H(t) = -\sum_{i,j} S_i(t+1) J_{ij} S_j(t) = -\sum_j S_j(t) \left[\sum_i J_{ji} S_i(t+1) \right]$$

$$= -\sum_j S_j(t) h_j(t+1) . \tag{9}$$

Subtracting this from $H(t+1) = -\sum_i |h_i(t+1)|$ we obtain, using $h_j(t+1) = |h_j(t+1)|\text{sgn}[h_j(t+1)]$ in tandem with $\text{sgn}[h_j(t+1)] = S_j(t+2)$,

$$\Delta H(t) = H(t+1) - H(t) = -\sum_i |h_i(t+1)|\left[1 - S_i(t)S_i(t+2)\right] \leq 0 \qquad (10)$$

as $|S_i(t)S_i(t+2)| \leq 1$. So H is a Lyapunov function. For finite N, H is bounded from below. After finitely many steps ΔH therefore has to vanish identically. This can be realized only if for all i we have $S_i(t)S_i(t+2) = 1$, i.e., $S_i(t+2) = S_i(t)$. Accordingly, the dynamics converges either to a fixed point or to a period-two limit cycle. In passing we note that in contrast to sequential dynamics nonvanishing diagonal terms J_{ii} do not modify the above argument. We refer to van Hemmen and Kühn [27, pp 38-40] for a treatment of the finite-temperature case using duplicate spins.

To see that for small enough α a pattern gives rise to a minimum of H we introduce the *overlap* with pattern ν,

$$m_\nu = N^{-1}\sum_{i=1}^{N} \xi_i^\nu S_i \;. \qquad (11)$$

For $S_i = \xi_i^\nu$ we get $m_\nu = 1$ whereas for an 'arbitrary' (=random) configuration we obtain $m_\nu \approx 0$. Suppose now that we offer the system a *noisy* pattern ν. Then at time $t = 0$ we have $m_\nu < 1$ and $h_i(0)$ may be written [cf. (5)]

$$h_i(0) = \xi_i^\nu m_\nu + N^{-1}\sum_{j(\neq i),\mu(\neq\nu)} \xi_i^\mu \xi_j^\mu S_j(0) \;. \qquad (12)$$

If m_ν is too small and/or $\alpha > \alpha_c$, the second term in (12) dominates. If, however, $\alpha < \alpha_c$ and m_ν exceeds a critical overlap m_c, then it is more advantageous to 'flip' all those neurons with $S_i(0) = -\xi_i^\nu$, which gives $m_\nu(1) = 1$ since the signal term now dominates the noise and, thus, this flipping will lower H; cf. (8).

This finishes our treatment of the Hopfield model with parallel dynamics, which is also called the Hopfield-Little model. Little studied a synchronous updating, Hopfield found a Lyapunov function for asynchronous zero-temperature dynamics, viz., $H = -\sum_{i,j} J_{ij} S_i S_j$, and estimated the storage capacity α_c numerically but neither of them found the Lyapunov function for parallel dynamics. A general theoretical context for this kind of system has been provided by Herz et al [28].

3 The Sublattice Idea

Assuming the simple Glauber dynamics (1) for a while and taking for granted the Hebbian learning rule (6) for spatio-temporal patterns, we want to get a more 'direct' feeling for what is going on. Suppose the network should perform the cycle $\xi^1 \to \xi^2 \to \cdots \to \xi^q \to \xi^1 \to \cdots$, and stay in each pattern ξ^μ, say, Δ ms. A rule that does this is

$$h_i(t) = \sum_{i=1}^{N} J_{ij} S_j(t - \Delta) \quad \text{and} \quad J_{ij} = N^{-1} \sum_{\mu=1}^{q} \xi_i^{\mu+1} \xi_j^\mu \tag{13}$$

where μ is taken to be mod q. In passing we note [36] that instead of the sum over μ itself we could also have taken the sign of J_{ij}. To see why the prescription (13) works, we employ a self-consistency argument. Let us assume that the cycle existed in the past and made the transition $\mu - 1 \to \mu$ at time $t = 0$. By the very definition (11) of the overlap,

$$h_i(t) = \sum_{\nu=1}^{q} \xi_i^{\nu+1} m_\nu(t - \Delta) \tag{14}$$

and by assumption $m_\nu(t - \Delta) = \delta_{\nu,\mu-1}$ as long as $t < \Delta$. At $t = \Delta$, however, $h_i(t)$ jumps from ξ_i^μ to $\xi_i^{\mu+1}$, i.e., the system makes the transition from μ to $\mu + 1$, as advertized.

Before proceeding we generalize (13) and write $J_{ij} = N^{-1} Q_\Delta(\xi_i; \xi_j)$ where ξ_i is the q-vector $(\xi_i^1, \xi_i^2, \ldots, \xi_i^q)$ representing the data at i and the subscript Δ denotes that a *delay* Δ is associated with J_{ij}. For the moment the delay Δ is the same for all bonds J_{ij}.

To analytically treat the dynamics we now introduce *sublattices* [23, 24, 26]

$$L(\mathbf{x}) = \{i \,|\, \xi_i = (\xi_i^1, \xi_i^2, \ldots, \xi_i^q) = \mathbf{x}\} \tag{15}$$

for all $\mathbf{x} \in \{-1, 1\}^q$. In words, the sublattice $L(\mathbf{x})$ consists of all i with $\xi_i = \mathbf{x}$; see also Fig. 1. We thus obtain a disjoint partition of $\{1, 2, \ldots, N\}$. Furthermore, we define the *sublattice activity*

$$m(\mathbf{x}) = |L(\mathbf{x})|^{-1} \sum_{i \in L(\mathbf{x})} \tilde{S}_i \tag{16}$$

where $|A|$ denotes the number of elements in the set A and $\tilde{S}_i = (1 + S_i)/2 \in \{0, 1\}$. We then obtain

$$m_\mu = (2/N) \sum_{i=1}^{N} \xi_i^\mu \tilde{S}_i = 2 \sum_{\mathbf{x}} (|L(\mathbf{x})| N^{-1}) |L(\mathbf{x})|^{-1} \sum_{i \in L(\mathbf{x})} \xi_i^\mu \tilde{S}_i$$

$$\equiv \sum_{\mathbf{x}} p_N(\mathbf{x}) x^\mu m(\mathbf{x}) . \tag{17}$$

In the present case $p_N(\mathbf{x}) \to p(\mathbf{x}) = \text{Prob}\{\xi_i = \mathbf{x}\}$ as $N \to \infty$ by the strong law of large numbers [33]. In view of (17) it suffices to determine $m(\mathbf{x}; t + 1)$, once we know all the $m(\mathbf{x}; t')$ for $t' \leq t$. Due to (17), the overlaps m_μ directly follow.

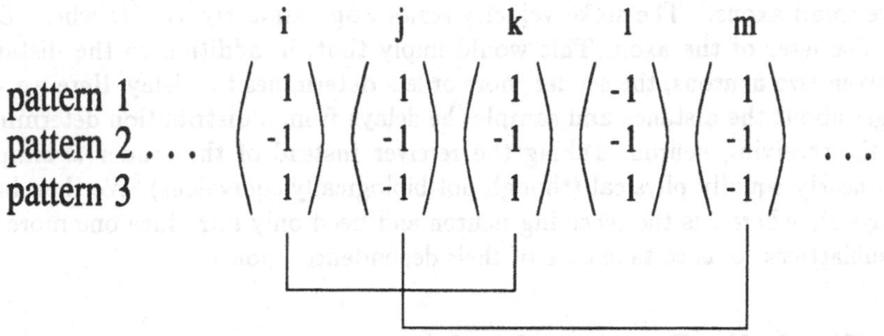

Fig. 1. *Sublattices.* A neural network has learned three different patterns $\{\xi_i^1\}, \{\xi_i^2\}, \{\xi_i^3\}$. Each pattern consists of a sequence of numbers $\{\xi_i^\mu = \pm 1; 1 \le i \le N\}$, out of which the configurations at 5 sites $(i, j, k, l,$ and $m)$ are shown. Each row contains the values of one pattern and each column the values for one neuron. Neurons i and k have been shown the same information and thus belong to the same sublattice. Neuron j and m belong to a different sublattice. Taken from [12].

Here we go. We have, as $N \to \infty$, by the very definition of sublattice and sublattice activity

$$
\begin{aligned}
h_i(t) &= \sum_{j=1}^{N} J_{ij} S_j(t - \Delta) \\
&= \sum_{\mathbf{y}} (|L(\mathbf{y})|N^{-1})|L(\mathbf{y})|^{-1} \sum_{j \in L(\mathbf{y})} Q_\Delta(\xi_i; \xi_j) S_j(t - \Delta) \\
&= \sum_{\mathbf{y}} p(\mathbf{y}) Q_\Delta(\xi_i; \mathbf{y}) m(\mathbf{y}; t - \Delta)
\end{aligned}
\tag{18}
$$

so that for all $i \in L(\mathbf{x})$

$$
h_i(t) = h(\mathbf{x}; t) = \sum_{\mathbf{y}} p(\mathbf{y}) Q_\Delta(\mathbf{x}; \mathbf{y}) m(\mathbf{y}; t - \Delta) .
\tag{19}
$$

That is to say, all $i \in L(\mathbf{x})$ "see" the *same* field $h(\mathbf{x}; t)$. The stochastic updating (1) is performed for each formal neuron independently of the others and thus we obtain in the limit $N \to \infty$, using the strong law of large numbers [33],

$$
\begin{aligned}
m(\mathbf{x}; t + 1) &= |L(\mathbf{x})|^{-1} \sum_{i \in L(\mathbf{x})} S_i(t + 1) \\
&= \langle S_i(t + 1) \rangle_\beta = \tanh[\beta h(\mathbf{x}; t)] .
\end{aligned}
\tag{20}
$$

So we have now arrived at our final result as a consequence of the *homogeneity* at each of the sublattices and the strong law of large numbers. The latter formalizes that (some) local processes are independent of each other.

We now want to drop the condition that all delays be equal. There exists the physiological rule of thumb that "big neurons have big axons and small neurons have small axons." The spike velocity scales approximately as $\sqrt{D}$ where D is the diameter of the axon. This would imply that, in addition to the distance between two neurons, the sender more or less determines the delay. Here we will forget about the distance and sample the delays from a distribution determined by the receiving neuron. Taking the receiver instead of the sender is simpler and nearly equally physical (though not biologically equivalent). We then have delays Δ_i where i is the receiving neuron and need only introduce one more set of sublattices so as to take care of their dependence upon i.

4 The Spike Response Model

4.1 Model of a Neuron

We now turn to the Spike Response Model (SRM). In so doing we follow Gerstner et al. [10, 11, 13, 14]. The spike generation process in a neuron is taken care of by three essential ingredients,

    (i)   a threshold ϑ,
    (ii) a refractory function $\eta(t - t^f)$,
    (iii) noise with 'strength' inversely proportional to β.

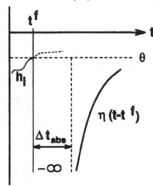

For the moment we follow the membrane potential h_i of neuron i (see figure). At time $t = t^f$ it reaches the threshold ϑ and the neuron fires, i.e., *emits a spike*. During a short period ($\approx 2\,\mathrm{ms}$) after t_f the neuron cannot spike at all; this is the absolute refractory period Δt_{abs}. Then it can only spike with some "extra" effort, the relative refractory periode. Both effects are taken into account through a refractory function η. We put $\eta(t) = -\infty$ for $0 < t < \Delta t_{\mathrm{abs}}$ and some negative value going to 0 for $t > \Delta t_{\mathrm{abs}}$; cf. the figure.

We now can write

$$h_i(t) = h_i^{\mathrm{s}}(t) + h_i^{\mathrm{r}}(t) \tag{21}$$

where h_i^{s} denotes the signal input at the soma, which will be discussed shortly, and for $\eta \in \mathcal{L}^1(\mathbb{N})$

$$h_i^{\mathrm{r}}(t) = \sum_f \eta(t - t_i^f) \tag{22}$$

where t_i^f ranges through the firing times of neuron i prior to t.

A neuron's firing is some kind of stochastic process which, in the present context, can be modeled conveniently as follows. A spike itself lasts for about 1 ms. We will argue (see below) that its detailed structure is not relevant to the collective phenomena we are interested in. So we *discretize time* in portions of size 1 ms. The state of a neuron can be characterized through

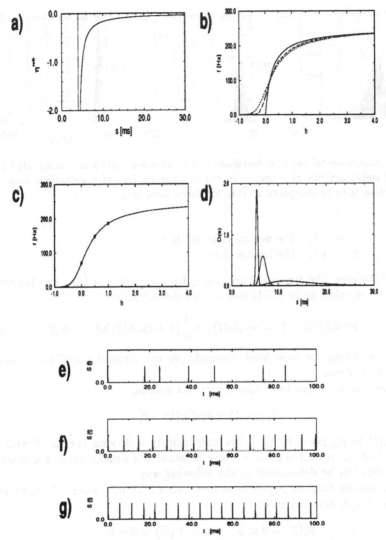

Fig. 2. *Standard model neuron.* **a)** Refractory function η. It is taken to be $-\infty$ during the absolute refractory period ($0 \le s \le 4\,\text{ms}$). **b)** Gain function with noise ($\beta = 8$ dotted, $\beta = 15$ dashed) and noiseless (solid). **c)** Gain function for $\beta = 8$. The 'bullets' indicate the values at which the interval distribution (d) and typical spike trains (e - g) have been calculated. **d)** Interval distribution with $\beta = 8$ and constant stimulus ($h^{\text{ext}} = 0$, relatively flat; $h^{\text{ext}} = 0.5$, middle; $h^{\text{ext}} = 1$, narrow distribution). It agrees well with experiment. **e)** - **g)** Typical *spike trains* with $\beta = 8$: **e)** $h^{\text{ext}} = 0$; **f)** $h^{\text{ext}} = 0.5$; **g)** $h^{\text{ext}} = 1.0$. All results are for a non-adaptive neuron. Taken from [10].

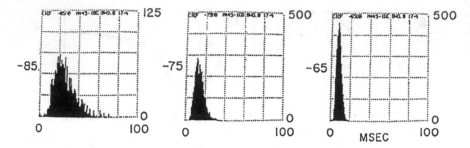

Fig. 3. *Experimental interval distribution* of spontaneous (left) and stimulated (center and right) activity of a neuron in the cochlear nucleus of anesthetized cats. The distribution is to be compared to Fig. 2d. Taken from [35].

$$S_i = -1, \quad \text{the neuron does not fire,}$$
$$S_i = +1, \quad \text{the neuron fires.}$$

As we have seen, the *transition probability* from the state at time t to the one et time $t + 1$ is taken to be a Glauber process described by

$$\text{Prob}\{S_i(t+1) = +1 | h_i(t)\} = \frac{1}{2}\{1 + \tanh[\beta(h_i(t) - \vartheta)]\} \tag{23}$$

with β indicating the noise level. Formally one can write $\beta = 1/T$ and interpret T as temperature.

In the limit $\beta \to \infty$ we obtain a noiseless neuron,

$$S_i(t+1) = \text{sgn}[h_i(t) - \vartheta] , \tag{24}$$

with $h_i(t)$ being given by (21). We now identify a driving current I with the input h^s of our model neuron. I being constant the neuron fires at a constant rate, which can be determined in the following way.

The neuron fires at regularly spaced times $t = nT$, where T is, to good approximation, determined by

$$h_i(T) - \vartheta = 0 \quad \Longrightarrow \quad h^r(T) = \vartheta - h^s .$$

A firing rate is also well–defined, if the neuron is noisy. We fix a time window $\mathbb{T}$ and count the number of spikes in this window for a given I. That is,

$$f(I) \equiv \#(\text{spikes in } \mathbb{T})/\mathbb{T} . \tag{25}$$

This is the *firing rate*. It is very popular among experimentalists. There are a number of problems, though

(i) How do we determine $\mathbf{T}$? In principle, $\mathbf{T} \to \infty$.
(ii) We should average, which is ok, only if $I = constant$.
(iii) A neuron is relevant only if it belongs to a *network*.
   But there $f \approx 40 - 60\,\text{Hz} \ll f_{max} = 500\,\text{Hz}$ of a single neuron.
   Thus the network is important.

Before treating the network behavior, however, we first make a detour.

Spike Trains of a Single Neuron. Given that a spike occured at $s = 0$ we can define various probabilities. For example, through the distribution of time intervals between two successive spikes or the probability that a neuron given it has fired at $s = 0$ has not spiked again yet, for obvious reason called the survivor function. So we define

– Interval distribution:

$$D(s, I)\Delta s = \text{Prob}\{\text{neuron fires between } s \text{ and } s + \Delta s\}$$

– Survivor function:

$$P(s, I) = \text{Prob}\{\text{neuron has not spiked yet at time } s\}$$

These two quantities are directly related by

$$D(s, I) = -\frac{d}{ds}P(s, I) \tag{26}$$

as can be seen by the following simple argument.

Proof. Prob{neuron fires in $[s, s + \Delta s)$}

$$\begin{aligned}
&= D(s, I)\Delta s \\
&= P(s, I) - P(s + \Delta s, I) \\
&= -\frac{d}{ds}P(s, I)\Delta s \ + \ O((\Delta s)^2) \ . \square
\end{aligned}$$

Using (26) we can now easily calculate the mean interval length

$$\begin{aligned}
\bar{s}(I) &= \int_0^\infty ds\ D(s, I)\ s \\
&= \int_0^\infty ds\ s\ [-\frac{d}{ds}P(s, I)] \\
&= -\int_0^\infty dP(s, I)]\ s \\
&= -sP(s, I)|_0^\infty \ + \int_0^\infty dsP(s, I)\ .
\end{aligned} \tag{27}$$

That is the quantity that is usually determined in experiments; cf. (25): $\bar{s}(I) = \lim_{\mathbf{T}\to\infty}[\mathbf{T}/\#(\text{spikes})]$. By abuse of language, this leads to the *mean firing rate* through $f(I) \equiv \bar{s}(I)^{-1}$. It differs from the 'mean firing frequency' in that

$$\int_0^\infty ds\ D(s,I)\ s^{-1} \geq f(I) = 1/\int_0^\infty ds\ D(s,I)\ s\ .$$

The inequality is due to the convexity of s^{-1} for $s > 0$ and Jensen [39].

The Spike Response Model itself is very flexible. It can reproduce *adaptation and bursts* in a natural (!) way; cf. Fig. 4 for the latter. In addition, we can faithfully simulate the behavior of various famous but complicated neuron models such as the one of Hodgkin and Huxley [31], who earned the Noble price by their careful fit of experimental data. We simply state as a *fait accompli* that this case can be represented by the refractory function

$$\eta(t) = \begin{cases} -\infty & \text{for } 0 \leq t \leq t_0 \\ \alpha[(t_m - t_0)^{-1} - (t - t_0)^{-1}] & \text{for } t_0 < t \leq t_m \\ 0 & \text{for } t > t_m \end{cases}$$

with $\alpha = 48, t_0 = 9\,\text{ms}, t_m = 19.1\,\text{ms}$.

4.2 Model of a Network

We now put the model neurons together. In so doing we take into account the following neurobiological facts. When j emits a spike, it takes some time Δ before it reaches a synapse on the dendritic tree of neuron i. All spikes have the same amplitude. It, therefore, suffices to specify the synaptic strength $J_{ij}(\Delta)$. Say, upon arrival of the spike $J_{ij}(\Delta)$ 'units' of charge are transferred to the dendritic tree of neuron i. We assume that the Maxwell equations govern the process of generating a response at the soma. Since the Maxwell equations are *linear* so is the response. That is, the response can be computed as the convolution with a response kernel ϵ,

$$\text{Response}(t) = \int_0^\infty ds\ \epsilon(s)\ \text{Input}(t - s)\ ,$$

temporal summation: linear *in time*

where ϵ can be determined experimentally. Here, we take an alpha function

$$\epsilon(s) = \frac{s}{(\tau_{\text{RC}})^2} \exp(-s/\tau_{\text{RC}})\ \text{so that}\ \int_0^\infty ds\epsilon(s) = 1\ . \tag{28}$$

In principle, ϵ is a function of i only but, for the sake of convenience, all response functions are taken to be equal, *provided* they correspond to neurons of the same type.

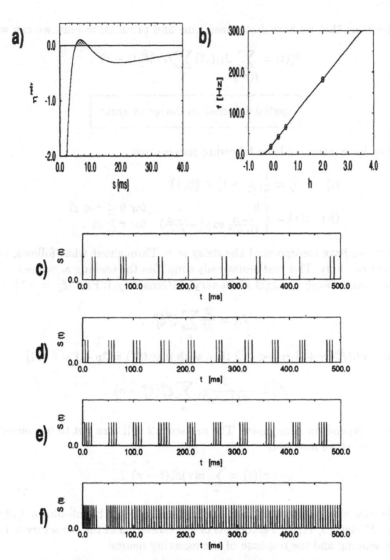

Fig. 4. *'Bursting' model neuron.* a) shows the refractory function η which can be described by three exponentially decaying components. One of them has a *depolarizing* effect (shaded). This yields a periodic bursting as shown in the spike trains c-f. In b) the gain function is plotted ($\beta = 15$). The values of h at which the spike trains c-f have been calculated have been marked by open circles. c) Noise induced firing ($h^{ext} = -0.1$). d) Periodic activity bursts of three spikes each ($h^{ext} = 0.2$). e) Periodic *bursts* of 4-5 spikes ($h^{ext} = 0.5$). f) Periodic spiking; no bursts can be discerned any more. ($h^{ext} = 2.0$). The present figure is to be compared with Fig. 2. Taken from [10].

Realizing that we have discretized time and taken $\Delta t = 1\,\text{ms}$, we can write

$$h_i^s(t) = \sum_{j(\neq i)} J_{ij}(\Delta) \sum_{\tau} \epsilon(\tau) \tilde{S}_j(t - \tau) . \tag{29}$$

> spatial summation: linear *in space*

Here we have introduced the following conventions:

(i) $\quad \tilde{S}_j = \frac{1}{2}(S_j + 1) \in \{0, 1\}$

(ii) $\quad \epsilon(\tau) = \begin{cases} 0 & \text{for } 0 \leq \tau < \Delta \\ \frac{\tau - \Delta}{(\tau_{RC})^2} \exp(-\frac{\tau - \Delta}{\tau_{RC}}) & \text{for } \tau \geq \Delta . \end{cases}$

That is, we have incorporated the delay in ϵ. Throughout what follows, Δ is a function of i only. This mathematically simplifies the ensuing arguments.

(iii) The network is taught *stationary* patterns only. If $\text{Prob}\{\xi_i^\mu = +1\} = 1/2$, then

$$J_{ij} = \frac{2}{N} \sum_{\mu} \xi_i^\mu \xi_j^\mu , \tag{30}$$

and if $\text{Prob}\{\xi_i^\mu = +1\} = p < 1/2$ then with $a \equiv \langle \xi_i^\mu \rangle = 2p - 1 < 0$ [21]

$$J_{ij} = \frac{2}{N(1 - a^2)} \sum_{\mu} \xi_i^\mu (\xi_j^\mu - a) , \tag{31}$$

does not depend on Δ anymore. The prefactor 2 will turn out to be convenient. In passing we note that

$$h_i^r(t) = \sum_{\tau} \eta(\tau) \tilde{S}_i(t - \tau) . \tag{32}$$

We now study the dynamics of the above network. The full-blown Spike Response Model [10, 11, 13, 14] consists of a *combination* of the refractory behavior of the sending and the response of the receiving neuron.

5 Network Dynamics

It is convenient to introduce *overlaps* in a slightly different way than in (11)

$$m_\mu(t) = \frac{2}{N} \sum_{i=1}^{N} \xi_i^\mu \tilde{S}_i(t) \tag{33}$$

or, if $a \neq 0$,

$$m_\mu(t) = \frac{2}{N(1 - a^2)} \sum_{i=1}^{N} (\xi_i^\mu - a) \tilde{S}_i(t) . \tag{34}$$

As before, $\tilde{S}_i = (1 + S_i)/2 \in \{0,1\}$. Let us first check that (34) is a sensible definition. To this end we put $S_i(t) = \xi_i^\mu$ and find ($a \neq 0$ and $N \to \infty$)

$$m_\mu = \frac{2}{N(1-a^2)} \sum_{i=1}^{N} (\xi_i^\mu - a) \cdot (1 + \xi_i^\mu)/2$$

$$\approx \frac{\langle (\xi^\mu - a)(1 + \xi^\mu) \rangle}{1 - a^2}$$

$$= [\langle \xi^\mu - a \rangle + \langle \xi^\mu(\xi^\mu - a) \rangle]/1 - a^2$$

$$= 1$$

whereas for $S_i(t) = \xi_i^\nu$ with $\mu \neq \nu$

$$m_\mu \approx \frac{\langle (\xi^\mu - a)(1 + \xi^\nu) \rangle}{1 - a^2}$$

$$= \frac{\langle \xi^\mu - a \rangle \langle 1 + \xi^\nu \rangle}{1 - a^2}$$

$$= 0 .$$

The approximate equality $\approx$ has been obtained through the strong law of large numbers and becomes exact as $N \to \infty$. Since $a \neq 0$ can be incorporated in the overlap there is no harm in taking $p = \text{Prob}\{\xi_i^\mu = +1\}$ as a free parameter.

In terms of the overlaps $h_i(t)$ reads

$$h_i(t) = h_i^s(t) + h_i^r(t)$$

$$= \sum_{\mu=1}^{q} \xi_i^\mu \sum_\tau \epsilon_i(\tau) m_\mu(t - \tau) + \sum_\tau \eta(\tau) \tilde{S}_i(t - \tau) \qquad (35)$$

where $\tau \geq 0$ ranges through the natural numbers. Note that this representation hinges on the axonal delays depending on i; cf. model B in [29, 30].

Equation (35) tells us that the *local* dynamics (23) depends on $(\xi_i^\mu; 1 \leq \mu \leq q) = \boldsymbol{\xi}_i$. Since $\boldsymbol{\xi}_i$ varies from neuron to neuron, solving the equations of motion seems prohibitively difficult. This need not be the case, however, once we realize that neuronal nets presumably have a *low* loading ($\alpha = q/N \ll 1$). We, therefor, take q to be finite and invoke the *sublattice* idea of Sect. 3, which we summarize here for the sake of convenience.

We know that $J_{ij} = N^{-1} Q_\Delta(\boldsymbol{\xi}_i; \boldsymbol{\xi}_j)$ for some synaptic kernel Q_Δ on $\mathbb{R}^q \times \mathbb{R}^q$. The vector $\boldsymbol{\xi}_i$ varies as i travels from 1 to N but it is *always* in a corner of $[-1, 1]^q$. Let $\mathbf{x} \in \{-1, 1\}^q$ be such a synaptic corner. Then the sublattice belonging to $\mathbf{x}$ is

$$L(\mathbf{x}) = \{i | \boldsymbol{\xi}_i = \mathbf{x}\} . \qquad (36)$$

In words, $L(\mathbf{x})$ consists of all i with $\boldsymbol{\xi}_i = \mathbf{x}$. Thus we obtain a disjoint partition

$$\{1, 2, 3, \ldots, N\} = \bigcup_{\mathbf{x}} L(\mathbf{x}) \qquad (37)$$

and induce a homogeneity on each sublattice. Explicitly, $\xi_i = \mathbf{x}$ for *all* $i \in L(\mathbf{x})$. Secondly,

$$J_{ij} = N^{-1}Q_\Delta(\mathbf{x}; \mathbf{y}) \quad \text{for } all \ i \in L(\mathbf{x}) \text{ and } all \ j \in L(\mathbf{y}) \ .$$

These simple observations are the indispensible key to several analytic results below. We now focus on the low-activity limit $a \to -1$.

In the models we are interested in, the ξ_i are independent random vectors that assume $\mathbf{x}$ with propability $p(\mathbf{x})$. Then we obtain by the strong law of large numbers

$$N^{-1}|L(\mathbf{x})| = N^{-1}\sum_{i=1}^{N} \mathbb{1}_{\{\xi_i = \mathbf{x}\}} \to \langle \mathbb{1}_{\{\xi_i = \mathbf{x}\}} \rangle = p(\mathbf{x}) \tag{38}$$

that is $|L(\mathbf{x})| \propto p(\mathbf{x})N$ as $N \to \infty$.

The notion of sublattice is associated with an order parameter, the *sublattice activity*

$$m(\mathbf{x}) = \frac{1}{|L(\mathbf{x})|} \sum_{i \in L(\mathbf{x})} \tilde{S}_i \ . \tag{39}$$

Let us now rewrite m_μ in terms of the $m(\mathbf{x})$. The argument below is typical. We have

$$
\begin{aligned}
m_\mu [2/(1-a^2)]^{-1} &= N^{-1}\sum_{i=1}^{N}(\xi_i^\mu - a)\tilde{S}_i \\
&= N^{-1}\sum_{\mathbf{x}} \sum_{i \in L(\mathbf{x})} (\xi_i^\mu - a)\tilde{S}_i \\
&= N^{-1}\sum_{\mathbf{x}}(x^\mu - a)|L(\mathbf{x})| \cdot \frac{1}{|L(\mathbf{x})|} \sum_{i \in L(\mathbf{x})} \tilde{S}_i \\
&\overset{N\to\infty}{=} \sum_{\mathbf{x}}(x^\mu - a)p(\mathbf{x})m(\mathbf{x}) \\
&\equiv \langle (x^\mu - a)m(\mathbf{x}) \rangle \ .
\end{aligned}
\tag{40}
$$

Before returning to our problem we analyze a simpler one, viz., the dynamics of a model with $J_{ij} = N^{-1}Q(\xi_i; \xi_j)$. We suppress the dependence upon the delay Δ and start by computing

$$
\begin{aligned}
h_i(t) &= \sum_j J_{ij}\tilde{S}_j(t) \\
&= N^{-1}\sum_j Q(\xi_i; \xi_j)\tilde{S}_j(t) \\
&= \sum_{\mathbf{y}} N^{-1} \sum_{j \in L(\mathbf{y})} Q(\xi_i; \mathbf{y})\tilde{S}_j(t) \\
&= \sum_{\mathbf{y}} N^{-1}Q(\xi_i; \mathbf{y}) \sum_{j \in L(\mathbf{y})} \tilde{S}_j(t)
\end{aligned}
$$

$$= \sum_{\mathbf{y}} N^{-1}|L(\mathbf{y})|Q(\boldsymbol{\xi}_i;\mathbf{y}) \cdot \frac{1}{|L(\mathbf{y})|} \sum_{j \in L(\mathbf{y})} \tilde{S}_j(t)$$

$$\stackrel{N \to \infty}{=} \sum_{\mathbf{y}} p(\mathbf{y})Q(\boldsymbol{\xi}_i;\mathbf{y})m(\mathbf{y};t) \ .$$

Now $i \in L(\mathbf{x})$ for some $\mathbf{x}$ and for *all* $i \in L(\mathbf{x})$ we obtain

$$h_i(t) = h(\mathbf{x};t) = \sum_{\mathbf{y}} Q(\mathbf{x};\mathbf{y})m(\mathbf{y};t)p(\mathbf{y}) \ . \tag{41}$$

That is, $h_i(t)$ does not depend on i *as long as $i \in L(\mathbf{x})$*.

For parallel dynamics a parallel updating is performed according to the rule (23). Hence we find, because of (41) and (38),

$$m(\mathbf{x};t+1) = |L(\mathbf{x})|^{-1} \sum_{i \in L(\mathbf{x})} \tilde{S}_i(t+1)$$

$$= \langle \tilde{S}_i(t+1) \rangle_\beta$$

$$= \frac{1}{2}\{1 + \tanh[\beta h(\mathbf{x};t)]\} \tag{42}$$

as $N \to \infty$. By (40), m_μ follows. We are done. The set of equations (42) has to be iterated. No approximation is involved.

We now return to our original problem, viz., determining the dynamics for (35). If i belongs to $L(\mathbf{x})$, then

$$h_i(t) = \sum_\mu x^\mu \sum_\tau \epsilon(\tau)m_\mu(t-\tau) + \sum_\tau \eta(\tau)\tilde{S}_i(t-\tau) \ . \tag{43}$$

Here $\epsilon(\tau)$ might depend on i and, thus, would induce a further partition of $L(\mathbf{x})$. To avoid that, we assume $\epsilon(\tau)$ does not depend on i at all. The reader is invited to check the theory below for the modifications which are necessary, if we drop this assumption.

There is, however, another complication, which we must take care of: the refractory field $h_i^r(t)$ steadily changes with time, even though i is fixed. There is no harm in assuming that the support of η is finite, so that

$$h_i^r(t) \in \{h^0, h^1, \ldots, h^r, \ldots, h^n\}$$

for some finite n. The h^r with $1 \le r \le n$ induce a *partition of* $L(\mathbf{x})$ and we therefore define

$$L(\mathbf{x}, h^r; t) = \{i | \boldsymbol{\xi}_i = \mathbf{x} \wedge h_i^r(t) = h^r\} \ . \tag{44}$$

Furthermore, in view of (38) we also put

$$|L(\mathbf{x}, h^r; t)| = N \ p(\mathbf{x}, h^r; t) \ . \tag{45}$$

At each time step all neurons perform the stochastic rule (23) *in parallel* and *in*dependently of each other. Thus we find, using (23),(45), and the strong law of large numbers [33],

$$m_\mu(t+1) = \lim_{N\to\infty} \frac{2}{1-a^2} \cdot N^{-1} \sum_{i=1}^{N} (\xi_i^\mu - a)\tilde{S}_i(t+1)$$

$$= \lim_{N\to\infty} \frac{2}{1-a^2} \cdot \sum_{\mathbf{x},r} \frac{|L(\mathbf{x},h^r;t)|}{N} \cdot \frac{x^\mu - a}{|L(\mathbf{x},h^r;t)|} \sum_{i\in L(\mathbf{x},h^r;t)} \tilde{S}_i(t+1)$$

$$= \frac{1}{1-a^2} \sum_{\mathbf{x},r} p(\mathbf{x},h^r;t)(x^\mu - a)\{1 + \tanh[\beta(h(\mathbf{x},r;t) - \vartheta)]\} \quad (46)$$

with

$$h(\mathbf{x},r;t) = \sum_\mu x^\mu \sum_\tau \epsilon(\tau)m_\mu(t-\tau) + h^r(t) \ . \quad (47)$$

In passing we note that here too we could have formulated the dynamical evolution in terms of the sublattice activities $m(\mathbf{x},r;t)$, which indicate fractions of active neurons.

Due to (46) and (47) the dynamical evolution is determined completely once we know the $p(\mathbf{x},r;t)$. These depend on the neuronal behavior, more precisely, on the neuronal state one time step earlier. A neuron in $L(\mathbf{x},r;t)$ can do only two things: At time $t+1$ it *either* fires and jumps to the state h^n, say, *or* it keeps quiet and moves to

$$h^{r'} = \sum_\tau \eta(\tau)\tilde{S}_i(t+1-\tau) \ . \quad (48)$$

If it never fires, it will end up in the category h^0 which has vanishing refractory field. Then we obtain in (fairly) selfexplanatory notation

$$p(\mathbf{x},h^n;t+1) = \sum_{r=0}^{n} p(\mathbf{x},h^r;t) \cdot \frac{1}{2}\{1 + \tanh[\beta(h(\mathbf{x},r;t) - \vartheta)]\}$$

$$p(\mathbf{x},h^{r'};t+1) \stackrel{(r\neq n)}{=} p(\mathbf{x},h^r;t) \cdot \frac{1}{2}\{1 - \tanh[\beta(h(\mathbf{x},r;t) - \vartheta)]\}$$

$$p(\mathbf{x},h^0;t+1) = p(\mathbf{x},h^0;t) \cdot \frac{1}{2}\{1 - \tanh[\beta(h(\mathbf{x},0;t) - \vartheta)]\}$$

$$+ p(\mathbf{x},h^1;t) \cdot \frac{1}{2}\{1 - \tanh[\beta(h(\mathbf{x},1;t) - \vartheta)]\} \ . \quad (49)$$

The choice of h^1 is somewhat arbitrary. If not only r but several states map onto r', then these have to be included as well. Degeneracy is not expected to be the rule, though. This we assume throughout what follows.

The Eqs. (46)–(49) completely describe the dynamical evolution. At any instant

$$\sum_{r=0}^{n} p(\mathbf{x},h^r;t) = p(\mathbf{x}) \quad (50)$$

which is evident by the very definition (45) and consistent with (49) as a consequence of the conservation of probability. Equation (47) explicitly tells us that h^r refers to the spiking history of *specific* neurons whereas h^s takes care of the spiking history of all the *other* neurons. Another stochastic process (instead of Glauber) gives rise to the same equations except for the fact that one has to change $1/2\{1 + \tanh[\beta(h - \vartheta)]\}$ into a firing probability $P_F(h)$ for this specific process. Plainly, both $P_F(h)$ and $[1 - P_F(h)]$ occur in (49).

5.1 Universality

An interesting special solution is a stationary one: $m_\mu(t) \equiv m_\mu$ for all t. Then (47) simplifies to

$$h(\mathbf{x}, r; t) = \sum_\mu x^\mu m_\mu + h^r \equiv h(\mathbf{x}, r) , \qquad (51)$$

one drops t from (49) and ends up with a set of fixed-point equations. The activity $m(\mathbf{x}, h^r)$ equals $P_F[h(\mathbf{x}, r)]$. The equation (46) for m_μ allows a simple interpretation,

$$\begin{aligned} m_\mu &= \frac{2}{1 - a^2} \sum_{\mathbf{x}} (x^\mu - a) \sum_r p(\mathbf{x}, h^r) P_F(h(\mathbf{x}, h^r)) \\ &\overset{(49)}{=} \frac{2}{1 - a^2} \sum_{\mathbf{x}} (x^\mu - a) \cdot p(\mathbf{x}, h^n) , \quad 1 \le \mu \le q , \end{aligned} \qquad (52)$$

with $p(\mathbf{x}, h^n)$ being the fraction of *active* neurons in $L(\mathbf{x})$ for a given constant input $h(\mathbf{x})$. That is, m_μ is a weighted sum of neuronal activities.

Let us try to (re)interpret this result. Due to (50), we immediatly get $p(\mathbf{x}, h^n) \le p(\mathbf{x})$. More precisely, we have

$$p(\mathbf{x}, h^n) = N^{-1} \sum_{i \in L(\mathbf{x})} \tilde{S}_i(t) = p(\mathbf{x})[|L(\mathbf{x})|^{-1} \sum_{i \in L(\mathbf{x})} \tilde{S}_i(t)] .$$

Since the process is ergodic with respect to the stationary state it is in, the expression in the braces equals

$$\lim_{\mathbb{T} \to \infty} \frac{1}{\mathbb{T}} \sum_{t=0}^{\mathbb{T}} \tilde{S}_i(t) = \lim_{\mathbb{T} \to \infty} \frac{1}{\mathbb{T}} \#(\text{spikes in } [0, \mathbb{T}]) = f[h(\mathbf{x})]$$

where $f[h(\mathbf{x})]$ is the *mean firing rate* for constant input $h(\mathbf{x})$. With the benefit of hindsight, this result is quite reasonable since $p(\mathbf{x}, h^n)$ incorporates all the neurons on $L(\mathbf{x})$ for a given input $h(\mathbf{x})$ that have just fired. Thus we can separate the network-dependent part from the process-dependent one and obtain

NETWORK-DEPENDENT   PROCESS-DEP.

$$m_\mu = \frac{2}{1-a^2} \sum_{\mathbf{x}} p(\mathbf{x})(x^\mu - a) \boxed{f[h(\mathbf{x})]} \qquad (53a)$$

with

$$h(\mathbf{x}) = \sum_{\mu=1}^{q} x^\mu m_\mu. \qquad (53b)$$

Note that the firing rate in (53a) is not in [Hz] yet as our unit of time is 1 ms so that $f[h(\mathbf{x})] \leq 1$; in [Hz], we get $10^3 f[h(\mathbf{x})]$.

Equations (53a) and (53b) constitute a *universality theorem* in that in the low-loading limit the stationary overlaps are determined *by the mean firing rate only* [11]. This result is *in*dependent of the specific firing mechanism. So it is universal. Phrased differently: As to stationary activity, all firing models whether Glauber, or Hodgkin-Huxley, or integrate-and-fire, or whatever, they all belong to the same universality class. This ceases to be true, however, as soon as time-dependent, e.g. coherent oscillatory, activity appears. In passing we note that the overlap can be stationary only if the neurons fire *in*coherently (why?). We will study oscillations in more detail. Before doing so, however, we study a practical application of (53a) and (53b).

Model with absolute refractory period only. We turn to a simple model that has an absolute refractory period Δt_{abs}, no relative refractory behavior, and assume that only a single activity pattern ν is involved. Then $m_\mu = m\delta_{\mu\nu}$ and $h(\mathbf{x})$ reduces to mx^ν. We first have to verify that this ansatz is consistent with the fixed-point equation (53a). For $\mu \neq \nu$ the left-hand side of the fixed-point equation reads $m_\mu = 0$ and for the right-hand side we require

$$m_\mu = \left(\frac{2}{1-a^2}\right) \sum_{\mathbf{x}} p(\mathbf{x})(x^\mu - a)f[h(\mathbf{x})] = 0 , \qquad (54)$$

which is indeed the case. To wit, we note that the patterns are sampled *independently* so that $p(\mathbf{x}) = p(x_1, \ldots, x_\mu, \ldots, x_q) = \prod_\mu p(x_\mu)$ with $\sum_\mu p(x_\mu) = 1$ and thus

$$\sum_{\mathbf{x}} p(\mathbf{x})(x^\mu - a)f[h(\mathbf{x})] = \left[\sum_{x^\mu = \pm 1} p(x^\mu)(x^\mu - a)\right]\left(\sum_{x^\nu = \pm 1} f(mx^\nu)\right) = 0 . \quad (55)$$

In a similar vein we obtain for $\mu = \nu$ a fixed-point equation for m,

$$m = \left(\frac{2}{1-a^2}\right) \sum_{x^\nu = \pm 1} p(x^\nu)(x^\nu - a)f(J_0 mx^\nu) \qquad (56)$$

where we have introduced an extra scaling factor J_0 which is supposed to be a common factor to all the J_{ij} .

For the model with absolute refractory period Δt_{abs} and input h one finds $f(h) = 1/\bar{s}(h)$ where in the noiseless case $\bar{s}(h) = \Delta t_{\text{abs}}/\Theta(h - \vartheta)$ is the mean interval length and Θ is the Heaviside function. In plain English, if $h > \vartheta$, then $\bar{s}(h) = \Delta t_{\text{abs}}$; otherwise $\bar{s}(h) = \infty$. For finite noise, i.e., β, the expressions are more involved and it is advantageous to take a continuum limit [10], which will not be considered here. So we end up with the fixed-point equation

$$m = \left[\frac{2}{\Delta t_{\text{abs}}(1 - a^2)} \right] \sum_{x^\nu = \pm 1} p(x^\nu)(x^\nu - a)\Theta(J_0 m x^\nu - \vartheta) , \qquad (57)$$

and we are done.

5.2 Oscillations

Given a network which we have taught *stationary* patterns only, e.g., $J_{ij} = \frac{2}{N}\sum_\mu \xi_i^\mu \xi_j^\mu$, can it perform an oscillatory response? Well, we have discussed the stationary solution (if any) but as yet we did not ask whether it was *stable*. We will see shortly that it need not be. For the sake of convenience we assume $a = 0$.

We return to the dynamical evolution as given by Eqs. (46) and (47), take the limit $\beta \to \infty$ and put $a = \vartheta = 0$ so that the dynamics itself is fully deterministic: $S_i(t + 1) = \text{sgn}[h_i(t)]$. We now check whether oscillatory activity is possible. That is, we check the self-consistency of the ansatz that there exists a *coherent* oscillation with period T, which is to be determined [10].

We assume that only pattern μ is involved. Then (46) and (47) read

$$m_\mu(t + 1) = \sum_{\mathbf{x},r} p(\mathbf{x}, h^r; t)x^\mu \{1 + \text{sgn}[h(\mathbf{x}, r; t)]\}$$

with

$$h(\mathbf{x}, r; t) = x^\mu \sum_\tau \epsilon(\tau)m_\mu(t - \tau) + h^r(t) .$$

Since the oscillation should be coherent all neurons that are active in pattern μ and, hence, are associated with $x^\mu = 1$, have fired coherently at times $t = 0, -T, -2T, \ldots$ whereas the 'off' neurons with $x^\mu = -1$ stay quiescent. Accordingly we obtain

$$m_\mu(t) = (2/N) \sum_{i=1}^N \xi_i^\mu \tilde{S}_i(t)$$

$$= \sum_{n=-\infty}^{+\infty} \delta_{t,nT} \qquad (58)$$

while for $0 < t < T$

$$h(\mathbf{x}, r; t) = x^\mu \sum_{\tau \geq 0} \epsilon(\tau) \sum_{n=0}^{-\infty} \delta_{t-\tau,nT} + h^r(t)$$

with

$$h^r(t) = \sum_{\tau \geq 0} \eta(\tau) \sum_{n=0}^{-\infty} \delta_{t-\tau, nT} \tag{59}$$

for the 'on' neurons and $h^r(t) \equiv 0$ for the 'off' neurons. That is

$$h(\mathbf{x}, r; t) = \sum_{n=0}^{+\infty} [\epsilon(t + nT) + \eta(t + nT)] \quad \text{for } x^\mu = 1 \ ,$$

$$h(\mathbf{x}, r; t) = -\sum_{n=0}^{+\infty} \epsilon(t + nT) < 0 \quad \text{for } x^\mu = -1 \ . \tag{60}$$

We now verify what happens at time $t = T$,

$$x^\mu = 1 : \quad h(\mathbf{x}, r; T) = \sum_{n=1}^{+\infty} [\epsilon(nT) + \eta(nT)] \overset{!}{>} 0$$

where T is to be the first time that the sum exceeds the threshold $\vartheta = 0$. For an oscillation we then need $h(\mathbf{x}, r; t) < 0$ for $0 < t < T$. We determine T by requiring

$$T = \inf_{t>0} h(\mathbf{x}, r; t) > 0 \tag{61}$$

and fix T *approximately* through

$$\boxed{h(\mathbf{x}, r; T) = \sum_{n=1}^{+\infty} [\epsilon(nT) + \eta(nT)] = 0 \ .} \tag{62}$$

Usually, $\sum_{n=2}^{+\infty} \epsilon(nT) \ll \epsilon(T)$ and $\sum_{n=2}^{+\infty} \eta(nT) \ll \eta(T)$ so that, if we reintroduce the threshold ϑ,

$$\boxed{\epsilon(T) + \eta(T) \doteq \vartheta} \quad \Rightarrow T \ . \tag{63}$$

One can solve $\epsilon(T) = \vartheta - \eta(T)$ graphically; e.g. for $\vartheta = 0$; cf. Fig. 5.

Once we know T we can determine the stability of this coherent oscillation. The key question is: How does it manage to keep the coherence intact? Suppose one of the neurons, say neuron j, has fired Δt too late, i.e., later than $t = 0$. Is there a 'force' pulling j back into the coherent oscillation? If so, the difference Δt must be smaller after the next cycle. We determine this time difference.

Neuron j fires again at time t' given by

$$h^s(t') + h^r(t' - \Delta t) = \vartheta \tag{64}$$

so that by a Taylor-expansion around T, we get

$$h^s(T) + \partial_t h^s|_T (t' - T) + h^r(T) + \partial_t h^r|_T (t' - T - \Delta t) = \vartheta \ .$$

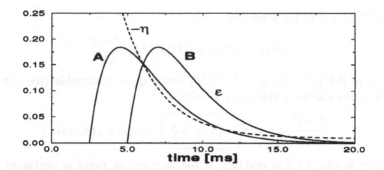

Fig. 5. *Network oscillations.* The negative of the refractory function $-\eta(\tau)$ (dotted line) and the synaptic response $\epsilon(\tau)$ (solid lines) have been plotted as a function of the time τ after a coherent spiking event in the network at $\tau = 0$. We consider two different axonal delay times, short delays (A, $\Delta = 2.5$ ms) and long delays (B, $\Delta = 5$ ms). The first intersection point $\epsilon(\tau) + \eta(\tau) = 0$ yields the oscillation period T. Note that in both models the oscillation period is approximately the same, but the oscillation is stable only if the slope of ϵ is *positive* at the intersection point, as in B. Taken from [10].

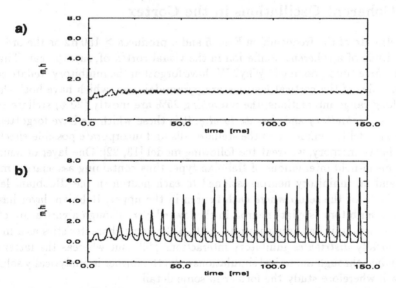

Fig. 6. *Dependence upon the axonal delay Δ.* The overlap $m(t)$ (solid lines) and the postsynaptic field $h^s(t)$ (dotted lines) have been plotted as a function of time. The network A with short delays ($\Delta = 2.5$ ms) settles into a stationary state (a), while system B with long delays ($\Delta = 5$ ms) goes into a state of synchronized spiking with period $T \approx 6$ ms (b). Note that the spiking in B occurs when $\frac{\partial}{\partial t} h^s(t) > 0$, which is a condition for stable oscillations (noise $\beta = 8$); cf. Fig. 5. Taken from [10].

Since $h^s(T) + h^r(T) = \vartheta$ we find

$$\partial_t h^r|_T \cdot (\partial_t h^s|_T + \partial_t h^r|_T)^{-1} = \frac{t' - T}{\Delta t} \ . \tag{65}$$

Assuming $\partial_t h^r|_T = \partial_t \eta|_T > 0$, which seems quite reasonable (see Fig.5), we arrive at the following stability condition

$$\frac{t' - T}{\Delta t} < 1 \Longleftrightarrow \boxed{\partial_t h^s|_T > 0} \quad \text{STABLE LOCKING} \ . \tag{66}$$

In other words, we find stability, if the intersection point is contained in the *ascending* part of the response (alpha) function ϵ. It is fulfilled for B but not for A. Hence a stable oscillation is possible for B (longer axonal delays Δ) whereas A only allows stationary activity patterns, where the neurons fire incoherently; cf. Fig. 6. Furthermore, the points 1, 2, and 3 all are bound to represent unstable oscillations. It turns out that the condition (66) is not only necessary but also sufficient for stability.

6 Coherent Oscillations in the Cortex

An estimate of the frequency in Figs. 5 and 6 produces $\geq 150\,\text{Hz}$ as the order of magnitude of a coherent oscillation in the visual cortex of, say, the cat. That is about *three* times too much. Why? We have forgotten the inhibitory stellate cells. About 75% of the cortical neurons are pyramidal cells, which have both short- and long-range interactions, the remaining 25% are mostly (20%) stellate cells, which are *inhibitory* and operate *locally*. It is these which we have forgotten.

To model the influence of the stellate cells and incorporate possible effects of associative memory, we treat the following model [13, 22]: One layer of neurons with mean-field interactions of Hebbian type, thus containing associative memory, and an inhibitory neuron assigned to each neuron in the 'Hebbian' layer. See Fig. 7 for the setup where each neuron in the upper, Hebbian, layer has its private inhibitory (stellar) neuron in the lower layer. Though a caricature of local interaction, it has been shown to contain the *same* characteristics as a model with locally *distributed* inhibitory interactions [44]. But whereas the latter can be studied through numerical simulations only, the former is analytically soluble. We will wherefore study the former in some detail.

In addition to the synaptic and refractory field used in (21) we also introduce an inhibitory field

$$h_i^{\text{inh}}(t) = \sum_\tau^{\tau_{\max}} \eta^{\text{inh}}(\tau)\tilde{S}_i(t - \tau - \Delta_i^{\text{inh}}) \ . \tag{67}$$

We now have an additional delay Δ^{inh} and the response kernel $\eta^{\text{inh}}(\tau)$ modeling the local inhibition. The summation over the past is stopped at an upper bound $\tau_{\max}$ which is large enough to include saturation effects. To be specific, we take

$$\eta^{\text{inh}}(t) = J^{\text{inh}} \exp(-t/\tau_\eta) \tag{68}$$

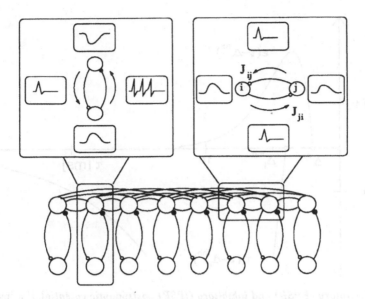

Fig. 7. *Network structure:* The network consists of two layers of 4000 neurons (only 8 are shown) which communicate through the exchange of spikes. The neurons in the top layer ('pyramidal' neurons) are fully connected by Hebbian synapses J_{ij} that store a fixed set of patterns. Each 'pyramidal' neuron is also connected to an inhibitory partner neuron (bottom). If one of the neurons fires, a spike is transmitted along the axon and, after some delay, evokes an excitatory (EPSP) or inhibitory (IPSP) postsynaptic potential at the receiving neuron (inset top left and top right). Taken from [13].

where J^{inh} is the strength and τ_η the decay constant of the local inhibitory interaction. For the sake of convenience, the refractory field only incorporates absolute refractory behavior, so we take

$$h_i^{\text{r}}(t) = \begin{cases} -\infty & \text{for } t_i^f \leq t \leq t_i^f + \tau_{\text{ref}} \\ 0 & \text{otherwise} \end{cases} \tag{69}$$

where t_i^f denote the firing times of neuron i and the absolute refractory period is set to $\tau_{\text{ref}} = 1\,\text{ms}$. The upshot is that, if we allow an external signal h_i^{ext}, we end up with four contributions to the local field of neuron i, namely

$$h_i(t) = h_i^{\text{s}}(t) + h_i^{\text{r}}(t) + h_i^{\text{inh}}(t) + h_i^{\text{ext}}(t) \tag{70}$$

describing the synaptic, refractory, local inhibitory, and external (inter)action, respectivly.

Suppose an external signal $h_i^{\text{ext}}(t) = \gamma(\xi_i^\mu + 1)/2$ is introduced at $t = t_0$. Neurons with $(\xi_i^\mu + 1) > 0$ want to fire and, for the first time, they all will. What happens thereafter is determined by the axonal delays, the response function ϵ and the refractory function η – in short, by the hardware of the network. Associative memory, if any, is determined by the 'software' in the synapses. We now consider three scenarios.

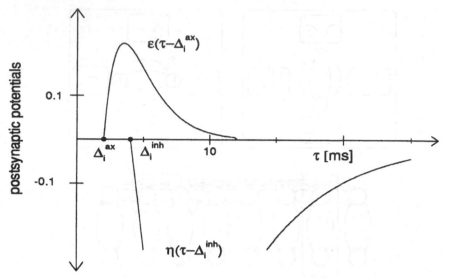

Fig. 8. *Excitatory (EPSP) and inhibitory (IPSP) postsynaptic potential.* If a 'pyramidal' neuron i fires at $\tau = 0$, a spike is transmitted to all other 'Hebbian' neurons to which it is connected. There it evokes – after some transmission delay Δ_i^{ax} – an EPSP. At the same time a signal is transmitted to the inhibitory partner neuron, which in turn sends a spike back so as to induce – after a delay Δ_i^{inh} – a strong IPSP; cf. Fig. 7. The time course of an EPSP is modeled by the alpha function $\epsilon(t) = (t/\tau_\epsilon^2)\exp(-t/\tau_\epsilon)$ with $\tau_\epsilon = 2$ ms (top). Inhibition is described by a sharp rise and an exponential decay with a time constant $\tau_\eta = 6$ ms (bottom). It lasts *much longer* than an EPSP. Taken from [13].

I. *Scenario I:* relatively short axonal delays in the range of 1–4 ms, delays in the inhibitory loop ranging between 3 and 6 ms. As compared to the previous discussion we have a true generalization since we include a *distribution* of delays for both the axons and the inhibitory loop. By construction, the IPSP can be incorporated in the refractory function η, which then explicitly depends on i due to the delay. As Fig. 9 shows, this scenario allows an osillation with the right timing, viz, 40–60 Hz. For the cat, the above delays are quite typical, so scenario I seems natural. Furthermore, it is a *weak locking* scenario [4, 13], which makes it even more realistic.

II. *Scenario II:* slightly longer axonal delays, viz., in the range 6–8 ms, and/or an EPSP which is smeared out to some extent, suffice to block a coherent oscillation and let the system relax into a state of *in*coherent firing, though at a slightly higher level than usual.

III. *Scenario III:* delays which are even longer than the ones in scenario II, viz., between 12 and 15 ms, lead again to coherent oscillations but now these *persist also after* the stimulus has been switched off. That is, the neurons are firing spontaneously. The reason is that $-\eta$ cuts ϵ, which is translated by a long Δ_{axon}, at a *positive* slope so that the resulting spontaneous oscillation is stable.

Plainly, this is no good. The period of the oscillation is shorter than the one of scenario I (why?). In addition, the oscillation is strongly locked, which is no good either.

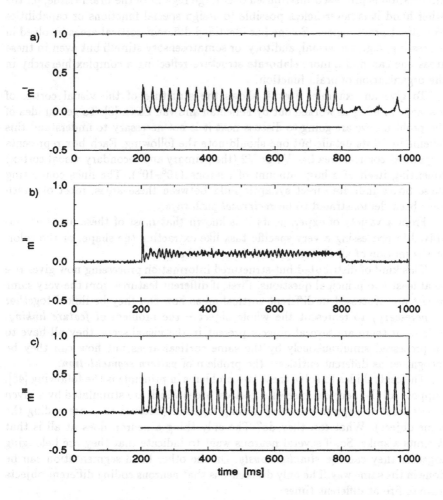

Fig. 9. *Response of a network to a stimulus:* Overlaps with a specific pattern are shown as a function of time for the three different scenarios. The pattern is supported by a weak external signal for a time between 200 and 800 time steps (ms) as indicated by the horizontal bar. Top: scenario I, oscillations occur only *while* an external stimulus is applied. Middle: scenario II, transient behaviour to a stationary retrieval state where the neurons fire incoherently. Bottom: scenario III, oscillations remain even after the stimulus has been turned off. Taken from [13].

7 Application to Binding and Segmentation

7.1 Experimental Evidence

During the last decades quite a lot of knowledge has been obtained concerning the organization of brain function. One of the main results is that on the one hand information is processed distributed over large regions of the brain tissue, on the other hand it is nevertheless possible to assign special functions or capabilities to distinct brain regions. So one has identified different cortical areas involved in processing, e.g., the visual, auditory, or somatosensory stimuli but even in these areas one can find a more elaborate structure reflecting a complex hierarchy in the organization of brain function.

To give an example, Fig. 10 shows the structure of the visual cortex of macaque monkey as worked out by Felleman and van Essen [8]. To get an idea of the problems we are going to discuss next it is not necessary to understand this scheme in all its details but one should note the following. Each box represents a specified cortical area like V1 or V2 (the primary and secondary visual cortex) consisting itself of a huge amount of neurons (10^8–10^9). The lines connecting these boxes indicate direct synaptic links between these areas, most of which have been demonstrated to be *reciprocal* pathways.

From a variety of experiments it is known that most of these areas are involved in processing a very specific task like extracting the shape, or the color, or the motion of all objects present in a visual scene.

This kind of distributed but structured information processing now gives rise to at least two principal questions. First, if different features from the very same object are processed in different cortical areas how can they be linked together (if necessary) to represent the whole object – the problem of *feature linking*. Second, if there are several objects present in the visual scene they all have to be processed simultaneously by the same cortical areas but how can they be recognized as different entities – the problem of *pattern segmentation*.

The central idea to solve these two problems in principle is the following [45]. Suppose all the neurons in several cortical areas which are stimulated by a given object want to express that they belong together (because they are coding the same object). What can they do? The only thing a neuron does at all is that it emits a spike. So, if several neurons want to indicate that they are belonging together they can fire *simultaneously*. On the other hand segmentation can be done in the same way. The only difference is that neurons coding different objects have to fire at different times.

As we have now seen it is at least in principle possible to solve the segmentation and binding task by the use of the time structure of the neuronal signal, but is this really the way the brain does it? In view of this concept a great deal of attention was payed to the discovery of coherent oscillations in the cortex in the late eighties [2, 18]. We would like to stress the word 'coherent'. Many efforts have been made to check the role of these oscillations which were found simultaneously at different sites of the same brain if the neurons were involved in processing the same object [17].

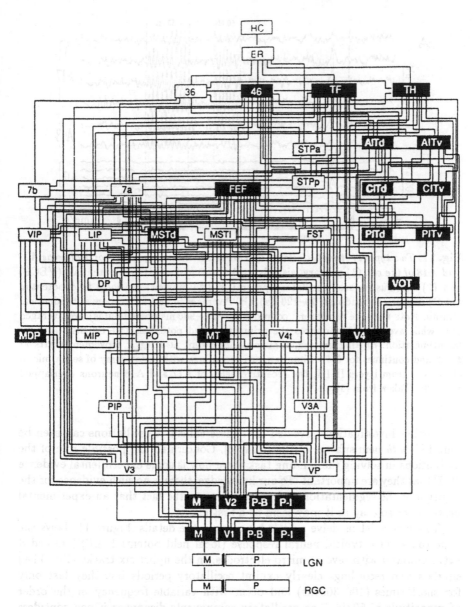

Fig. 10. *Hierarchy of visual areas.* This hierarchy shows 32 visual cortical areas, 2 subcortical visual stages (the retinal ganglion cell layer and the LGN), plus several nonvisual areas (area 7b of somatosensory cortex, perirhinal area 36, the ER, and the hippocampal complex). The areas are connected by 187 linkages, most of which have been demonstrated to be it reciprocal pathways. (Reproduced after Fig. 4 from Felleman and van Essen [8].)

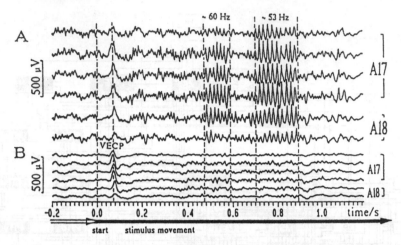

Fig. 11. *Two types of stimulus-related synchronizations among visual cortical areas A17 and A18 of the cat:* (1) primary, visually evoked (event-locked) field potentials (VECP) and (2) stimulus induced oscillatory potentials. (A) Single-sweep local field potential (LFP) responses (bandpass: 10–120 Hz). (B) Averages of 18 LFP responses to identical stimuli. Note that the oscillatory components (480–900 ms poststimulus) are averaged out, while averaging pronounces the primary evoked potentials (50–120 ms). Binocular stimulation: drifting grating; 0.7 cycles/deg swept at 4 deg/s; movement starts at $t = 0$ and continues for 3.4 s. Simultaneous recordings with linear array of seven microelectrodes from layers II/III; the receptive fields of A17 and A18 neurons overlapped partially. Taken from [5].

Further investigations revealed that these collective oscillations can even be found in *both* hemispheres of the brain [6]. Concerning the possible role of the oscillations in solving the binding task there now is some experimental evidence [3, 7] that they are important. To our knowledge, there is no such evidence for the solution of the segmentation task, mainly due to the fact that an experimental proof is not so easy as it may seem at a first glance.

To see this, let us delve a little bit more in the details. Figure 11 shows the time course of a typical neural response (local field potential, LFP) recorded extracellularly with several micro-electrodes. In the upper six tracks (Fig. 11A) single sweep recordings clearly exhibit oscillatory periods but they last only for short times (100–300 ms) and occur with variable frequency in the order of magnitude of 50 Hz. The oscillatory components disappear if one considers an average over several runs under the same stimulus condition. This clearly indicates that there is a variability either in the onset of the oscillatory periods or in their frequencies or both.

The first experimental evidence that these coherent oscillations may be involved in binding was given by Gray et al [17]. The main result of their experiment is a reflection of global stimulus properties by long-range synchronization achieved in the primary visual cortex of anesthetized cats; cf. Fig. 12 for details.

We finish here our review on experimental findings and now turn to the mod-

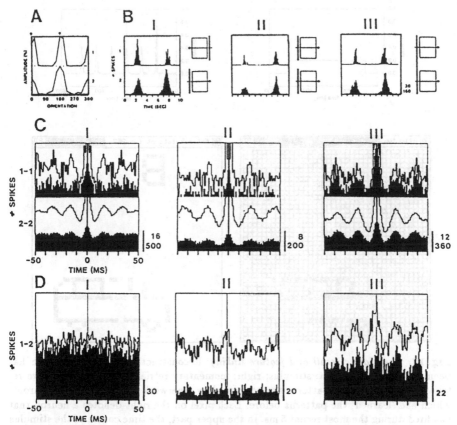

Fig. 12. *Long-range synchronization* in area 17 of the visual cortex of the cat reflects global stimulus properties. (A) Orientation tuning curves of multi-unit responses recorded at two sites separated by 7 mm. At both sites, the responses were tuned to vertical orientations (arrows). (B) Peri-stimulus-time histograms of the responses recorded at each site for three different stimulus conditions: (I) two light bars moved in opposite directions; (II) two light bars moved in the same direction; and (III) one long light bar moved across both receptive fields. A schematic plot of the receptive fields and the stimulus configuration is displayed to the right of each peri-stimulus-time histogram. (C) Auto-correlograms (1-1, 2-2) computed for the responses at both sites for each of the three stimulus conditions (I-III). (D) Cross-correlograms computed for the same responses. Note that the strongest response synchronization is observed with a continuous long light bar. For each pair of correlograms except the two displayed in C (I, 1-1) and D (I), the second direction of stimulus movement is shown with unfilled bars. Taken from [17].

eling and simulation of the described phenomena. We do not want to reproduce all the details but we will demonstrate how a solution of the binding and segmentation task can be achieved in a neural network in the same way as proposed in this section.

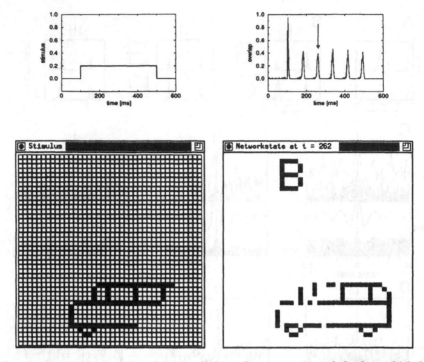

Fig. 13. *Associative recall and feature linking:* Reconstruction of the bus which has been 'learned' before (left: stimulus; right: momentary retrieval state). Associative retrieval and 'labeling' of patterns are performed by a network of 40x42 *spiking* neurons, which had 'learned' the patterns before. Each pixel on the right denotes a neuron that has fired during the most recent 5 ms. In the upper part, the time course of the stimulus and overlap with the pattern 'bus' are shown. The momentary time step is indicated by an arrow. Note that the network evolves to a weakly locked oscillatory state which is interpreted as linking the different neurons coding the bus. Neurons not touched by the stimulus but belonging to the 'bus' are excited via synaptic inputs from the rest of the network. This leads to associative retrieval of the *full* pattern including the label B. Taken from [38].

7.2 Associative linking

Now we choose the regime of weak locking which seems to be the state which is biologically most relevant and also most powerful in handling the binding problem – as we will show in this section. A network of 40x42 neurons is trained with $q = 7$ patterns, each pattern consisting of a simple contour labelled by a letter; cf. Fig 13 for an example. The fraction of neurons which are active in a pattern varied in a range between 4% and 7% (pattern-to-pattern variation). The mean activity averaged over all neurons and all patterns is $a = -0.9$. Taking many more patterns is allowed but computationally expensive. We have therefore refrained from doing so.

First of all, it is worth noticing that the associative capabilities of the Hopfield model can be found in this network of *spiking* neurons as well. The neurons are represented by pixels to allow an easy interpretation to the human reader. This does not mean that the network has been arranged according to some two–dimensional topology. Neurons which have fired one or more spikes during the most recent five time steps are coded by black pixels while the rest is white. Figure 13 shows on the left-hand side the stimulus which is applied to the network. It is switched on at $t = 100$ and kept constant up to $t = 500$. After some time, the network has reconstructed the pattern, as seen on the right-hand side of the figure. We also note that the system has accomplished some sort of hetero-association task by assigning a 'label', here a B, to the retrieved pattern. We would like to stress that this kind of pattern retrieval is time–dependent, as can be seen in Fig. 13 top right, because, due to inhibition, the neurons are not able to fire all the time. So, feature *linking* is done through the *simultaneous* activity of all neurons coding the same object.

7.3 Pattern segmentation

On the other hand, intermittent firing opens the possibility of exciting several patterns one after the other. This is exactly what is done when a superposition of *several* patterns is presented to the network. Figure 14 (top left to bottom right) shows the stimulus, then the network 100 time steps after the onset of the stimulus and finally the system as it has evolved to a weakly locked oscillatory behavior where the four simultaneously presented patterns are active *one after the other*. The network has *not* been trained to perform this oscillatory behavior; it has only learned the different patterns which constitute the scene as separate entities. The frequency of the oscillation mainly depends on the strength J_{inh} and duration τ_η of the local inhibition as can be seen in the following argument. A first estimate for the frequency of the bursts can be given in the noiseless case for a neuron which is only connected to its inhibitory partner neuron and receives a constant input higher than the activation threshold ϑ. Then Eq. (1) reduces to an exact firing condition. With the local field given in (67) and (68) one gets for a single neuron i

$$t_i^{\mathrm{burst}} = \Delta_i^{\mathrm{inh}} + \tau_\eta \ln \frac{|J^{\mathrm{inh}}|}{\gamma - \vartheta} \quad \text{for } \gamma > \vartheta \ . \tag{71}$$

Here γ denotes the strength of the external stimulus ($\gamma > \vartheta$). So, choosing $J^{\mathrm{inh}} = -2.5$ and $\tau_\eta = 25\,\mathrm{ms}$ we end up with a frequency about $11\,\mathrm{Hz}$ which is lower than the frequencies found in the cortex but convenient for the simulation. Note that parts of the patterns hidden in the stimulus are reconstructed and 'labeling' is performed by assigning different letters to the different patterns (T-B-H-L). Due to noise, locking is not perfect in that some neurons drop out of the picture as they come a little too early or too late. It is fair to say, though, that the network can perform both the segmentation of the presented scene and the completion of the objects therein. We also note that in principle a superposition of *arbitrarily* many low-activity patterns can be segmented by choosing the 'hardware' suitably. We have done so [37].

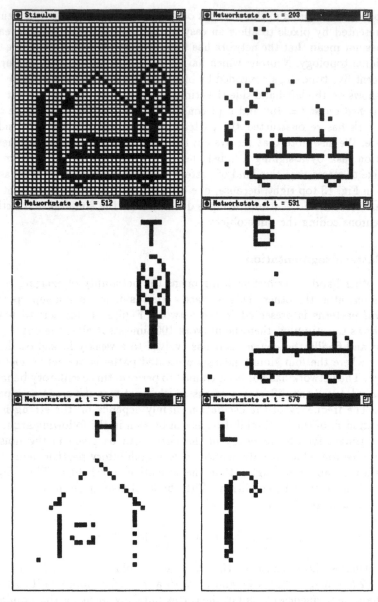

Fig. 14. *Pattern segmentation:* Four patterns which have been taught to the network as separate entities are now presented in a superposition (top left). After an intermediate phase the network nicely completes and separates the original patterns (T-B-H-L) *through temporally disjoint, coherent spiking* – in short, through coherent oscillations. Taken from [38].

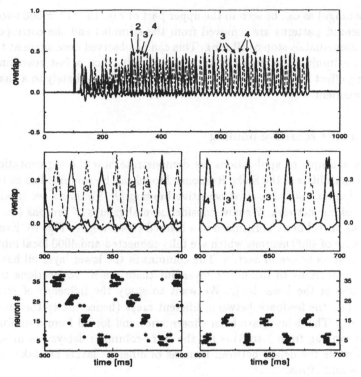

Fig. 15. *Overlap and spike raster during pattern segmentation:* A superposition of four patterns has been presented to a network of 4000 neurons. The overlap with four different patterns during a segmentation run with varying stimulus is shown as a function of time (top). All microscopic parameters are in a biologically realistic range. After 200 ms the four patterns have been separated completely. The network reacts even faster, if patterns 1 and 2 are removed from the stimulus at t = 500 ms (middle). Some typical spike rasters (bottom) before and after t = 500 ms have also been shown. Since we have *weak locking* there is no strict periodicity. Taken from [38].

7.4 Switching between patterns

The next picture (Fig. 15) shows the time structure of a segmentation run in more detail. The overlaps with the presented patterns have been plotted during the whole simulation run (top) while in the middle two sections of their time course can be seen at a better time resolution. We also show a spike raster of some representative neurons from each of the stimulated patterns (bottom). In contrast to the example given above, all microscopic parameters are now in a biologically reasonable range ($J_{inh} = -1$ and $\tau_\eta = 6$ ms). We have verified that the maximum number of patterns that can be distinguished at the same time is restricted to *four* which is supported by psychophysical data [40, 42]. In principle this number is arbitrary and depends mainly on the strength and duration of the local inhibition.

In addition, the network is able to change its global behavior on a fast time scale (ms range) as can be seen in the upper part of Fig. 15. At $t = 500$ two of the four presented patterns are removed from the stimulus and the corresponding overlaps immediately stop oscillating. This can be observed once again at $t = 900$ when the stimulus is removed completely. Plainly, such a fast reaction is an interesting effect since it allows the network to react immediately to a change in the environment.

7.5 Context sensitive binding

As a last example of applications we demonstrate how the segmentation and binding capabilities of the Spike Response Model can be combined so as to solve the even harder task of context sensitive binding [9]. To do so, we modify the structure of our network. We now consider several groups of neurons (let us call them areas or 'columns') organized in a layered structure; see Fig. 16. Every column consists of 4000 neurons which are fully connected and 4000 local inhibitory neurons, just as before in Sect. 6. The columns in the lower layer all have feedforward connections to the upper layer but there are no connections between the columns of the lower layer. We want to study the influence of *structural feedback*, i.e., the feedback between different areas (hence the epitheton ornans 'structural'). The delays have been chosen once and for all from a uniform distribution ranging from 1 to 3 ms for the intra-columnar delays (as in scenario I before) while the delays between neurons of different layers are taken to vary between 8 and 10 ms.

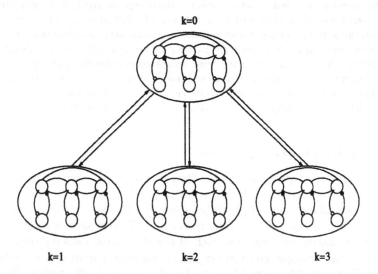

Fig. 16. *Structure of the layered network:* Four columns are organized in two layers. Each column consists of 4000 fully connected neurons and 4000 inhibitory local neurons. Every neuron of the lower layer is connected to every neuron of the upper layer (and vice versa if we allow feedback connections), but there are no connections between neurons of different columns of the same layer. Taken from [38].

The patterns have been learned in the following way. The training set of each column consists of static low-activity patterns and results in intra-columnar couplings of the same type as before [cf. Eq. 31]

$$
J_{ki,kj} = g_k \frac{2}{N(1-a^2)} \sum_{\mu=1}^{q} \xi_{ki}^{\mu}(\xi_{kj}^{\mu} - a), \quad \begin{matrix} 0 \leq k \leq K , \\ 1 \leq i,j \leq N \end{matrix} \tag{72}
$$

where k labels the columns and i the neurons in each column. In contrast to this, the feedforward and feedback couplings connecting the two layers store combinations of patterns in the lower layer by associating them with a specific pattern in the upper layer [43]. This results in inter-columnar couplings for the feedforward connections

$$
J_{0i,kj} = g_0 \frac{2}{N(1-a^2)} \sum_{(\mu,\mu')} \xi_{0i}^{\mu}(\xi_{kj}^{\mu'} - a) , \quad \begin{matrix} 1 \leq k \leq K, \\ 1 \leq i,j \leq N \end{matrix} \tag{73}
$$

and the feedback connections, if available, are given by

$$
J_{ki,0j} = g_k \frac{2}{N(1-a^2)} \sum_{(\mu',\mu)} \xi_{ki}^{\mu'}(\xi_{0j}^{\mu} - a) , \quad \begin{matrix} 1 \leq k \leq K, \\ 1 \leq i,j \leq N \end{matrix} . \tag{74}
$$

Here $\langle \mu, \mu' \rangle$ denotes a combination of patterns μ' in the different columns of the lower layer that is connected to pattern μ in the upper layer. The normalization g_k is taken so that N/g_k equals the number of connections to every neuron in column k, so

$$
g_k = \begin{cases} (1+K)^{-1} & \text{for } k = 0 \\ 1 & \text{for } 1 \leq k \leq K \text{ without feedback} \\ 0.5 & \text{for } 1 \leq k \leq K \text{ with feedback.} \end{cases} \tag{75}
$$

Putting everything together we get for the local field of neuron i in column k

$$
h_{ki}^{\text{syn}} = g_k \sum_{\mu=1}^{q} \xi_{ki}^{\mu} \overline{m}_{\mu}^{k}(t - \Delta_{ki}) + g_k \sum_{(\mu',\mu)} \xi_{ki}^{\mu'} \overline{m}_{\mu}^{0}(t - \Delta_{ki}^{\text{ar}}) \tag{76}
$$

with $1 \leq k \leq K$ and

$$
h_{0i}^{\text{syn}} = g_0 \sum_{\mu=1}^{q} \xi_{0i}^{\mu} \overline{m}_{\mu}^{0}(t - \Delta_{0i}) + g_0 \sum_{(\mu,\mu')} \sum_{k=1}^{K} \xi_{0i}^{\mu} \overline{m}_{\mu'}^{k}(t - \Delta_{0i}^{\text{ar}}) , \tag{77}
$$

where we have introduced the convention

$$
\overline{m}_{\mu}^{k}(t) = \frac{2}{N(1-a^2)} \sum_{\tau=0}^{\infty} \epsilon(\tau) \sum_{j=1}^{N} (\xi_{kj}^{\mu} - a) S_{kj}(t-\tau) , \quad 0 \leq k \leq K . \tag{78}
$$

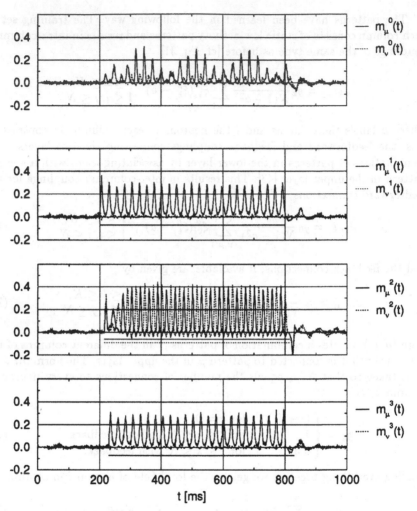

Fig. 17. *Without feedback the binding task cannot be solved.* During 600 ms (horizontal bar) the lower layer is stimulated by a learned combination of patterns but column $k = 2$ gets an additional pattern as input. The onset of the stimulation varies from column to column. We plot the time evolution of the overlaps with the stimulated patterns for columns 1 to 3 (lower three parts). On top, the overlap with the associated pattern in the upper layer (column $k = 0$) is shown; note the beat. Without feedback the activities in the lower layer do not synchronize. Thus the network is not able to solve the binding task. Taken from [38].

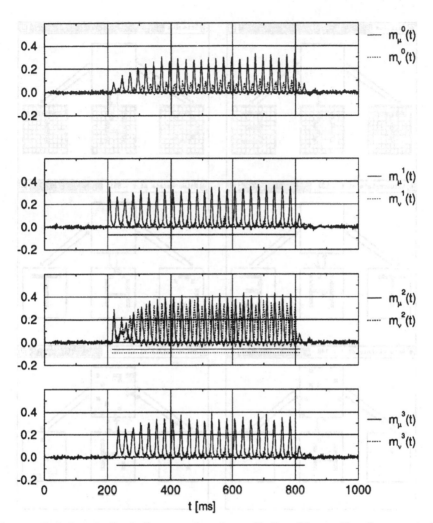

Fig. 18. *Including feedback the network performs binding.* We consider the same task as before (cf. Fig 17) but we now allow feedback connections from the upper to the lower layer. This results in synchronizing the corresponding overlaps in the lower layer. A stationary oscillation arises in the upper layer in contrast to the beat seen before. Taken from [38].

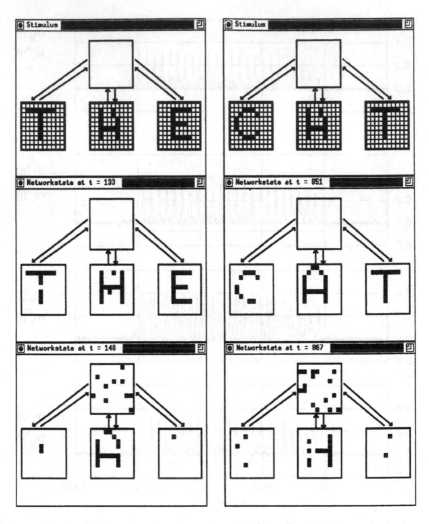

Fig. 19. *Illustration of context-sensitive binding:* Column 2 of our network is confronted with a mix of the two letters 'A' and 'H'. This results in synchronous firing of all neurons representing the same letter but phase shifted with respect to the neurons representing the other letter. When the 'T' and the 'E' is stimulated in the other columns, then the 'H' synchronizes with them to represent 'THE' (left-hand side), while in case of stimulating 'C' and 'T' in columns 1 and 3 the 'A' synchronizes thus recalling 'CAT' (right-hand side). So the very same stimulus (for column 2) is interpreted in two different ways, depending on the stimulation of the other columns. Therefore we call this *context-sensitive binding*. Taken from [38].

To demonstrate the capabilities of this network we consider the following task. The lower layer gets patterns as input that correspond to a learned combination in the upper layer but column 2 gets an additional pattern as input so that it is confronted with a superposition of two patterns. The onset of the stimulus is taken to vary for the different columns. Due to the weak locking scenario which we have chosen, this results in oscillating overlaps with the presented patterns. In column 2 the patterns are separated through a phase shift in the activity of the corresponding neurons. If we do not allow any feedback, the oscillations in the different columns will not synchronize and the oscillation frequency in column 2 will be slightly lower due to the two concurrent patterns. This results in a beat in the time evolution of the overlap with the associated pattern in the upper layer, as is shown in Fig. 17.

If we allow *structural feedback*, however, the situation changes. Now the activity of the recalled *combination* of patterns synchronizes throughout the lower layer, while the overlap with the additional pattern that is stimulated in column 2 oscillates phase-shifted by π with respect to the others. In the upper layer the overlap oscillates with stationary amplitude; cf. Fig. 18.

To get a better idea of how this can be interpreted and used for pattern recognition we finally consider the same task as before but with a partially ambivalent stimulus. In Fig. 19 we plot the stimuli and selected states of the network. For convenience only, the patterns have been taken to represent 'letters', so that one may identify the learned combinations of patterns with 'words'. In column 2 an incomplete superposition of 'A' and 'H' is given as stimulus. Depending on the stimulus in the other columns the 'H' can be completed and synchronized with the 'T' and the 'E' so as to represent the learned word 'THE' while the 'A' is still present but phase-shifted by π with respect to the other letters. On the other hand, if the stimulus in columns 1 and 3 is changed to 'C' and 'T' (in the very same simulation run), the network now combines the 'A' with the other patterns to represent the learned word 'CAT', while the 'H' is activated with some delay.

So it is fair to say that the network is able to do context sensitive binding in a way that the very same stimulus (here for column 2) is interpreted in two different ways, depending on the stimulation of the other columns, the context. In addition, this gives an idea what feedback connections that are found almost everywhere in the cortex may be good for.

8 Outlook

Because the temporal resolution of a spike is 1 ms, for practical problems spiking neurons have a new degree of freedom: *time*. Due to the refractory behavior of a neuron spikes have an excellent temporal resolution, due to the width of the ascending part of the response (alpha) function the system can perform some tuning and generate a finite domain of attraction for an emergent limit cycle, and both aspects are taken care of by the Spike Response Model.

Coherent spiking may signal that neurons belong to the same set, the same 'concept', spiking of different subsets at different times enables the network to perform a pattern segmentation. We have seen that associative feature linking and pattern segmentation or figure-ground segregation are performed by the very same network that has learned the constituents of a scene (stationary patterns) as separate entities. Though it was never trained to do a pattern segmentation, it did so when confronted with a superposition of several patterns. In a network with biologically realistic time constants the maximum number of segments of a scene that can be discerned appears to be *four* – in agreement with psychophysical experiments [40, 42]. In technical applications, however, we can easily perform a segmentation of ten *low*-activity patterns through a suitable hardware, which is not available to a biological network. It is a open problem to realize similar ideas for spatio-*temporal* patterns.

Structural feedback is a dominating feature of biological networks that consist of many 'areas', which perform specific tasks, and have both feedforward and feedback connections. What the latter may be good for has been illustrated by our *context-sensitive binding*, which provided a kind of 'attention' for an ambivalent segment of a composite pattern.

The underlying learning rule was taken to be 'Hebbian', that is, spatio-temporal patterns can be learned by correlating the delayed presynaptic and the momentary postsynaptic [41] behavior of the neurons that 'sandwich' a synapse. We so combine correlations in space (pre- and postsynaptic neuron) and time (through the delay). Also for applications it is important to realize that this learning rule is strictly local and unsupervised. Any local rule, however, has the disadvantage that it cannot cope with *global* correlations in the network. This is a serious problem. It is also quite urgent since varying activities – as they exist in reality – let the storage capacity shrink to zero. This is the *correlation catastrophe*. We expect that, e.g., unlearning [25, 20, 46], a Hebbian *anti*-learning, can remove the correlation catastrophe and greatly improve the network performance. It may well be that other local procedures will also solve the problem of global correlations, but suitable candidates have not been announced yet.

Summarizing, neural coding through spiking neurons provides us with new possibilities to process information. These possibilities have not been fully understood and exploited yet, but we hope the reader has gained a first impression.

Acknowledgments

This work was supported by the Deutsche Forschungsgemeinschaft (grant number: HE 1729/2-1). RR holds a scholarship of the Freistaat Bayern. It is a great pleasure to thank Wulfram Gerstner for his help and advice as well as for the many stimulating discussions we had with him concerning our common research. We are also most grateful to N. S. Sutherland (Brighton) for drawing our attention to the work of Sperling.

References

1. Amit, D.J., Gutfreund, H., Sompolinsky, H.: Statistical mechanics of neural networks near saturation. Ann Phys (NY) **173** (1987) 30–67
2. Eckhorn, R., Bauer, R., Jordan, W., Brosch, M., Kruse, W., Munk, M., Reitboeck, H.J.: Coherent oscillations: A mechanism of feature linking in the visual cortex? Biol. Cybern. **60** (1988) 121–130
3. Eckhorn, R., Brosch, M.: Synchronous oscillatory activities between areas 17 and 18 in the cat's visual cortex. J. Neurophysiol. (1994) in press.
4. Eckhorn, R., Obermueller, A.: Single neurons are differently involved in stimulus-specific oscillations in cat visual cortex. Exp. Brain Res. **95** (1993) 177–182
5. Eckhorn, R., Reitboeck, H.J., Arndt, M., Dicke, P.: Feature linking via synchronization among distributed assemblies: Simulations of results from cat visual cortex. Neur. Comp. **2** (1990) 293–307
6. Engel, A.K., König, P., Kreiter, K., Singer, W.: Interhemispheric synchronization of oscillatory neural responses in cat visual cortex. Science **252** (1991) 1177–1179
7. Engel, A.K., König, P., Singer, W.: Direct physiological evidence for scene segmentation by temporal coding. Proc. Natl. Acad. Sci. USA **88** (1991) 9136–9140
8. Felleman, D.J., van Essen, D.C.: Distributed hierarchical processing in the primate cerebral cortex. Cerebral Cortex 1 (1991) 1–47
9. Fuentes, U.: Einfluß der Schicht- und Arealstruktur auf die Informationsverarbeitung im Cortex. Diplomarbeit, Technische Universität München, 1993.
10. Gerstner, W., van Hemmen, J.L.: Associative memory in a network of 'spiking' neurons. Network **3** (1992) 139–164
11. Gerstner, W., van Hemmen, J.L.: Universality in neural networks: The importance of the mean firing rate. Biol. Cybern. **67** (1992) 195–205
12. Gerstner, W., van Hemmen, J.L.: Coding and information processing in neural networks. In E. Domany, J.L. van Hemmen, and K. Schulten, Editors, *Models of neural networks II*. Springer, New York, 1994. Chap. 1.
13. Gerstner, W., Ritz, R., van Hemmen, J.L.: A biologically motivated and analytically soluble model of collective oscillations in the cortex: I. Theory of weak locking. Biol. Cybern. **68** (1993) 363–374
14. Gerstner, W., Ritz, R., van Hemmen, J.L.: Why spikes? Hebbian learning and retrieval of time-resolved excitation patterns. Biol. Cybern. **69** (1993) 503–515
15. Goles, E., Olivos, J.: Comportement périodique des fonctions à seuil binaires et applications. Discr. Appl. Math. **3** (1981) 93–105
16. Goles, E., Vichniac, Y.: Lyapunov functions for parallel neural networks. In J.S. Denker, Editor, *Neural networks for computing*, pp. 165–181. American Institute of Physics, New York, 1986.
17. Gray, C.M., König, P., Engel, A.K., Singer, W.: Oscillatory responses in cat visual cortex exhibit inter-columnar synchronization which reflects global stimulus properties. Nature **338** (1989) 334–337
18. Gray, C.M., Singer, W.: Stimulus-specific neuronal oscillations in orientation columns of cat visual cortex. Proc. Natl. Acad. Sci. USA **86** (1989) 1698–1702
19. Hebb, D.O.: *The organization of behavior*. Wiley, New York, 1949
20. van Hemmen, J.L.: Hebbian learning and unlearning. In *Neural networks and spin glasses*, pp. 91–114. World Scientific, Singapore, 1990.
21. van Hemmen, J.L, Gerstner, W., Herz, A.V.M., Kühn, R., Sulzer, B., Vaas, M.: Encoding and decoding of patterns which are correlated in space and time. In

G. Dorffner, Editor, *Konnektionismus in Artificial Intelligence und Kognitionsforschung*, pp. 153–162. Springer, Berlin, Heidelberg, New York, 1990.

22. van Hemmen, J.L, Gerstner, W., Ritz, R.: A 'microscopic' model of collective oscillations in the cortx. In J.G. Taylor et al., Editors, *Perspectives in neural computing*, pp. 250–257. Springer, Berlin, Heidelberg, New York, 1992.

23. van Hemmen, J.L., Grensing, D., Huber, A., Kühn, R.: Elementary solution of classical spin glass models. Z. Phys. B **65** (1986) 53–63

24. van Hemmen, J.L., Grensing, D., Huber, A., Kühn, R.: Nonlinear neural networks I and II. J. Stat. Phys. **50** (1988) 231–257 and 259–293

25. van Hemmen, J.L., Ioffe, L.B., Kühn, R., Vaas, M.: Increasing the efficiency of a neural network through unlearning. Physica A **163** (1990) 386–392

26. van Hemmen, J.L., Kühn, R.: Nonlinear neural networks. Phys. Rev. Lett. **57** (1986) 913–916

27. van Hemmen, J.L., Kühn, R.: Collective phenomena in neural networks. In E. Domany, J.L. van Hemmen, and K.Schulten, Editors, *Models of neural networks*. Springer, Berlin, Heidelberg, New York, 1991.

28. Herz, A.V.M., Li, Z., van Hemmen J.L.: Statistical mechanics of temporal association in neural networks with transmission delays. Phys. Rev. Lett. **66** (1991) 1370–1373

29. Herz, A.V.M., Sulzer, B., Kühn, R., van Hemmen, J.L.: The Hebb rule: Representation of static and dynamic objects in neural nets. Europhys. Lett. **7** (1988) 663–669

30. Herz, A.V.M., Sulzer, B., Kühn, R., van Hemmen, J.L.: Hebbian learning reconsidered: Representation of static and dynamic objects in associative neural nets. Biol. Cybern. **60** (1989) 457–467

31. Hodgkin, A.L., Huxley, A.F.: A quantitative description of ion currents and its applications to conduction and excitation in nerve membranes. J. Physiol. (London) **117** (1952) 500–544

32. Hopfield, J.J.: Neural networks and physical systems with emergent collective computational abilities. Proc. Natl. Acad. Sci. USA **79** (1982) 2554–2558

33. Lamperti, J.: *Probability*. Benjamin, New York, 1966. Sects. 7 and 15.

34. Little, W.A., Shaw, G.L.: Analytical study of the memory storage capacity of a neural network. Math. Biosc. **39** (1974) 281–290

35. Pfeiffer, R.R., KiangY.S.: Spike discharge patterns of spontaneous and continously stimulated activity in the cochlea nucleus. Biophys. J. **5** (1965) 301–316

36. Riedel, U., Kühn, R., van Hemmen, J.L.: Temporal sequences and chaos in neural nets. Phys. Rev. A **38** (1988) 1105–1108

37. Ritz, R., Gerstner, W., van Hemmen, J.L.: A biologically motivated and analytically soluble model of collective oscillations in the cortex: II. Application to binding and pattern segmentation. Biol. Cybern. (1994) submitted.

38. Ritz, R., Gerstner, W., van Hemmen, J.L.: Associative binding and segregation in a network of spiking neurons. In E. Domany, J.L. van Hemmen, and K. Schulten, Editors, *Models of neural networks II*. Springer, New York, 1994. Chap. 5.

39. Rudin, W.: *Real and Complex Analysis*. McGraw-Hill, New York, 1974. p. 63.

40. Sperling, G.: The information available in brief visual presentations. Psychol. Monogr. **74**(11 Whole No. 498) (1960) 1–29

41. Stuart, G.J., Sakmann, B.: Active propagation of somatic action potentials into neocortical pyramidal cell dendrites. Nature **367** (1994) 69–72

42. Sutherland, S.: Only four possible solutions. Nature **353** (1991) 389–390

43. Sutton, J.P., Beis, J.S., Trainor, L.E.H.: Hierarchical model of memory and memory loss. J. Phys. A **21** (1988) 4443–4454
44. Trefz, T.: Oszillationen im Cortex. Diplomarbeit, Technische Universität München, 1991.
45. von der Malsburg, C.: The correlation theory of brain function. Internal Report 81-2, MPI für Biophysikalische Chemie, Göttingen, 1981; reprinted in E. Domany, J.L. van Hemmen, and K. Schulten, Editors, *Models of neural networks II.* Springer, New York, 1994. Chap. 2.
46. Wimbauer, S., Klemmer, N., van Hemmen, J.L.: Universality of unlearning. Neural Networks **7** (1994) in press

Hebbian Unlearning

Stefan Wimbauer and J. Leo van Hemmen

Physik-Department, Technische Universität München

D-85747 Garching bei München, FR Germany

Abstract

Unlearning, a reverse process to learning according to Hebb's rule, is a local and unsupervised procedure that gives rise to a substantial improvement of the retrieval properties of an associative neural network: (i) an enhancement of both the storage capacity and the domains of attraction, (ii) the possibility to store correlated patterns, and (iii) the capability to distinguish between patterns and non-retrieval states. Three different versions of this type of algorithm are reviewed and the basic properties of these algorithms are investigated. We demonstrate that the same microscopic mechanism underlies all three of them. Furthermore, unlearning is applied succesfully to the storage of *temporal* sequences of *correlated* patterns which have been learned in a purely Hebbian way.

1 Introduction

Talking about the *storage* of patterns in an associative neural network the notion of *unlearning* seems to be a bit paradoxical. The name has its origins in the fact that during unlearning a reverse process to Hebb's learning [1] takes place, and we will argue, and show, that unlearning greatly improves the network performance.

Unlearning shares the two main advantages of learning according to Hebb's rule. Learning and unlearning are perfectly unsupervised and local procedures. Unsupervised means that once the algorithm has been formulated and the patterns have been presented, the network runs without making any contact to the outside world. Locality on the other hand describes the property that only information of "neighbouring" neurons is required to adjust a synapse that connects them. Both the locality and the capability to do without an external "teacher" are necessary prerequisites for an algorithm to be biologically plausible or to be realizable in an electrical or optical hardware implementation of a neural network.

An associative network, however, that employs the Hebb's rule suffers from two severe shortcomings. First, its critical storage capacity $\alpha_c = \max(q/N) = 0.138$ [2] lies far below the theoretical upperbound of $\alpha_c = 2$ (or $\alpha_c = 1$ for symmetric bonds), as calculated by Gardner [3]. Furthermore, only patterns that are uncorrelated can be stored and retrieved correctly.

Unlearning helps to overcome the above difficulties. An enhancement of the storage capacity, even reaching the theoretical upperbound of $\alpha_c = 1$ for symmetric bonds in one formulation, is achieved and the storage of correlated patterns becomes possible. At the same time unlearning shows other favorable properties, e.g., the capability of distinguishing between patterns and non-retrieval states of a network with synchronous dynamics.

Unlearning has been introduced by Hopfield, Feinstein and Palmer [4], who implemented an idea of Crick and Mitchinson [5] about the function of dream sleep. Another interesting contribution to the field has been made by Plakhov and Semenov [6]. This article reviews work about unlearning done by van Hemmen, Ioffe, Kühn, Vaas, Klemmer, and Wimbauer [7],[8],[9]. In particular, we aim at demonstrating that unlearing works in a wide range of possible implementations and explaining the common underlying mechanisms that give rise to this property.

In Section 2 we present a short overview of associative networks and alternative algorithms to improve the retrieval properties of these networks are depicted. In Section 3 we introduce three different unlearning Scenarios and describe the basic properties of these formulations. We consider the three unlearning procedures as prototypes of a wider class of learning algorithms that are quadratic in the couplings (see below) and hence nonlocal and that can be transformed into local rules by introducing stochastic initial conditions. In Section 4 we calculate the temporal development of the couplings during unlearning. From this calculation in conjunction with a discussion of the stability parameters an explanation of the effects of unlearning can be deduced. In Section 5 the unlearning procedure is applied to networks which have to store temporal sequences of *correlated* patterns and the features of this model are deduced numerically. We close our considerations with a number of conclusions in Section 6.

2 Learning in an associative network

The prototypes of an associative network were described by Hopfield and Little [10],[11]. N formal twostate neurons $S_i = \pm 1$ are connected all-to-all through symmetric synapses $J_{ij} = J_{ji}$. In order to store q static patterns $\{\xi_i^\mu; 1 \leq i \leq N\}$ the synapses are adjusted according to Hebb's rule

$$J_{ij} = \frac{1}{N} \sum_{\mu=1}^{q} \xi_i^\mu \xi_j^\mu. \tag{1}$$

For timedependent patterns Hebb's rule has to be modified so as to read [12]

$$N\Delta J_{ij} = \frac{1}{T} \sum_{0 \leq t \leq T} S_i(t+1)S_j(t), \tag{2}$$

where T is the learning time. That is to say, the presynaptic signal emitted by j arrives at i at time t and is correlated with i's ensuing postsynaptic behaviour at time $t+1$. The rule is *local* in that the change ΔJ_{ij} depends only on the states of the neurons at i and j.

During the retrieval phase a new pattern $\{\zeta_i; 1 \leq i \leq N\}$ that resembles one of the stored patterns is presented to the network as an input. The degree of similarity between the presented and one of the stored patterns is expressed by the socalled overlap

$$m_\mu = \frac{1}{N} \sum_{i=1}^{N} S_i \zeta_i^\mu. \tag{3}$$

The network relaxes to a final state, which is then the output of the network. Intuitively, one expects the network to relax to that stored pattern as a final state which the presented pattern resembles most.

After each time step of the relaxation procedure the local fields

$$h_i(t) = \sum_j J_{ij} S_j(t). \tag{4}$$

of all neurons are to be calculated. During the next time step the neurons will take the values

$$S_i(t+1) = \text{sgn}\left[h_i(t)\right]. \tag{5}$$

This zero temperature parallel dynamics will be used throughout the article. It is called parallel since *all* neurons are updated at the same time.

As was already mentioned above, a faithful retrieval of patterns within the Hopfield network is only possible as long as the number of patterns lies below the critical storage capacity of $\alpha_c = \max(q/N) = 0.138 \ll \alpha_c^{max} = 1.0$ and the patterns are not correlated.

There have been many attempts to overcome the difficulties of overlearning and correlation. The most straightforward way would be to use a projector onto the subspace $\mathcal{L}$ of $\mathbf{R}^N$ spanned by the q stored patterns as a coupling matrix. The patterns have to be assumed linearly independent, which is not a severe restriction. The upshot is the so called pseudo-inverse matrix or projector matrix

$$J_{ij} = \frac{1}{N} \sum_{\mu\nu} \xi_i^\nu (C^{-1})_{\nu\mu} \xi_j^\mu, \tag{6}$$

where C is the correlation matrix of the patterns

$$C_{\nu\mu} = \frac{1}{N} \sum_{i=1}^N \xi_i^\nu \xi_i^\mu. \tag{7}$$

How this coupling matrix works can be seen by putting a pattern $\{\xi_j^k; 1 \le j \le N\}$, or a linear combination of patterns, into the local potential

$$
\begin{aligned}
h_i &= \sum_i J_{ij}\xi_j^\kappa = \sum_j \frac{1}{N} \sum_{\mu\nu} \xi_i^\mu (\hat{C}^{-1})_{\mu\nu} \xi_j^\nu \xi_j^\kappa \\
&= \sum_{\mu\nu} \xi_i^\mu (\hat{C}^{-1})_{\mu\nu} \hat{C}_{\nu\kappa} = \xi_i^\kappa.
\end{aligned} \tag{8}
$$

The local potential is the stored pattern itself. If the presented pattern, however, lies in the subspace $\mathcal{L}_\perp$ that is perpendicular to the one spanned by the stored patterns, then the local potential will be zero.

Hence the projector coupling matrix overcomes both shortcomings of Hebb's rule. As long as the patterns are linearly independent a correlation among the patterns does not cause any problem. Furthermore, the theoretical upper bound of the storage capacity for symmetric bonds of $\alpha_c = 1$ can be obtained. However, it should be noted, that $\alpha_c = 1$ can be reached only if one sets the self-interaction terms J_{ii} equal to zero. Otherwise α_c is limited to $\alpha_c = 0.5$, since the basins of attraction vanish above this value [13].

The main problem with the projector matrix, however, is that we have to invert the correlation matrix C in order to derive (6) directly. This is a highly *nonlocal* operation.

Therefore algorithms would be desirable that are local and that converge towards the projector matrix. One of these algorithms is the socalled Adaline learning rule; for an overview see [14]. During each iteration step one has to adjust the synapse according to the following rule:

$$J_{ij} \rightarrow J_{ij} + \frac{\gamma}{N} \sum_{\nu} [1 - \xi_i^\nu h_i^\nu] \xi_i^\nu \xi_j^\nu \qquad (9)$$

Adaline learning can be viewed as a sort of weighted Hebbian learning. If ξ_i^ν and h_i are of the same sign, the pattern will be retrieved correctly. In this case the factor $[1 - \xi_i^\nu h_i^\nu]$ and hence the correction of the coupling is small. The opposite is true, if ξ_i^ν and h_i are of different sign. The algorithm converges towards the projector matrix, the number of iteration steps scaling as N^3 with the size N of the network [14].

Adaline learning fulfills one of the two requirements which we have set for a learning rule that is expected to be biologically plausible or realizable in an hardware implementation. It is a local rule in that only information of neighbouring neurons is required to adjust a synapse. However, Adaline learning is still a supervised procedure. During each iteration step all the patterns have to be presented anew. In a way a second external memory device would be necessary to fulfill this task.

Now unlearning enters the scene. It is a local and unsupervised procedure. In one of the three formulations presented in the following a convergence towards the projector matrix takes place. The other two formulations approximate the projector matrix and a perfect storage can be reached.

3 Three unlearning procedures

The starting point for unlearning are couplings learnt according to Hebb's rule

$$J_{ij} = \frac{1}{N} \sum_{\mu=1}^{q} \xi_i^\mu \xi_j^\mu. \qquad (10)$$

Starting with these or similiar J_{ij} a single loop of the unlearning procedure consists of three stages [10]: (i) Random shooting, i.e., we pick an initial state at random. (ii) Relaxation according to the deterministic dynamics

$$S_i(t+1) = \text{sgn}\,[h_i(t)] \qquad (11)$$

where

$$h_i(t) = \sum_j J_{ij} S_j(t). \qquad (12)$$

(iii) Unlearning, a change of the J_{ij} as will be specified more precisely below. Loosely formulated, it is learning with a *minus* sign. The unlearning loop is repeated D times. In the present paper we analyze three rather different scenarios.

- **Scenario I**: As proposed by Plakhov and Semenov [6], the local fields for a random initial configuration are calculated and all the J_{ij} are updated according to

$$J_{ij} \rightarrow J_{ij} - \frac{\epsilon}{N} h_i h_j. \qquad (13)$$

ϵ is a small number of the order of 10^{-2}. As will be demonstrated in the next section this procedure converges towards the projector matrix. However, for the convergence to take place it is crucial that the self-interaction terms J_{ij} are included. As has already been mentioned above, the storage capacity of a projector matrix including self-interaction terms is limited to $\alpha_c = 0.5$, which lies below the storage capacity achievable by the next two algorithms that will be introduced in the following.

Additionally, the number of operations one hase to perform until the couplings converge towards the projector matrix scales as N^5 with the size of the network. This lies by a factor of N^2 or $N^{1.5}$ above the number of operations needed in the two other algorithms.

However, the great advantage of this formulation lies in the fact, that it can be treated analytically, because it does not contain any nonlinearities. So the purpose of this formulation is not so much to use it in practical applications but to clarify what is going on during unlearning.

- **Scenario II**: Starting with random initial configurations the local fields are calculated and a single update of the neurons is performed

$$\eta_i = \text{sgn}\left(\sum_j J_{ij} S_j\right). \tag{14}$$

Then the new couplings are calculated according to the rule

$$J_{ij} \rightarrow J_{ij} - \frac{\epsilon}{N}\eta_i\eta_j \tag{15}$$

As was deduced numerically a storage capacity of $\alpha_c \approx 0.55$ can be achieved in this scenario. The self-interaction terms have to be excluded in Scenario II, since J_{ii} would decrease linearly with the number of unlearning steps D. This would finally cause negative stability parameters and hence impede a faithful retrieval.

The main advantage of this learning rule is that the number of operations until an optimal storage is reached scales only as N^3 with the system size. This is the same value as one obtains for Adaline learning.

Only in Scenario I does a real convergence towards the projector matrix take place. As will be derived in the next Section, in Scenario II and Scenario III the coupling matrix approximates the projector matrix for a while only and deviates from it again as unlearning proceeds. This is illustrated by Fig. 1 where we have plotted the mean final overlap of a network relaxing from the patterns to a stationary point in dependence upon the total number of unlearning steps D multiplied by the quotient of the unlearning parameter ϵ and the number of patterns q, i.e., $D\epsilon/q$. D_{opt} denotes the optimal number of unlearning steps. At this point the basins of attraction of the network are maximal, that is, the initial overlap (between the presented and the stored pattern) necessary to recognize a pattern is *minimal*. On the other hand, D_c is the critical number of unlearning steps at which the retrieval performance starts deteriorating again. Both values can be derived numerically for Scenario II,

$$D_{opt} = 1.3q/\epsilon \tag{16}$$
$$D_c = q/\epsilon(2 + 1.7\alpha). \tag{17}$$

In all three scenarios one obtains *perfect* retrieval (final overlap $m_f = 1$), for $D_{opt} \leq D \leq D_c$.

• **Scenario III**: Starting from a random initial configuration the network *relaxes to a final state*. Because the parallel dynamics is used this need not be a stationary one but it can also be a two-cycle. In order to incorporate the effects of two-cycles properly, i.e., à la (2) but now with T large, the unlearning rule has to be formulated in the following way [8],

$$J_{ij} \to J_{ij} - \frac{\epsilon}{2N}(\eta_i^1\eta_j^2 + \eta_i^2\eta_j^1), \tag{18}$$

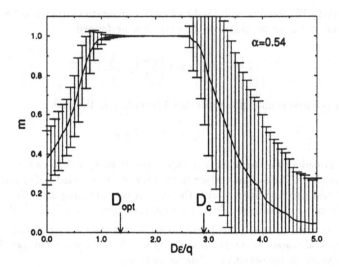

Figure 1: Average final overlap m versus $D\epsilon/q$ for a system with $N = 128$ formal neurons, $q = 70$ stored patterns, an unlearning parameter $\epsilon = 0.01$, and Scenario II. D_{opt} and D_c denote the optimal and critical number of unlearning loops at which the retrieval performance is optimal (maximal basins of attraction) and starts deteriorating again.

where η_i^1 and η_i^2 denote the two states of the two-cycle. For stationary points $\eta_i^1 = \eta_i^2$ and (18) reduces to (15).

The optimal and critical number of unlearning steps can again be derived numerically. One has to distinguish the cases that self-interaction terms are ex- or included.

Without self-interactions:	With self-interactions:
$D_{opt} = q/(2\epsilon)(1 + \alpha)$	$D_{opt} = q/\epsilon$
$D_c = q/(2\epsilon)(3 - \alpha)$	$D_c = q/\epsilon(1.14 - 0.53 \log \alpha)$

$$(19)$$

The number of operations necessary to achieve an optimal retrieval is proportional to $N^{3.5}$ and thus slightly higher than the one of Scenario II. This is because the network has to relax to a stationary state during each unlearning loop with the relaxation time scaling as $\sqrt{N}$ as can be deduced numerically.

The critical storage capacity of the network without self-interaction is $\alpha_c \approx 0.59$. In contrast to Scenario II the inclusion of self-interaction terms leads to a substantial improvement of the retrieval properties in that the *maximal* storage capacity of $\alpha_c = 1.0$ for symmetric bonds can be reached.

Stepping back for an overview, for uncorrelated patterns we obtain a storage capcity $\alpha_c = 0.50$ for Scenario I, $\alpha_c = 0.55$ for Scenario II, and $\alpha_c = 0.59$ for Sce-

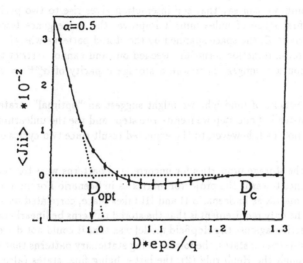

Figure 2: The arithmetic mean of the self-interaction $\langle J_{ii} \rangle$ versus $D\epsilon/q$ for Scenario III with $N = 200$ formal neurons and $q = 100$ ($\alpha = 0.5$).

nario III without self-interaction and $\alpha_c = 1.0$ with self-interaction. The results for Scenario III seem quite astonishing as compared to Scenario I and II. Whereas self-interaction terms improve the convergence towards the projector matrix and therefore have to be included in Scenario I, they limit the storage capacity to $\alpha_c = 0.5$. In Scenario II, self-interaction terms even have to be excluded because they cause a complete loss of memory shortly after having reached the $m = 1$ plateau. So one may wonder why the effect of self-interactions is so beneficial to Scenario III that α_c saturates the theoretical upperbound $\alpha_c^{max} = 1.0$: *One cannot do better.* Why is that so?

In order to understand the effects of self-ineraction in Scenario III, the evolution of the J_{ii} during unlearning is examined; see Figure 2.

One has to realize that the development of the self-interaction terms mainly depends on whether stationary patterns or two-cycles are the final states of the relaxation during an unlearning loop that has started with a random initial configuration. When unlearning starts and $D = 0$, the coupling matrix is positive semi-definite as a consequence of the self-interaction. Hence $H = -2^{-1} \sum_{ij} S_i J_{ij} S_j$ is a Lyapunov function and it is easy to show that the network relaxes *to stationary states only.* One therefore obtains, cf. (18),

$$J_{ii}(t + 1) = J_{ii}(t) - \frac{\epsilon}{2N}(\eta_i^1 \eta_i^1 + \eta_i^2 \eta_i^2) = J_{ii}(t) - \frac{\epsilon}{N}. \tag{20}$$

So J_{ii} decreases linearly and from a certain D onwards the coupling matrix is not positive semi-definite any longer. The network now can relax to two-cycles again, which means $\eta^1 = -\eta^2$ for a portion of the neurons – and it does so since in contrast to Scenario II the procedure is not finished after one time step. For the "blinking" neurons the self-interaction will increase again. After a slight "overshooting" to *small* but negative J_{ii} the mean self-interaction will approach zero, as is brought out by Figure 2.

Summerizing, we can say that self-interaction gives rise to two positive effects: during a first stage of unlearning it improves the convergence towards a coupling matrix in $\mathcal{L}$, the space spanned by the stored patterns whereas in a second stage the self-interaction terms are weeded out and cannot restrict the domains of attraction any longer. Therefore a storage capacity of $\alpha_c^{max} = 1.0$ becomes possible.

With the benefit of hindsight we might suggest an "optimal" strategy for Scenario II: Instead of one step we iterate *two* steps and use the unlearning rule (18). This does not lead, however, to the expected result since two-cycles do not occur that early.

So far only the storage properties for uncorrelated patterns were treated. However, since a convergence towards the projector matrix as in Scenario I or an approximation of the projector matrix as in Scenario II and III takes place, correlated patterns can be stored as well. The only requirement is that the stored patterns be linearly independent.

A main objection against the Hopfield model was that it could not discern between retrieval and non-retrieval states, the former being stationary patterns that were taught the network through the Hebb rule (2), the latter being final states (also stationary) that usually are not wanted. Now it is a suprising discovery that in Scenario II and III with and without selfinteraction the appearance of two-cycles as final states of the relaxation, i.e., blinking neurons, can be used as a criterion to distinguish between retrieval and non-retrieval states. That is, the network *either* relaxes to a stored pattern *or* to a final state that blinks and in this way signals that it has not been learnt.

In order to unambiguously demonstrate this capability of the network after unlearning the retrieval rate R that describes the fraction of patterns that reach a final overlap $m_f > 0.9$ when starting from an initial overlap m_i, and the fraction c of final states that are two-cycles, are plotted versus the initial overlap m_i for various system sizes N; see Figure 3.

Both the R and c curves are sigmoid functions that approach a step function as N increases without bound, what is equivalent to a first-order phase-transition. The *common* discontinuity m_c, the critical overlap, seperates the retrieval from the nonretrieval states and is at the intersection of all R and c curves. The behaviour of c and R is exactly complementary. If a state of the network is not retrieved, what happens for an initial overlap $m_i < m_c$, this is signaled by the appearance of a two-cycle. On the other hand, if it is retrieved for $m_i > m_c$, the final state is a stored pattern and, hence, not a two-cycle.

4   Unlearning as a quadratic stochastic process

In the following the common mechanisms that underly all three unlearning procedures will be investigated. To this end, we go back to Scenario I that can be treated analytically.

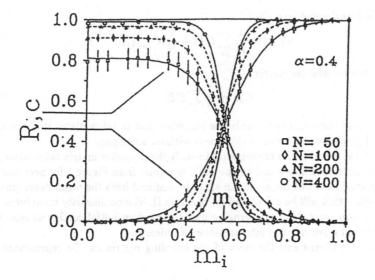

Figure 3: Retrieval rate R and fraction c of trials converging to a two-cycle for Scenario II and various system sizes. The critical overlap m_c is marked by an arrow. It turns out to be at the intersection of all curves. In the bulk limit $N \to \infty$ both R and c converge to indicator functions with their discontinuity at m_c.

In order to analyze the effect of unlearning on the synaptic weights J_{ij} it would be desirable to follow each of them seperately. In view of Eq. (12), which is an average over all j, it is simpler to *average* the change of a J_{ij} over the initial, random configuration. We then obtain [6] for the mean change,

$$J_{ij} \to J_{ij} - \frac{\epsilon}{N} \sum_k J_{ik} J_{kj}, \tag{21}$$

which can be formulated as a matrix equation,

$$\hat{J}(t+1) = \hat{J}(t) - \frac{\epsilon}{N} \hat{J}^2(t). \tag{22}$$

Calculating the square of the matrix of the couplings $\hat{J}$ is a nonlocal operation since we have to sum over all synapses k. This is to be contrasted with the only biologically plausible sum one employs in evaluating the local field h as in (4). Equation (10), however, is nothing but an averaged version of the local equation (5). Accordingly unlearning represents a possibility to transform a quadratic and nonlocal rule into a local but stochastic one.

In the formal limit "1 → 0", Eq. (22) is equivalent to a differential equation of the form

$$\frac{d\hat{J}}{dt} = -\frac{\epsilon}{N} \hat{j}^2 \tag{23}$$

with the solution

$$J_{ij}(t) = \sum_k J_{ik}^{(0)} (1 + \frac{\epsilon}{N} t \hat{J}^{(0)})_{kj}^{-1} \tag{24}$$

This can be transformed into

$$J_{ij}(t) = \frac{1}{N} \sum_{\mu\nu} \xi_i^\mu (1 + \frac{\epsilon}{N} t\hat{C})^{-1}_{\mu\nu} \xi_j^\nu, \tag{25}$$

where $\hat{C}$ is the correlation matrix

$$C_{\mu\nu} = \frac{1}{N} \sum_i \xi_i^\mu \xi_i^\nu. \tag{26}$$

Therefore a true convergence towards the projector matrix takes place. Plakhov and Semenov [6] give an exact proof of this result without averaging.

In Scenarios II and III no convergence towards the projector matrix takes place, as was already mentioned above and as one could recognize from Figure 1 for Scenario II.

An explanation of this behavior can again be deduced from the mean development of the couplings that will be calculated for Scenario II. The nonlinearity introduced by the sign function precludes the averaging over the random initial configurations. We therefore present a geometrical interpretation of unlearning.

The stored patterns and the rows of the coupling matrix can be represented as vectors in $\mathbf{R}^N$,

$$\boldsymbol{\xi}^\mu = (\xi_i^\mu | 1 \le i \le N), \tag{27}$$

$$\mathbf{J}_i = (J_{ij} | 1 \le j \le N). \tag{28}$$

The couplings learnt through Hebb's rule (2) are restricted to the subspace $\mathcal{L}$ of $\mathbf{R}^N$ that is spanned by the pattern vectors $\boldsymbol{\xi}^\mu$, which are assumed to be linearly independent,

$$\mathbf{J}_i(0) = \frac{1}{N} \sum_{\mu=1}^q \xi_i^\mu \boldsymbol{\xi}^\mu. \tag{29}$$

Unlearning in the sense of Scenario II can now be formulated in the following way,

$$\begin{aligned} J_{ij}(t+1) &= J_{ij}(t) - \frac{\epsilon}{N} \mathrm{sgn}\left(\sum_{k=1}^N J_{ik} S_k\right) \mathrm{sgn}\left(\sum_{l=1}^N J_{jl} S_l\right) \\ &= J_{ij}(t) - \frac{\epsilon}{N} \mathrm{sgn}\left(\mathbf{J}_i \cdot \mathbf{S}\right) \mathrm{sgn}\left(\mathbf{J}_j \cdot \mathbf{S}\right). \end{aligned} \tag{30}$$

We imagine the space $\mathbf{R}^N$ to be subdivided by two hyperplanes perpendicular to $\mathbf{J}_i$ and $\mathbf{J}_j$ as illustrated in Figure 4. The sign of the correction $\pm\epsilon/N$ depends only on the projection of $\mathbf{S}$ on $\mathbf{J}_i$ and $\mathbf{J}_j$ and, therefore, on the question whether or not $\mathbf{S}$ lies in the shaded region. $\mathbf{S}$ being a random vector the mean correction of J_{ij} is only a function of the angle between $\mathbf{J}_i$ and $\mathbf{J}_j$

$$\begin{aligned} \langle J_{ij}(t+1) \rangle &= J_{ij}(t) - \frac{\epsilon}{N}\left(\frac{2(\pi - \phi_{ij})}{2\pi} - \frac{2\phi_{ij}}{2\pi}\right) \\ &= J_{ij}(t) + \frac{2\epsilon}{\pi N} \arccos\left(\frac{\mathbf{J}_i \cdot \mathbf{J}_j}{|\mathbf{J}_i||\mathbf{J}_j|}\right) - \frac{\epsilon}{N} \\ &\equiv J_{ij}(t) + \langle \Delta J_{ij} \rangle, \end{aligned} \tag{31}$$

where the angular brackets denote an average over the stochastic initial conditions $\mathbf{S}$. A Taylor expansion of the correction term $\langle \Delta J_{ij} \rangle$ which becomes increasingly better for growing system size N gives

$$\langle \Delta J_{ij} \rangle = -\frac{2\epsilon}{\pi N}\left[\frac{\mathbf{J}_i \cdot \mathbf{J}_j}{|\mathbf{J}_i||\mathbf{J}_j|} + \frac{1}{6}\left(\frac{\mathbf{J}_i \cdot \mathbf{J}_j}{|\mathbf{J}_i||\mathbf{J}_j|}\right)^3 + \dots\right]. \tag{32}$$

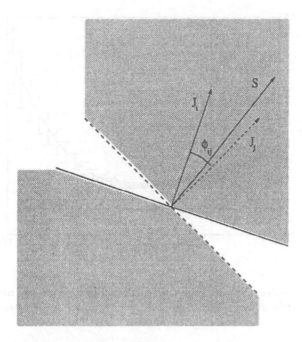

Figure 4: Geometrical interpretation of unlearning. $\mathbf{J}_i$ and $\mathbf{J}_j$ are normal vectors to two hyperplanes (here indicated by a solid and a dashed line, respectively). Depending on whether or not $\mathbf{S}$ falls into the shaded region the correction $\langle \Delta J_{ij} \rangle$ is negative or positive; cf. (31).

The first-order term differs from the correction term of Scenario I by the prefactor $2/(\pi |\mathbf{J}_i||\mathbf{J}_j|)$ that has to be examined more closely.

Plotting $|\mathbf{J}_i|$ vers $D\epsilon/q$ (see Figure 5) reveals that the $|\mathbf{J}_i|$ show a comparitively narrow distribution. Hence the following firstorder approximation seems to be reasonable,

$$\Delta J_{ij} \approx -\frac{2\epsilon}{\pi N} \frac{1}{|\mathbf{J}(D)|^2} \left(\hat{J}^2\right)_{ij} \tag{33}$$

Starting with couplings that lie within the subspace $\mathcal{L}$ spanned by the stored patterns and were learnt through Hebb's rule one easily verifies that to first order the couplings will remain in $\mathcal{L}$ during unlearning.

If one made the additional assumption that $|\mathbf{J}_i|$ does not depend on D, which seems reasonable in view of Figure 5 within the neighbourhood of the optimal number of unlearning steps $D_{opt} \approx 1.3q/\epsilon$, a convergence towards the projector matrix as in Scenario I would take place. The projector matrix is a unit matrix in $\mathcal{L}$ and vanishes in $\mathcal{L}^\perp$, the subspace perpendicular to $\mathcal{L}$.

What are the effects of higher-order terms on the approximation? It is to be noted that for $n > 1$

$$(\mathbf{J}_i \cdot \mathbf{J}_j)^n = \left(\left(\hat{J}^2\right)_{ij}\right)^n \neq \left(\hat{J}^{2n}\right)_{ij}. \tag{34}$$

Therefore, higher-order terms in the correction ΔJ_{ij} are no longer restricted to the subspace $\mathcal{L}$ and cause an increasing deviation from a convergence towards the projector matrix, that vanishes in $\mathcal{L}^\perp$. This finally causes the breakdown of the retrieval properties for $D > D_c$.

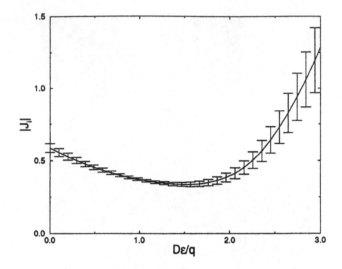

Figure 5: Mean value and width of the length-distribution of the row vectors of the coupling matrix $\hat{J}$ versus the normalised number of unlearning steps $D\epsilon/q$ with $D_{opt}\epsilon/q \approx 1.3$

The mean development of the couplings is thus dominated by two counteracting effects: an approximate movement towards the scaled unit matrix in $\mathcal{L}$ caused by the first-order term and an increase of components of $\hat{J}$ in $\mathcal{L}^\perp$ as a result of higher-order terms in $\Delta\hat{J}$.

In order to check this result numerically we have investigated the stability parameters. They are defined in the following way,

$$\text{stab}_i^\mu = \xi_i^\mu h_i(\boldsymbol{\xi}^\mu) = \xi_i^\mu \left(\sum_j J_{ij}\xi_j^\mu \right). \tag{35}$$

The patterns ξ_i^μ are stationary points of the dynamics, if ξ_i^μ and $h_i(\boldsymbol{\xi}^\mu)$ are of the same sign for all i and μ. Figure 6 shows a plot of the mean stability parameter. The distribution is characterized by a small width within the retrieval range of $1 \leq D\epsilon/q \leq 3$. So one comes close to a situation where to excellent approximation

$$\sigma\xi_i^\mu = \sum_j J_{ij}\xi_j^\mu \tag{36}$$

for some fixed σ. This would be equivalent to $\hat{J}$ being a scaled unit matrix within the subspace $\mathcal{L}$ of the stored patterns as deduced from the mean development of the couplings. However, it is impossible to draw any conclusions about the matrix elements of J_{ij} in $\mathcal{L}^\perp$ from an analysis of the stability parameters.

Since in Scenario III we allow the system to relax to a final state, which introduces correlations, it has not been possible to calculate the mean development of the couplings for this case. The similarity between the storage properties of Scenario II and III that are deduced numerically in the next section, leads to the conclusion that in principle the same microscopic mechanisms underlie both Scenario II and Scenario III.

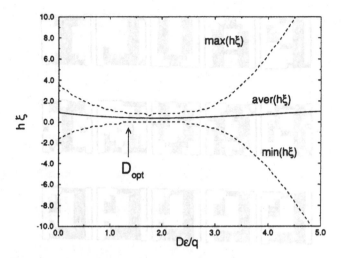

Figure 6: Distribution of the stability parameters for a system with size $N = 128$, $q = 60$, and Scenario II; $h\xi$ has been defined in (35) and max, min, aver are the maximum, minimum, and the arithmetic mean of $\xi_i^\mu h_i(\xi^\mu)$ with respect to i and μ. The arrow marks D_{opt} to be derived in the next section.

5   Unlearning of temporal sequences

Most everyday information is not presented as a static pattern but as a temporal sequence of patterns (e.g., a word consisting of letters). A succesful algorithm for storing temporal sequences, i.e., spatio-temporal patterns, in a purely Hebbian way has been proposed by Herz et al. [12],[15],[16]. It uses a modified Hebb rule that incorporates the effects of *delays* omnipresent in the brain - or also an electrical circuit. With each connection $j \to i$ we associate a delay τ_{ij}. The latter is taken to be a random variable, chosen from a given distribution and assigned to every synapse J_{ij}. Generalizing the Hebb rule (2) to the present situation we obtain [12]

$$J_{ij}(\tau_{ij}) = \frac{1}{N} \sum_{\mu=1}^{q} \frac{1}{T_\mu} \sum_{t_\mu=1}^{T_\mu} S_i(t_\mu + 1) S_j(t_\mu - \tau_{ij}) \qquad (37)$$

where T_μ is the duration of spatio-temporal pattern μ. As in (2), the presynaptic signal emitted by j at time $t - \tau_{ij}$ arrives at i at time t and is correlated with i's ensuing postsynaptic behaviour at time $t + 1$.

During retrieval the local fields at time t are

$$h_i(t) = \sum_j J_{ij}(\tau_{ij}) S_j(t - \tau_{ij}) \qquad (38)$$

where $J_{ij}(\tau_{ij})$ includes the very same delay τ_{ij}, that also prevailed during the learning session.

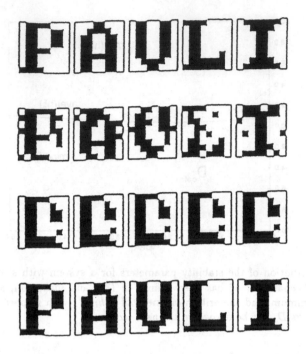

Figure 7: The word **PAULI** is stored as a *temporal* sequence P-A-U-L-I where the single letters follow each other as patterns lasting one time step. The 8×8 network has an uniform distribution of delays between $0 \leq \tau_{ij} \leq 4$.

1st line: The stored sequence. 2nd line: Four consequent letters of the word, e.g., **AULI** with 12.5% noise are presented to the network. 3^{d} line: Before unlearning, the network is not able to discern the letters and relaxes to a stationary state without a significant overlap with any one of the stored letters. 4th line: After unlearning the sequence is retrieved completely, despite the noisy initial conditions.

Unlearning now proceeds in a similiar way to the learning process (37). Suppose unlearning should start at time $t = 0$. Since the system has a memory dating back into the past for τ_{max} time steps we specify a "history" of random states during the time interval $[-\tau_{max}, 0]$. Then the network is allowed (i) to *relax* during $t_{relax} > T_\mu$ time steps, and (ii) to continue its dynamical evolution and *unlearn* during another $t_{unlearn} = T_\mu$ time steps. During this phase unlearning is performed after each time step according to the rule

$$J_{ij}(\tau_{ij}) \rightarrow J_{ij}(\tau_{ij}) - \frac{\epsilon}{N} S_i(t+1) S_j(t - \tau_{ij}) \tag{39}$$

with $t_{relax} + 1 \leq t \leq t_{relax} + t_{unlearn}$. As before, the total unlearning loop is repeated D times.

The plot of the final overlap m versus the normalized number of unlearning steps $D\epsilon/q$ starting from an unperturbed sequence shows an equivalent behaviour to the one of static patterns; cf. Fig. 1. After having reached a plateau of *perfect* retrieval with $m = 1$ the retrieval properties deteriorate again from a critical number of unlearning steps onwards and the final overlaps vanish for D large enough.

The process of unlearning of sequences depends on a wide range of different parameters, especially the length of the stored sequences T_μ and the distribution of the delays τ_{ij} of the network. For an overview what storage capacity can be obtained by unlearning and how it depends on the network architectures, we refer to [9].

As in the case of static patterns, unlearning also allows the storage of sequences of *correlated* patterns. This is demonstrated by a network that has stored a sequence of letters as an 8×8 pixel image. The results of retrieving the sequence before and after unlearning can be compared in Figure 7.

6 Conclusions

Unlearning works in a wide range of implementations with respect to relaxation times and transferfunctions as we have demonstrated by examining the three Scenarios I, II, III and by extending the unlearning procedure to temporal sequences. Central to unlearning is the introduction of stochastic initial conditions that allow the transformation of a learning rule that is quadratic in the couplings, and therefore nonlocal, into a local one. In this sense unlearing is universal. Furthermore, not only do the geometrical arguments of Section 3 apply to an exact sign function but they are also equally valid for any sigmoid functions. This makes unlearning applicable to a lot of different biological and hardware contexts. The high biological plausibility of unlearning that is caused by its locality and its unsupervised character may be enhanced even more by replacing the formal McCulloch-Pitts neurons by coding through spiking neurons.

For stationary patterns in a network of size N, the mechanism underlying unlearning can be explained as follows. The space $\mathbf{R}^N$ can be split up into two subspaces: $\mathcal{L}$ that is spanned by the patterns, and its orthogonal complement $\mathcal{L}^\perp$. In Scenario I, the coupling matrix $\hat{J}$ converges towards the projector matrix in that we obtain a limit that is equivalent to a scaled unit matrix in $\mathcal{L}$ and zero in $\mathcal{L}^\perp$. In Scenarios II and III, both without self-interaction, only a partial convergence towards the unit matrix in $\mathcal{L}$ occurs. Because of higher-order effects, matrix elements connecting $\mathcal{L}$ with $\mathcal{L}^\perp$ appear and, as the number D of unlearning steps proceeds, the retrieval properties deteriorate and are destroyed beyond D_c. In Scenario III *with* self-interaction a subtle compensation occurs and the algorithm saturates the theoretical upper bound for the storage capacity of a network with unbiased patterns. That is to say, one cannot do better.

References

[1] D.O. Hebb, *The organization of behavior*. Wiley, New York (1949)

[2] D. Amit, H. Gutfreund, H. Sompolinsky, *Statistical mechanics of neural networks near saturation*. Ann. Phys. N.Y. **173**, 30-67 (1987)

[3] E. Gardner, *The space of interactions in neural network models*. J. Phys. A: Math. Gen. **21**, 257-270 (1988)

[4] J.J. Hopfield, D.I. Feinstein, R.G. Palmer, *Unlearning has stabilizing effects in collective memories*, Nature **304**,158-9 (1983)

[5] F. Crick, G. Mitchinson *The function of dream sleep*, Nature **304**, 111-114 (1983)

[6] A. Plakhov, S. Semenov, *Unlearning-type procedures for reaching a perfect storage in neural networks*. Moscow, preprint (1992)

[7] J.L. van Hemmen, L.B. Ioffe, R. Kühn, M. Vaas, *Increasing the efficiency of a neural network through unlearning*. Physica A **163**, 386-392 (1990)

[8] J.L. van Hemmen, N. Klemmer, *Unlearning and its relevance to REM sleep: Decorrelating correlated data*. In: J.G. Taylor, E.R. Caianiello, R.M.J. Cotterill and J.W. Clark (Eds.) *Neural network dynamics*. London:Springer, 30-43 (1992)

[9] S. Wimbauer, N. Klemmer, J.L. van Hemmen, *Universality of unlearning*, Neural Networks (to appear)

[10] J.J. Hopfield, *Neural Networks and physical systems with emergent collective computational abilities*. Proc. Natl. Acad. Sci. USA **79**, 2554 - 2558 (1982)

[11] W.A. Little, *The existence of persistent states in the brain*. Math. Biosci. **19**, 101-120 (1974)

[12] A.V.M. Herz, Z. Li, J.L. van Hemmen, *Statistical mechanics of temporal association in neural networks with transmisson delays*. Phys. Rev. Lett. **66**, 1370-73 (1991)

[13] I. Kanter, H. Sompolinsky, *Associative recall of memory without errors*. Phys. Rev. A **35**, 380-392 (1987)

[14] W. Kinzel, M. Opper, *Dynamics of learning*, in: E. Domany, J.L. van Hemmen, K. Schulten (Eds.) *Models of neural networks*. Springer, Berlin, 149-172 (1991)

[15] A.V.M. Herz, B. Sulzer, R. Kühn, J. L. van Hemmen, *The Hebb rule: Storing static and dynamic objects in an associative neural network*. Europhys. Lett. **7**, 663-669 (1988)

[16] A.V.M. Herz, B. Sulzer, R. Kühn, J. L. van Hemmen, *Hebbian learning reconsidered: Representation of static and dynamic objects in associative neural nets*. Biol. Cybern. **60**, 457-467 (1989)

Mapping Discounted and Undiscounted Markov Decision Problems onto Hopfield Neural Networks

Alexandru Murgu

Department of Mathematics, University of Jyväskylä
FIN-40351 Jyväskylä, Finland

Abstract. This paper presents a framework for mapping the value-iteration and related successive approximation methods for Markov Decision Problems onto Hopfield neural networks, for both discounted and undiscounted versions of the finite state and action spaces. We analyse the asymptotic behaviour of the control sets and we give some estimates on the convergence rate for the value-iteration scheme. We relate the convergence properties on an energy function which represents the key point in mapping Markov Decision Problems onto Hopfield networks. Finally, an application from queueing systems in communication networks is taken into consideration and the results of computer simulation of Hopfield network running for the equivalent Markov Decision Problem are presented, together with some comments on possible developments.

1 Introduction

We consider some results related on the asymptotic behaviour of the value-iteration scheme

$$v_i(n+1) = \min_{k \in K(i)} \left[q_i^k + \beta \sum_{j=1}^{N} P_{ij}^k v_j(n) \right] \quad , \quad 1 \le i \le N \quad , \quad n = 0,1,2,\ldots \tag{1}$$

which arises in the finite-state ($N < \infty$) and finite action ($1 \le |K(i)| < \infty$) Markov Decision Processes [2, 15].

Here q_i^k and $P_{ij}^k \ge 0$ denote, respectively, the one-step expected reward and transition probability to state j when action k is chosen in state i ($\mathbf{P}^k$ is a stochastic matrix satisfying the relation $\sum_j P_{ij}^k = 1$, $i = 1,\ldots,N$). The starting point $\mathbf{v}(0)$ (initial value vector) is arbitrary and $v_i(n)$ denotes the minimum possible expected n-period reward starting from state i.

Asymptotic results are of interest because they show the relation between the finite-horizon and infinite-horizon models where the use of the latter case is justified if the planning horizon is large, although possibly not (exactly) known. Two types of asymptotic results are presented. One type involves the asymptotic behaviour of the value function, i.e.,

(1) $v(n)$ in the *discounted* case where the discount factor β satisfies $0 \le \beta < 1$, or

(2) $v(n) - ng^*$, where g^* is the maximal gain rate vector, in the *undiscounted* case where $\beta = 1$.

The other type of asymptotic result concerns the behaviour of the sequence of the sets of optimizing policies $S(n)$ for the stage n, defined as

$$S(n) = \prod_{i=1}^{N} K(n,i) \quad , \quad n = 1,2,3,\ldots \tag{2}$$

where

$$K(n,i) = \left\{ k \in K(i) \middle| v_i(n) = q_i^k + \beta \sum_{j=1}^{N} P_{ij}^k v_j(n-1) \right\} \quad , \quad i = 1,2,\ldots,N \tag{3}$$

is the optimal action set associated with stage n and state i.

Also, of great interest are so-called initially stationary or periodic optimal or ε-optimal strategies. The following notation will be employed. We let

$$S = \prod_{i=1}^{N} K(i) \tag{4}$$

denote the finite set of *policies*, where $K(i)$ does not depend on the current stage. A strategy $\pi = (\ldots, A^{(l)}, \ldots, A^{(1)})$ is an *infinite sequence of policies* where applying strategy π means using policy $A^{(l)}$ when there are l periods to go.

A strategy is said to be *stationary* if it uses the same policy at each period, i.e., if $A^{(l)} = A$ for all $l = 1,2,\ldots$ Note that each policy specifies a stationary strategy. Likewise, a strategy $\pi = (\ldots, A^{(l)}, \ldots, A^{(1)})$ is called *initially stationary* if there exists an integer $n_0 \ge 1$ and a policy A such that $A^{(l)} = A$ for all $l \ge n_0$. A strategy is *optimal* (or ε-optimal for any $\varepsilon > 0$) if for each $n = 1,2,\ldots$, each component of its expected n-period reward vector equals (comes within ε of) the minimal vector $v(n)$. Observe that a strategy $\pi = (\ldots, A^{(l)}, \ldots, A^{(1)})$ is optimal if and only if $A^{(l)} \in S^{(l)}$ for all $l = 1,2,\ldots$

For each $\varepsilon > 0$ and $n = 1,2,\ldots$, we define $S(n,\varepsilon)$:

$$S(n,\varepsilon) = \left\{ A \in S \middle| q_i^A + \beta \left[P^A v(n-1) \right]_i \ge v_i(n) + \varepsilon \quad , \quad i = 1,\ldots,N \right\}. \tag{5}$$

Associated with each policy $A = (A(1), A(2),..., A(N)) \in S$, where $A(i)$ is the action chosen in state i, are the reward vector $\mathbf{q}^A = \left[q_i^{A(i)} \right]$ and the transition probability matrix $\mathbf{P}^A = \left[P_{ij}^A \right]$. Thus (1) may be written as:

$$v(n+1) = \mathbf{T}v(n) = \mathbf{T}^{n+1}v(0) \quad , \quad n = 0, 1, 2, 3, ... \tag{6}$$

where $\mathbf{T}$ is an operator defined by

$$\mathbf{T}x = \min_{A \in S} \left[\mathbf{q}^A + \beta \mathbf{P}^A x \right] \tag{7}$$

and the minimization is performed component by component.

Separate treatment will be given for the discounted and undiscounted cases. In both models, the geometric rate of convergence of the value function $v(n)$ or $v(n) - n\mathbf{g}^*$ plays a central role.

In section 6 we show that elementary data transformations turn both discounted and undiscounted Markov Renewal Programs [8] into the discrete case Markov Decision Problems which are equivalent in the sense that they have the same state and policy space as well as the total discounted return or gain rate vector for each policy.

In section 7 we present the mapping of the optimization problem involved by the Markov Decision Process onto a Hopfield network and in section 8, some numerical results of simulation on Hopfield network for a queueing system problem.

2 Discounted Case: Asymptotic Behaviour of $v(n)$

The discounted case possesses an elegant treatment because the $\mathbf{T}$ operator defined by (7) is a contraction operator with contraction constant less or equal to $\beta < 1$ (which is exactly the discount factor in the value iteration scheme) when the L_∞-norm is used:

$$\|\mathbf{T}x - \mathbf{T}y\| \le \beta \|x - y\| \quad \text{for all} \quad x, y \in R^N. \tag{8}$$

The classical theory of contraction operators (summarized for example in Denardo, [7]) provides the following immediate results:

(i) $\mathbf{T}$ has a unique fixed point $v^* = \mathbf{T}v^*$.

(ii) For any starting point x, $\mathbf{T}^n x$ converges geometrically to the fixed point:

$$\|\mathbf{T}^n x - v^*\| \le \beta^n \|x - v^*\|, \quad n = 1, 2, 3, ... \tag{9}$$

(iii) An upper bound on the distance between $T^n x$ and v^* can be computed after just one iteration of T by

$$\left\| T^n x - v^* \right\| \leq \beta^n \frac{\left\| Tx - x \right\|}{1 - \beta} \quad , \quad n = 1, 2, \ldots \tag{10}$$

which is fairly sharp provided β is not too close to unity.

Additional properties follow from the fact that T is a *monotone operator* ($x \geq y$ implies $Tx \geq Ty$). The most important implication of this is given by:

$$v_i^* = \min_{A \in S} v_i^A \quad , \quad 1 \leq i \leq N, \tag{11}$$

where v^A is the total expected discounted return vector associated with policy A:

$$v^A = \left[I - \beta P^A \right]^{-1} q^A. \tag{12}$$

Observe that both v^* and v^A, $A \in S$, are independent of the initial value vector $v(0) \in R^N$. Thus, the unique fixed point of the T operator coincides with the minimal total discounted return vector, which in turn is achievable by some *stationary strategy*. As for the second implication, the monotonicity of the T operator shows that a stationary strategy A is optimal if and only if it satisfies the N minima on the right of the fixed point equation $v^* = Tv^*$. This in turn shows the existence of policies A that achieve the N minima in (11) simultaneously. Moreover,

$$x \geq v^* \ (x \leq v^*) \quad \text{implies} \quad T^n x \downarrow v^* \ (T^n x \uparrow v^*), \tag{13}$$

$$Tx \geq x \ (Tx \leq x) \quad \text{implies} \quad T^n x \downarrow v^* \ (T^n x \uparrow v^*). \tag{14}$$

Some of these results can be modified by using instead of (8) the following relation:

$$\beta(x - y)_{\min} \leq (Tx - Ty)_{\min} \leq (Tx - Ty)_{\max} \leq \beta(x - y)_{\max} \tag{15}$$

where

$$x_{\min} = (x)_{\min} = \min_{1 \leq i \leq N} x_i \text{ and } x_{\max} = (x)_{\max} = \max_{1 \leq i \leq N} x_i. \tag{16}$$

Thus (10) is replaced by

$$x_i + \frac{(Tx - x)_{\min}}{1 - \beta} \leq (Tx)_i + \frac{\beta(Tx - x)_{\min}}{1 - \beta} \leq \cdots \leq (T^n x)_i + \frac{\beta^n (Tx - x)_{\min}}{1 - \beta} \leq$$

$$v_i^{A^{(n)}} \leq v_i^* \leq (T^n x)_i + \frac{\beta^n (Tx - x)_{max}}{1 - \beta} \leq \cdots$$

$$(Tx)_i + \frac{\beta(Tx - x)_{max}}{1 - \beta} \leq x_i + \frac{(Tx - x)_{max}}{1 - \beta}, \quad n = 1, 2, \cdots \tag{17}$$

where $A^{(n)} \in S(n)$, with $v(0) = x$ and where $v^{A^{(n)}}$ is the associated total return vector. Note that the bounds in (17) are invariant to adding a constant c to each component of x. Additional improvements on the bounds as well as on the rate of convergence can be based upon data transformations [21, 26, 27] or Gauss-Seidel variants of the iterative scheme [11, 16, 22, 26], extrapolation and overrelaxation techniques [23, 29], delinearization schemes, as well as by removal self-transitions.

In terms of the original value-iteration scheme $v(n) = T^n x$ where $x = v(0)$, the above results are useful in at least four ways:

(1) $v(n)$ is shown to approach v^* geometrically fast.
(2) The $n = 1$ version of (17) with $x = v(n-1)$ gives computable bounds on the error between the fixed point v^* and the current best guess $v(n)$.
(3) Elimination via the bounds of alternatives which are not optimal for the infinite-horizon problem (MacQueen [18], Hastings [13]).
(4) Prior estimation of how many *additional* iterations $n(x)$ are required given that the current estimate of v^* is x, until the new estimate $T^n x$ lies within ε-neighbourhood of v^* or until a policy $A^{(n)} \in S(n)$ found at the end of these n iterations has a return vector $v^{A^{(n)}}$ which lies within ε-neighbourhood of v^*.

Bounds on $n(x)$ are obtained by setting

$$\left\| T^n x - v^* \right\| \leq \frac{\beta^n \| Tx - x \|}{1 - \beta} \leq \varepsilon, \tag{18}$$

or according to [10],

$$0 \leq \left\| v^* - v^{A(n)} \right\| \leq \frac{2\beta^n \| Tx - x \|}{1 - \beta} \leq \varepsilon, \tag{19}$$

with the result that at most

$$n(x) = \ln\left(\frac{2 \| Tx - x \|}{\varepsilon(1 - \beta)} \right) / \left| \ln \beta \right| \tag{20}$$

additional iterations are required. This has the property that $n(\mathbf{Tx}) \leq n(\mathbf{x}) - 1$, and so the number of remaining iterations to get accuracy ε decreases by at least unity with each iteration. Hence, the termination criterion will be met after a finite number of steps.

Unfortunately, $n(\mathbf{x})$ can be large if β is close to unity or if the initial guess $\mathbf{x}$ is far from $\mathbf{v}^*$. An encouraging feature is that $n(\mathbf{x})$ varies only logarithmically with ε such that it is practical to achieve high precisions as long as β is not too close to unity. We note that using (17) the upper bound for $n(\mathbf{x})$ in (20) may be replaced by:

$$n(\mathbf{x}) \leq \ln\left(\frac{\mathrm{sp}[\mathbf{Tx} - \mathbf{x}]}{\varepsilon(1 - \beta)} \right) / |\ln \beta| \tag{21}$$

where $\mathrm{sp}[\mathbf{x}] = x_{max} - x_{min}$ denotes the *span* of $\mathbf{x}$.

3 Discounted Case: Asymptotic Behaviour of $S(n)$ and the Existence of Initially Stationary ε-Optimal Strategies

The main question of interest is the relation of the sets $S(n)$ to the set S^* of policies which are optimal for the infinite horizon problem:

$$S^* = \left\{ A \in S \middle| \mathbf{v}^* = \mathbf{q}^A + \beta P^A \mathbf{v}^* \right\}. \tag{22}$$

Note that S^* is uniquely determined and has a Cartesian product structure. It follows directly from (19) and (20) that for each starting point $\mathbf{x} = \mathbf{v}(0)$, $S(n) \subseteq S^*$ holds for sufficiently large n, say $n \geq n_1(\mathbf{x})$. As an evaluation of $n_1(\mathbf{x})$, one may use (20) with

$$\varepsilon < \varepsilon_0 = \begin{cases} \min\left\{ v_i^A - v_i^* \middle| A \in S \text{ and } 1 \leq i \leq N \text{ such that } v_i^* < v_i^A \right\}, & \text{if } S^* \neq S, \\ \infty & , \text{if } S^* \neq S. \end{cases} \tag{23}$$

The threshold ε_0 is the minimal separation of v_i^* and the total return when starting in state i and employing a non-optimal stationary strategy. It is positive since the state and action spaces are finite. Thus value-iteration eventually settles upon optimal policies. Unfortunately this result can not be used in general while performing calculations because the lack of prior knowledge about $\mathbf{v}^*$ and the resulting inability to evaluate ε_0 makes it impossible to calculate $n_1(\mathbf{x})$ a priori. Estimation of $n_1(\mathbf{x})$ remains an outstanding problem. No ways are available to deduce whether a policy in $S(n)$ lies in S^*, except if S^* is a singleton and all non-optimal actions are eliminated. That is, a policy can appear during the first (say) fifty iterative steps yet fail to be optimal for the infinite horizon-model. Furthermore, a policy from S^* might appear in say $S(1)$, *not* appear in $S(2)$ and reappear in $S(4)$ (or never reappear), so that a policy which has "dropped out" of $S(n)$ cannot be eliminated as suboptimal. In the very special case where v^* is known, ε_0 may be estimated [28] from:

$$(v^* - v^A)_i = \left[I - \beta P^A\right]^{-1} \Gamma_i^A = \sum_{n=0}^{\infty} \sum_{j=1}^{N} (\beta P^A)_{ij}^n \Gamma_j^A \tag{24}$$

where

$$\Gamma^A = \left[q^A + \beta P^A v^* - v^*\right] \geq 0. \tag{25}$$

Namely, assuming that S^* is a proper subset of S, i.e., for all $\varepsilon_0 < \infty$, we can pick a pair (i, A) which achieve ε_0 in (23) and a state j and an integer $n \leq N$, such that $(\beta P^A)_{ij}^n > 0$ and $\Gamma_j^A > 0$. We thus find

$$\varepsilon_0 \geq (\beta \alpha)^N \delta_0 \tag{26}$$

where,

$$\alpha = \min\left\{P_{rs}^k \mid \text{for all } 1 \leq r, s \leq N \text{ and } k \in K(r) \text{ with } P_{rs}^k > 0\right\}, \tag{27}$$

$$\delta_0 = \min\left\{\Gamma_j^A \mid \text{for all } A \in S, \ 1 \leq j \leq N \text{ with } \Gamma_j^A > 0\right\} > 0, \tag{28}$$

the last inequality following from the assumption $S^* \neq S$. Hence it suffices to take $\varepsilon_0 = (\beta \alpha)^N \delta_0$ when computing $n_1(x)$.

Proposition 1. *Concerning the convergence of $S(n)$ the following properties hold for large enough n:*

1) *If S^* is a singleton, $S(n)$ must reduce to S^* for large enough n (i.e., for $n \geq n_1(x)$).*

2) *If S^* is not a singleton, $S(n)$ does not need to possess a limit as n tends to infinity.*

3) *Since $S(n)$, for large n, may oscillate or contain only a proper subset of S^*, the individual $S(n)$'s do not by themselves determine S^*.*

□

Remark 1. Related to property 2), Shapiro [28] has constructed a 2-state example where $S(n)$ oscillates with period 2 between the two members of S. His example suggests that the set $S(n)$ exhibits at least an ultimately *periodic* behaviour. However, an example which is similar to the one given in Bather [1] for the *undiscounted* case shows that the worst behaviour of $S(n)$ will be *non-periodic* oscillations.

Remark 2. For property 3) one may find the entire set S^* from ε -optimal policies, namely from

$$S^* = \lim_{n \to \infty} S(n, \varepsilon_n) \tag{29}$$

where $\{\varepsilon_n\}_{n=1}^{\infty}$ may be taken as an arbitrary $\{\varepsilon_n\}_{n=1}^{\infty}$ is slower than the one $\{v(n)\}_{n=1}^{\infty}$ exhibits, i.e., whenever $\varepsilon_n \beta^{-n} \to \infty$ as $n \to \infty$. One possible choice for $\varepsilon_n = n^{-1}$ and more generally, we can take ε_n^{-1} as a positive polynomial in n. To confirm (29), note that

$$S(n, \varepsilon_n) = \left\{ A \in S \middle| q^A + \beta P^A v(n-1) - v(n) \le \varepsilon_n 1 \right\}, \tag{30}$$

where 1 is the N-vector with all components unity. Inserting $v(n) = v^* + O(\beta^n)$ one obtains

$$S(n, \varepsilon_n) = \left\{ A \in S \middle| q^A + \beta P^A v^* - v^* \le \varepsilon_n 1 + O(\beta^n) \right\}. \tag{31}$$

Since $\varepsilon_n 1 + O(\beta^n)$ approaches zero when $\varepsilon_n \to 0$, there exists a number $\varepsilon_0 > 0$ such that $0 < \varepsilon_n 1 + O(\beta^n) < \varepsilon_0$ for all n sufficiently large. Therefore, $S(n, \varepsilon_n) = S^*$ for all n sufficiently large.

We finally turn to the issue of determining initially stationary optimal strategies. We observe before that an optimal strategy must lie in $\prod_{n=1}^{\infty} S(n)$. Shapiro's example [28] shows that in general there may be no (optimal) policy which is contained within all of the sets $S(n)$ for all n large enough. That is, none of the sequences of policies that may be generated by value-iteration needs to converge. Hence in general, no initially stationary optimal strategy may exist and the adaptation of example 1 in Bather [1], mentioned above, shows that in general no initially *periodic* optimal strategy needs to exist either. Only in the case where S^* is a singleton $(S^* = \{B\})$, we know that $S(n) = \{B\}$ for all $n \ge n_1(x)$, so that in this case every optimal strategy is initially stationary. In other words, B is the best choice of current policy if the planning horizon is *at least* $n_1(x)$ *additional* periods and this choice is optimal without knowing the exact length of the planning horizon (this type of behaviour is known as a *turnpike property*).

Fortunately, we observe that every policy in S^* comes closer and closer to being optimal at the nth stage, as n tends to infinity. This may be verified from the relation

$$\|v(n) - q^A - \beta P^A v(n-1)\| = \|v(n) - v^* - \beta P^A [v(n-1) - v^*]\|$$

$$\leq 2\beta^n \frac{|Tx - x|}{1-\beta}, \tag{32}$$

where $A \in S^*$ and $x = v(0)$, using (10). This in turn implies, for every $\varepsilon > 0$, the existence of an initially stationary strategy that is ε-optimal.

Proposition 2. *The following two properties hold:*

(i) *Any policy in S may be used in the initially stationary part of the ε-optimal strategy, i.e., the initially stationary part does not depend upon the scrap-value vector $v(0)$.*

(ii) *An upper bound for the length of the non-stationary tail of the ε-optimal strategy is given by*

$$m(x) \leq \ln\left(\frac{2\|Tx - x\|}{\varepsilon(1-\beta)^2}\right) / |\ln \beta| \quad , \quad x = v(0) \tag{33}$$

which varies again logarithmically with the accuracy ε .

□

4 Undiscounted Case: Asymptotic Behaviour of $v(n) - ng^*$

In the undiscounted case, $\beta = 1$ and the operator T is a non-expansive operator verifying the relation

$$(x - y)_{min} \leq (Tx - Ty)_{min} \leq (Tx - Ty)_{max} \leq (x - y)_{max} \quad \text{for all } x, y \in R^N. \tag{34}$$

Additionally, the T operator has the property

$$T(x + c1) = Tx + c1 \quad \text{for all } x \in R^N \text{ and scalars } c. \tag{35}$$

Note as a consequence of (35) that the T operator never has a unique fixed point and hence is never a contraction operator on R^N (and neither is any of its powers). Both (34) and (35) suggest choosing

$$sp[x] = x_{max} - x_{min} \tag{36}$$

as a quasi-norm [1]. However, the **T** operator (or any of its powers) is not necessarily *contracting* with respect to the sp-norm either. Hence, only under special conditions with respect to the (chain and periodicity) structure of the problem, there exist a number $0 \leq \alpha < 1$ and an integer $n \geq 1$ such that

$$\text{sp}\left[\mathbf{T}^n \mathbf{x} - \mathbf{T}^n \mathbf{y}\right] \leq \alpha \text{sp}\left[\mathbf{x} - \mathbf{y}\right] \ , \ \text{for all } \mathbf{x}, \mathbf{y} \in R^N . \tag{37}$$

As a consequence, the asymptotic behaviour of $\{v(n)\}_{n=1}^{\infty}$ requires an entirely different and more complicated analysis in the undiscounted case.

Define the gain rate vector $\mathbf{g}^A$ of policy $A \in S$ by

$$\mathbf{g}^A = \lim_{m \to \infty} \frac{1}{m} \left[\mathbf{I} + \mathbf{P}^A + (\mathbf{P}^A)^2 + \cdots + (\mathbf{P}^A)^{m-1}\right] \mathbf{q}^A \tag{38}$$

and we define the minimal gain rate vector $\mathbf{g}^*$ by

$$g_i^* = \min_{A \in S} g_i^A \ , \ 1 \leq i \leq N. \tag{39}$$

Howard [15] and Derman [9] have shown that policies exist which attain the N minima simultaneously, so the set

$$S_{MG} = \left\{A \in S \middle| \mathbf{g}^A = \mathbf{g}^*\right\} \tag{40}$$

of minimal gain policies is non-empty.

In contrast with the discounted case, a pair of *optimality equations* is needed in order to characterize the set of optimal (minimal gain) policies

$$g_i = \min_{k \in K(i)} \sum_{j=1}^{N} P_{ij}^k g_j \ , \ i = 1, \cdots, N, \tag{41}$$

$$v_i = \min_{k \in L(i)} \left[q_i^k - g_i + \sum_{j=1}^{N} P_{ij}^k v_j\right] \ , \ i = 1, \cdots, N, \tag{42}$$

where

$$L(i) = \left\{k \in K(i) \middle| g_i = \sum_{j=1}^{N} P_{ij}^k g_j\right\} \ , \ i = 1, \cdots, N. \tag{43}$$

Unlike the discounted case, v *is not* uniquely determined by (42) and then so does $v + c\mathbf{1}$ for any scalar c.

We consider the set $V = \{v | v \text{ satisfies (42)}\}$ and for each $v \in V$, we define

$$S^*(v) = \{A \in S | v = q^A - g^* + P^A v\} \tag{44}$$

i.e., $S^*(v)$ is the Cartesian product set of policies achieving the minima in (42) for the particular solution $v \in V$.

A policy A is *minimal gain* if for some $v \in R^N$ satisfying (42), $A(i)$ attains the minimum in (41) for all $i = 1, \cdots, N$ as well as in (42) for every state that is recurrent under P^A. *Conversely*, if A is minimal gain, then $A(i)$ satisfies (41) for all $i = 1, \cdots, N$, as well as (42) for any $v \in V$ and in all states $i \in \Omega$ that are recurrent under P^A [8].

In the case where each $P^A > 0$, Bellman [2] showed that $v(n)$ has the asymptotic behaviour $v(n) \approx ng^*$ for any $v(0) \in R^N$. Note that Bellman's condition is the strongest that one can make with respect to the chain and periodicity structure of the problem, implying, e.g., unichainedness and aperiodicity. Brown [6] showed in all generality that $\{v(n) - ng^*\}_{n=1}^{\infty}$ is bounded in n, permitting the interpretation of $g_i^* = \lim_{n \to \infty} v_i(n)/n$ as the minimal expected return per unit time starting from state i. Two cases can be distinguished concerning the asymptotic behaviour of $\{v(n) - ng^*\}_{n=1}^{\infty}$.

In the first case $\{v(n) - ng^*\}_{n=1}^{\infty}$ has a limit for any choice of $v(0)$. This corresponds roughly to the case of the discounted process. In the second case, $\{v(n) - ng^*\}_{n=1}^{\infty}$ has a limit for some, but not all choices of $v(0)$. These two cases are exhaustive since for each Markov Decision Process there exist $v(0) \in R^N$ such that $\lim_{n \to \infty} [v(n) - ng^*]$ exists, namely $v(0) = v^* + ag^*$ where v^* satisfies the optimality equation (42) and a is sufficiently large. Conditions determining the existence of $\lim_{n \to \infty} [v(n) - ng^*]$ are of importance for at least the following reasons:

(i) If $v(0)$ is such that $\lim_{n \to \infty} [v(n) - ng^*]$ exists, then $v(n) - v(n-1)$ converges to g^*, and $nv(n-1) - (n-1)v(n)$ converges to a solution $v \in V$. That is, both the minimal gain rate vector and a solution to the optimality equation (42) can be computed. Both limits are approached geometrically. However, if $\{v(n) - ng^*\}_{n=1}^{\infty}$ does not converge, then g^* cannot be estimated from $[v(n) - v(n-1)]$. It is still possible to estimate g^* from $v(n)/n$; however, the rate of convergence is now only of the order $O(n^{-1})$, whereas no estimates are available for $v \in V$.

(ii) Convergence of $\left\{v(n)-ng^*\right\}_{n=1}^{\infty}$ guarantees that $S(n)\subseteq S_{MG}$ for all n large enough, hence value iteration may be used to identify minimal gain policies [20]. However, if $v(0)$ is such that $\lim_{n\to\infty}\left[v(n)-ng^*\right]$ does not exist, then $S(n)\subseteq S_{MG}$ is not guaranteed to hold for all large n. Since value-iteration is the most practical computational method for finding minimal gain policies when n is very large, it is desirable to check whether $\lim_{n\to\infty}\left[v(n)-ng^*\right]$ is guaranteed to exist, or whether a data transformation should be performed on the original data so as to enforce the convergence.

(iii) Convergence of $\left\{v(n)-ng^*\right\}_{n=1}^{\infty}$ guarantees the existence of initially stationary ε-optimal strategies for any positive ε. Conversely, MDP's may be constructed in which $\left\{v(n)-ng^*\right\}_{n=1}^{\infty}$ fails to converge, no initially stationary strategy can be found which is ε-optimal for ε small enough.

Sufficient conditions for the convergence of $\left\{v(n)-ng^*\right\}_{n=1}^{\infty}$ were obtained by White [30] and others. There exists a positive integer J^* such that

$$\lim_{n\to\infty}\left[v(nJ^*+r)-(nJ^*+r)g^*\right]$$ (45)

exists for any $v(0)$ and any $r=0,\cdots,J^*-1$. Whenever $\left\{v(n)-ng^*\right\}_{n=1}^{\infty}$ converges, it can be shown that the approach to the limit is geometric in the sense that there exist numbers C and λ, with $0\le\lambda<1$ such that

$$sp\left[v(n)-ng^*-L(v(0))\right]\le C\lambda^n$$ (46)

where $L(v(0))=\lim_{n\to\infty}\left[v(n)-ng^*\right]\in V$. Note that λ, is independent of the starting point $v(0)$ and, in the special case where a unique minimal gain policy A exists, reduces to the subdominant eigenvalue of the matrix P^A, [19]. (46) is derived by showing that there exists an integer $M>1$ which is independent of $v(0)$, and an integer $n_0(v(0))\ge 1$ such that for all $n\ge n_0$

$$sp\left[v(n+M)-(n+M)g^*-L(v(0))\right]\le\mu(v(0))sp\left[v(n)-ng^*-L(v(0))\right]$$ (47)

where $0\le\mu(v(0))<1$. In (47), $n_0(v(0))<\infty$ indicates the number of steps after which the span-norm of $\left\{v(n)-ng^*-L(v(0))\right\}_{n=1}^{\infty}$ is monotonically non-increasing. The latter is guaranteed as soon as the T operator selects policies $A\in\prod_i L(i)$ exclusively, which is known to occur after a finite number of steps. The first n_0

steps of value-iteration thus constitute a *first* phase of the convergence process, during which the behaviour of $\left\{ v(n) - ng^* - L(v(0)) \right\}_{n=1}^{\infty}$ may be very irregular (e.g., it may be alternatively increasing and decreasing). The first phase is obviously *non-existent* whenever $L(i) = K(i)$ for all $i = 1, \cdots, N$ which in turn is guaranteed to hold whenever the minimal gain rate is independent of the initial state of the system (g^* is a multiple of 1). M indicates the number of steps needed for strict contraction in the *second* phase and is uniformly bounded in $v(0)$. Clearly one would like to obtain an upper bound for M as a function of N.

The bound $M \le N^2 - 2N + 2$ can be obtained [12] under the special condition

(H) $\qquad\qquad\qquad$ $v \in V$ is unique up to a multiple of 1, $\qquad\qquad\qquad$ (48)

which is equivalent to the existence of a *randomized* optimal policy which has R^* as its single subchain (closed, irreducible set of states).

The geometric convergence result is surprising since it was noted above that T is not necessarily contracting with respect to the sp[x]-"norm". In fact, it can be shown that no *uniform* m-step contraction factor needs to exist, for any $m = 1, 2, \cdots$, i.e., we may have for all $m = 1, 2, \cdots$ and $n \ge n_0$

$$\sup\left\{ \frac{\text{sp}\left[Tx - (n+m)g^* - L(x) \right]}{\text{sp}\left[Tx - ng^* - L(x) \right]} \middle| \text{ all x for which } L(x) \text{ exists and } Tx \notin V \right\} = 1. \quad (49)$$

An open question is obtaining a computationally tractable estimate of the size of λ. Although bounds on λ are unavailable, bounds on the minimal gain rate vector g^* have been proposed. In the case where $g^* = \langle g^* \rangle 1$, with $\langle g^* \rangle$ a scalar, these were obtained by Hastings [12] namely:

$$(Tx - x)_{\min} \le g_i^* \le \langle g^* \rangle \le (Tx - x)_{\max} \quad (50)$$

for all $x \in R^N$ and $i = 1, \cdots, N$, where policy A satisfies $Tx = q^A + P^A x$. In the context of value-iteration this becomes

$$(v(n+1) - v(n))_{\min} \le \langle g^* \rangle \le (v(n+1) - v(n))_{\max}. \quad (51)$$

The bounds move inward monotonically as n increases and if $\lim_{n \to \infty} \left[v(n) - ng^* \right]$ exists, the bounds both converge geometrically fast to $\langle g^* \rangle$. In the case where $g^* = \langle g^* \rangle 1$, it is common to avoid the linear divergence of $v(n)$ with n by using instead the variables $y(n) = v(n) - \|v(n)\| 1$ introduced by White [30] and employing the iterative scheme

$$y(n+1) = Ty(n) - \|Ty(n)\|1 \tag{52}$$

with the property that when $v^* = \lim_{n \to \infty}[v(n) - ng^*]$ exists, then the following relations hold

$$y(n) \to v^* - \|v^*\|1 \in V, \tag{53}$$

$$\|Ty(n)\| \to \langle g^* \rangle, \tag{54}$$

$$(Ty(n) - y(n))_{max} \downarrow \langle g^* \rangle, \tag{55}$$

$$(Ty(n) - y(n))_{min} \uparrow \langle g^* \rangle, \tag{56}$$

where all four limits are approached geometrically fast. The bounds in (50) have been generalized to semi-Markov Decision Processes where $g^* = \langle g^* \rangle 1$, [12]. Under (H) hypothesis, the bounds on the scalar gain rate $\langle g^* \rangle$ have been unaccompanied by corresponding bounds on the deviation of the current vector x from $v \in V$. The existence of such bounds is also useful for demonstrating *convergence* of this or related types of value-iteration schemes.

Zangwill [31] has shown that an iterative scheme $x(n+1) = Qx(n)$ will converge to x if the continuous operator Q and a continuous Lyapunov function $\Phi(x)$ satisfy:

$$\Phi(x) \geq 0 \quad \text{for all} \quad x \in R^N \tag{57}$$

$$\Phi(x) = 0 \quad \text{if and only if} \quad x = x^* \tag{58}$$

$$\Phi(Qx) \leq \Phi(x) \quad \text{for all} \quad x \in R^N \tag{59}$$

for some integer $m \geq 1$, $\Phi(Qx) < \Phi(x)$ for all x with $\Phi(x) > 0$. $\tag{60}$

One possible choice of a Lyapunov function, not computable until v^* is known, is

$$\Phi_1(x) = sp[x - v^*] \tag{61}$$

with

$$Qx = Tx - \|Tx\|1 \tag{62}$$

$$x^* = v^* - \|v^*\|1 \tag{63}$$

which confirms (52).

Another choice of Lyapunov function which may be computed while in the midst of the value-iteration process is

$$\Phi_2(x) = sp[Tx - x] \tag{64}$$

with the same choice of Q and x^*. The important *new* property is that the deviation of v^* from x may be deduced from $\Phi_2(x)$, just as (10) and (17) were used in the *discounted* case.

Proposition 3. *Under* (H) *hypothesis there exists a constant* $\rho \geq 0$ *such that*

$$\frac{1}{2}\Phi(x) \leq sp[x - v^*] \leq \rho\Phi(x) \tag{65}$$

for all x *if and only if there exists a randomized policy which has*

$$\hat{R} = \{i | i \text{ is recurrent under some policy} A \in S\} \subseteq R^* \tag{66}$$

as its single subchain.

$\square$

5 Undiscounted Case: Asymptotic Behaviour of $S(n)$ and the Existence of Initially Stationary or Periodic ε-Optimal Strategies

In the case where $v^* = \lim_{n \to \infty} [v(n) - ng^*]$ exists, then for large n the following relation holds

$$S(n) \subseteq S^*(v^*) \subseteq S_{MG} \subseteq \prod_{i=1}^{N} L(i) \tag{67}$$

Thus (67) shows that value-iteration settles upon minimal gain policies, provided that convergence is guaranteed. Even in the case where $\{v(n) - ng^*\}_{n=1}^{\infty}$ converges (in fact even in the case where each policy is unichained and aperiodic), $S(n)$ may have a very irregular behaviour, the worst case of which exhibits nonperiodic oscillations. As a consequence, we are guaranteed to have an initially stationary (or periodic) optimal strategy only if $S^*(v^*)$ is a singleton where $v^* = \lim_{n \to \infty} (v(n) - ng^*)$ exists. Using the geometric convergence result as discussed in section 4, we obtain that for all $\varepsilon > 0$, there exists an initially periodic strategy which is ε-optimal.

Finally, several difficulties appear when trying to find the set S_{MG}. First, for all $v \in V$, $S^*(v)$ can be a *strict* subset of S_{MG} so that value-iteration fails to yield all maximal gain policies. Indeed even $\bigcup_{v^* \in V} S^*(v^*)$ can be a strict subset of S_{MG} so that varying the starting point $v(0)$ of value-iteration will fail to identify all minimal-gain policies. The explanation is that for states that are transient under P^A, a minimal gain policy A is merely required to choose actions within $L(i)$ (section 4). The second difficulty is provided by the irregular behaviour of the sets $\{S(n)\}_{n=1}^{\infty}$ as described above. This difficulty can, however, be overcome by noting as in the discounted case that whenever $v^* = \lim_{n \to \infty} (v(n) - ng^*)$ exists

$$S(n, \varepsilon_n) = S^*(v^*) \tag{68}$$

for all n sufficiently large provided that $\{\varepsilon_n\}_{n=1}^{\infty}$ is taken as a sequence of positive numbers approaching 0, at a slower rate than the (geometric) convergence rate of $\{v(n) - ng^*\}_{n=1}^{\infty}$, i.e., whenever $\lim_{n \to \infty} \varepsilon_n / \lambda_n = \infty$, e.g., when taking $\varepsilon_n = n^{-1}$.

6 Data Transformations

In section 4 we observed that only if $\{v(n) - ng^*\}_{n=1}^{\infty}$ converges will value-iteration be guaranteed to ultimately settle upon minimal gain policies and only then can sequences be derived from $\{v(n)\}_{n=1}^{\infty}$ which converge to g^* and some $v \in V$. In case $J^* > 2$, i.e., in case $\{v(n) - ng^*\}_{n=1}^{\infty}$ may fail to converge, the following two alternatives are possible:

A. Use the discounted value-iteration scheme with discount factor β depending upon the index of the iteration stage, n, i. e .,

$$w_i(n+1) = \min_{k \in K(i)} \left[q_i^k + \beta_n \sum_{j=1}^{N} P_{ij}^k w_i(n) \right] , \quad i = 1, \cdots, N, \tag{69}$$

where $\beta_n \to 1$. In the case $g^* = \langle g^* \rangle 1$ then,

$$w(n) - \gamma_n g^* \to w^* \in V , \quad \text{as } n \to \infty, \tag{70}$$

where $\{\gamma_n\}_{n=1}^{\infty}$ is obtained recursively by

$$\gamma_{n+1} = 1 + \beta_n \gamma_n , \quad \gamma_0 = 0, \tag{71}$$

provided that

$$\beta_n \beta_{n-1} \cdots \beta_1 \rightarrow 0 \qquad (72)$$

$$\sum_{j=2}^{n} \beta_n \cdots \beta_{j+1} \left| \beta_j - \beta_{j-1} \right| \rightarrow 0. \qquad (73)$$

(72) and (73) are guaranteed to hold when choosing $\beta_n = 1 - n^{-b}$, $0 < b \le 1$. The numerical difficulty of divergence of $w(n)$ is again avoided by using instead the variables $\tilde{w}(n) = w(n) - \|w(n)\| 1$. The convergence rate is $O(n^{-b} \ln n)$ which is considerably slower than the geometric convergence, exhibited by ordinary undiscounted value-iteration.

Proposition 4. *The above scheme has the following properties:*

(1) *Convergence is guaranteed regardless of the periodicity structure of the problem.*

(2) $\left\{ w(n) - \gamma_n g^* \right\}_{n=1}^{\infty}$ *converges to the optimal bias-vector rather than to an arbitrary solution* $v \in V$.

$\qquad\qquad\qquad\qquad\qquad\qquad\qquad\qquad\qquad\qquad\qquad\qquad\qquad\qquad$ $\square$

B. The following data-transformation can be used:

$$\tilde{P}_{ij}^k = \tau(P_{ij}^k - \delta_{ij}) \quad , \quad \tilde{q}_i^k = q_i^k \quad , \quad 1 \le i, j \le N \ , \ k \in K(i), \qquad (74)$$

where $0 < \tau < 1$, and δ_{ij} is the Kronecker delta-function. The transformed problem has the nice property of *aperiodicity* for all of the transition probability matrices since all of the diagonal elements are positive. That is, the transformed problem has $J^* = 1$ and convergence of $\left\{ v(n) - ng^* \right\}_{n=1}^{\infty}$ is guaranteed for any $v(0)$. In addition the following relation exists between V and $\tilde{V}$, the set of solutions to the optimality equation (42) in the transformed model

$$V = \left\{ v \in R^N \middle| \tau v \in \tilde{V} \right\}. \qquad (75)$$

This data-transformation turns every undiscounted Markov Renewal Program into an equivalent undiscounted MDP, in which each transition probability matrix is aperiodic. Hence, undiscounted value-iteration when applied to the transformed model, is guaranteed to settle upon minimal gain policies, and sequences may be derived which converge to the minimal gain rate vector and a solution to the optimality equation for Markov Renewal Programs.

7 Hopfield Network Implementation

The operation of the continuous Hopfield network as formulated in [14] is governed by the following nonlinear differential equation:

$$\frac{du_i}{dt} = -\frac{u_i}{\tau} + \sum_{j=1}^{M} W_{ij} \psi_j(u_j) + \theta_i \ , \quad \text{for } i = 1, 2, \ldots, M \tag{76}$$

where $\mathbf{u} = [u_1, \ldots, u_M]^T$ represents the state of the M neurons, $\mathbf{W}$ is the weight connection matrix, $\psi(\cdot)$ is the nonlinear input-output function of a neuron and $\theta = [\theta_1, \ldots, \theta_M]^T$ is the bias vector. It has proven [17] that this network evolves towards stable equilibrium point minimizing the following energy function:

$$E = -\frac{1}{2} \xi^T W \xi - \theta^T \xi + \frac{1}{\tau} \sum_{i=1}^{M} \int_0^{\xi_i} \psi^{-1}(\xi) d\xi \tag{77}$$

where $\xi = [\xi_1, \ldots, \xi_M]^T$ denotes the the vector of the neuron outputs, and $\xi_i = \psi_i(u_i)$. In the above expression of the energy function, the term $(1/\tau) \sum_{i=1}^{M} \int_0^{\xi_i} \psi^{-1}(\xi) d\xi$ can be neglected because we can make $\|\mathbf{W}\|$ or τ enough large. Here, the nonlinearity of the network is defined as

$$\psi_i(u_i) = \begin{cases} 1 & , \text{ if } u_i > 1, \\ u_i & , \text{ if } 0 \leq u_i \leq 1, \\ 0 & , \text{ if } u_i < 0. \end{cases} \tag{78}$$

The energy function (77) behaves as a Lyapunov function [17] for the nonlinear dynamic system described by (76). The idea is to relate this property on the mentioned Lyapunov function of the Markov Decision Problem (61) or (64), by an appropriate selection of state variables of neural network. Indeed, we consider the output vector of the network as being the associator function between a given state and the optimal control which minimizes the cost (1). Equivalently, the selected control represents the asymptotical solver of the operatorial equations (6) (discounted case) or (41) (undiscounted case). This associator function is defined as

$$\xi_{ik} = \begin{cases} 1 & , \text{ if control } k \text{ is optimal for state } i, \\ 0 & , \text{ if control } k \text{ is not optimal for state } i. \end{cases} \tag{79}$$

and so, the Hopfield network represents a decision network in the sense that its outputs represents the optimal decision which has to be taken, when the system modelled by Markov chain arrived at a given state. The general this setting for this association state-control can be seen in Figure 1.

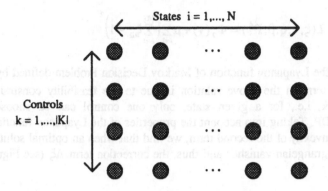

States i = 1,..., N

Controls
k = 1,...,|K|

Figure 1. Hopfield neural network for solving Markov Decision Problems

The Hopfield network operation is represented schematically in Figure 2. Here $P(k,t)$ represents a projection operator responsible for the dynamics of neural network, and $\Lambda(\xi, v, q, \beta, P^A)$ is a constraint feedback dealing with the optimization task formulated in equation (1), section 1.

Network operation

(1) Projection of **u** along the dynamics' trajectory according to

$$\mathbf{u}(t+1) = P(k,t)\mathbf{u}(t) \quad , \quad t = 1, 2, \ldots \tag{80}$$

(2) Passing the state vector **u** through the nonlinear threshold function in order to constraint ξ to belong to the unit hypercube.

(3) Change in ξ given by the gradient of the optimization energy term E^{op}, as

$$\Delta\xi = \Lambda(\xi, v, q, \beta, P^A) \tag{81}$$

Specifically, the projection operator $P(k,t)$ and the function Λ are defined as follows:

$$P(k,t) = (1-\eta)\mathbf{I} + \Delta t \mathbf{W}(k)\mathbf{D}_v \tag{82}$$

$$\Lambda(\xi, v, q, \beta, P^A) = \nabla_\xi L(\xi, v, q, \beta, P^A) \tag{83}$$

where $L(\xi, v, q, \beta, P^A)$ is the augmented Lagrangean [3] of the joint variables (ξ, v), the first variable being related to the dynamical behaviour of the decision neural network and the second representing the value-iteration variable for MDP:

$$L(\xi, v, q, \beta, P^A) = \Phi_2(v) + \mu \sum_{i=1}^{M} \left(\sum_{k=1}^{|K|} \xi_{ik} - 1 \right)^2 \tag{84}$$

Here, $\Phi_2(v)$ is the Lyapunov function of Markov Decision Problem defined by (64) and the second term in the above relation is due to the feasibility constraint on decision network, i.e., for a given state, only one control can be choosen as minimizer in MDP. Taking into account the properties of the Lyapunov function for MDP and the convexity of the second term, we find that when an optimal solution is achieved, the Lagrangeian vanishes and thus, the correction term $\Delta\xi$ (see Figure 2) does so as well.

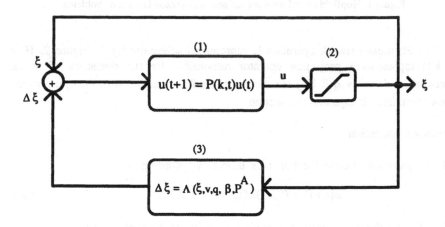

Figure 2. Schematic diagram of Hopfield network for MDP

In relation (82), Δt stands for the integration time step, $\eta = \Delta t / \tau$ with τ the time constant of the dynamical system (76). The weight connection matrix $W(k)$ is choosen to depend on the history of control over time

$$W(k) = (1 - r(k))W_0 + r(k)\frac{\eta}{\Delta t} D_v^{-1} \tag{85}$$

where W_0 is the arbitrary initial weight connection matrix and $r(k)$ is a learning measure of control defined by

$$r(k) = r(k,t) = 1 - \frac{1}{M} \sum_{i=1}^{M} \sum_{j=1}^{|K|} |k_i(t) - k_i(t-1)| \quad , \quad t = 1, 2, \ldots \tag{86}$$

considering that initially $r(k,0) = 0$. Clearly, when the policy became stationary, $r(k) = 1$ and so, the equilibrium state of the network does not depend on the initial weight connections between neurons.

$\mathbf{D}_\psi$ is a diagonal operator containing as entries the transfer functions of the neurons defined as follows:

$$\mathbf{D}_\psi = \text{diag}(\psi_1, \psi_2, \ldots, \psi_M) \tag{87}$$

8 Application to Closed Queueing Networks

As for an application of the developed theory we consider an cyclic closed queueing network as it appears in Figure 3. This diagram can represent a virtual circuit with a sliding-window control (for more details on the reason and practical significance of this communication network problem see [5]). The queueing system consists of M servers working for processing N packets which have to be cyclically circulated through the network.

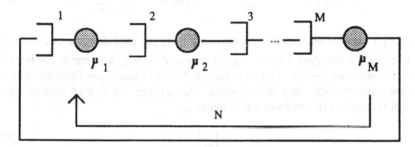

Figure 3. Cyclic queueing network

Here, the vector $\mu = [\mu_1, \mu_2, \ldots, \mu_M]^T$ represents the processing rates of the M servers. In words, the optimization problem is that of minimization the total length of time interval needed for a complete circulation of the packets through the network. The state of this system is described by the vector of the packet numbers on queues at each time step, so

$$\mathbf{n} = [n_1, n_2, \ldots, n_M]^T \tag{88}$$

where n_i is the number of packets in the i queue, and $i = 1, 2, \ldots, M$. The following constraint is imposed on the state variables:

$$\sum_{i=1}^{M} n_i = N \tag{89}$$

From the above constraint, we can see that the number of packets in a given queue can range from 0 to N. As a matter of technical conditions, in order to avoid further complications, we assume that the queueing capacity is infinity for each server [5]. The dynamics modelling of this system is done in a quite standard way [36], namely, we assume the product form solution of the closed queueing networks with an exponential distributed service time (with parameter μ_i for the ith queue):

$$p(\mathbf{n}) = p(n_1, n_2, \ldots, n_M) = \prod_{i=1}^{M} \left(\frac{\lambda_i}{\mu_i}\right)^{n_i} p(0) = \prod_{i=1}^{M} \rho_i^{n_i} p(0) \tag{90}$$

where $p(\mathbf{n})$ stands for the probability of system to be in a state $\mathbf{n}$ and $\rho_i = \lambda_i / \mu_i$ represents the traffic intensity [36] of the ith queue. The free parameters of this system according to our choice are the processing rates μ_i which must be selected among a finite set of available rates in order to minimize the following cost function:

$$E[T_w] = \sum_{i=1}^{M} \sum_{j=1}^{N} \int_0^{\infty} e^{-\alpha t} P_{ij} \frac{\lambda_i n_i(t)}{\mu_j} dt \approx h \sum_{i=1}^{M} \sum_{j=1}^{N} \sum_{t=0}^{\infty} e^{-\alpha t h} P_{ij} \frac{\lambda_i n_i(th)}{\mu_j} \tag{91}$$

where $E[\cdot]$ denotes the expectation operator, T_w is the total duration of a complete cycle, $\alpha > 0$ is a damping factor over the time and finally, h is stands for the time step of the queueing system. Thus, the state and control spaces are finite and discrete sets (we have N states and $|K|$ controls). The uniform version of Markov chain problem is based on the uniform traffic intensity

$$\rho = \max_{1 \leq i \leq M} \{\rho_i\} \tag{92}$$

because of unichain Markov property [37]. Then the discount factor of the dynamic programming equation can be computed from

$$\beta = \frac{\rho}{\rho + \alpha} \tag{93}$$

where we see that taking $\alpha = 0$, we arrive at an undiscounted problem. The cost per stage, when the system is in state i and we select control k is given by

$$q_i^k = \frac{h}{\rho + \alpha} \sum_{j=1}^{N} P_{ij} \frac{\lambda_j n_j}{\mu_j},$$

(94)

and the transition probability matrix $\mathbf{P}^k$ can be easily calculated from (90) using the Bayes formula. With these settings, the above optimization problem falls into the category of Markov Decision Problems.

Simulation results

As a numerical simulation of Hopfield network corresponding to the closed queueing system problem, we took, the following data:

$$M = 10 \ , \quad N = 10 \ , \quad |K| = 3,$$

and

$$\lambda_1 = \lambda_2 = \cdots = \lambda_{10} = \lambda = 1200 \text{ bps},$$

$$\alpha_1 = 0.2 \text{ (discounted) and } \alpha_2 = 0 \text{ (undiscounted)}.$$

The packets were considered as having an exponential distribution of the length with mean value $l = 1000$ bits, and the rates of processing were chosen to be 2400, 4800 and 9600 bps. The behaviour of the optimal solution found by running dynamics of neural network and learning factor $r(k)$ defined by (86) appear in Figures 4-13 and 14, respectively. We considered for graphical representation, the evolution of controls (processing rates) over a 10 time steps period. We remark the achievement of stationary policy under 5 time steps for the discounted case and in maximum 8 steps for undiscounted case, no matter of the length of packets. This simulation can be useful for designing the time horizon of a controller overheading the closed queueing system. Possible developments of this framework are related on the continuous state variable case (in application, the Markov process was a discrete time counting process of number of packets under queueing system).

Conclusions

We presented a possible mapping the value-iteration and related successive approximation methods for Markov Decision Problems onto Hopfield neural networks in both discounted and undiscounted versions of the finite state and action spaces. The asymptotical behaviour of the control sets has been considered and estimates on the convergence rate for the value-iteration scheme have been presented. The energy function of the Hopfield network was related to Lyapunov function of the Markov Decision Problem. Finally, an application from queueing systems in communication networks has been considered and the results of computer simulation on Hopfield network displayed, mentioning the possibility of solving the continuous state variable Markov Decision Problem.

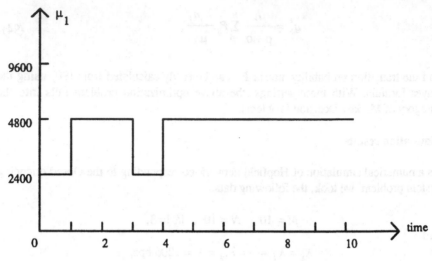

Figure 4. Evolution of processing rate μ_1 for 10 time steps

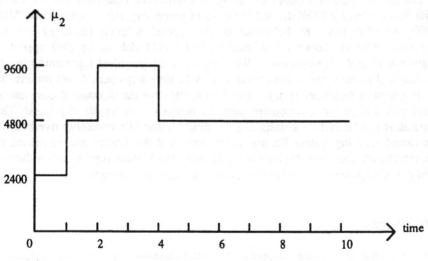

Figure 5. Evolution of processing rate μ_2 for 10 time steps

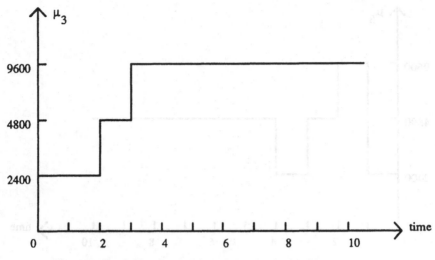

Figure 6. Evolution of processing rate μ_3 for 10 time steps

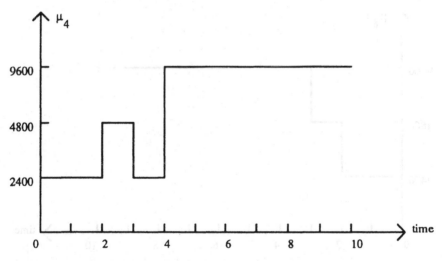

Figure 7. Evolution of processing rate μ_4 for 10 time steps

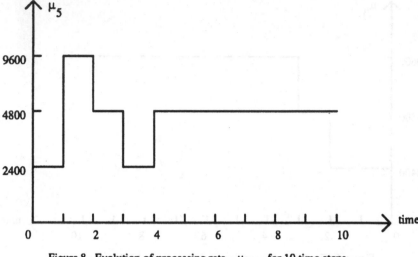

Figure 8. Evolution of processing rate μ_5 for 10 time steps

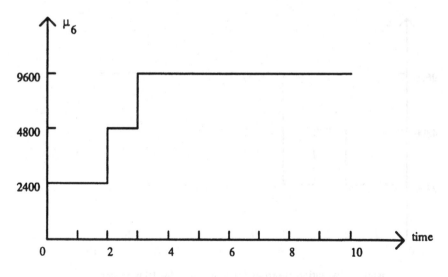

Figure 9. Evolution of processing rate μ_6 for 10 time steps

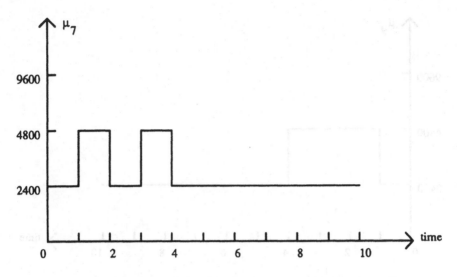

Figure 10. Evolution of processing rate μ_7 for 10 time steps

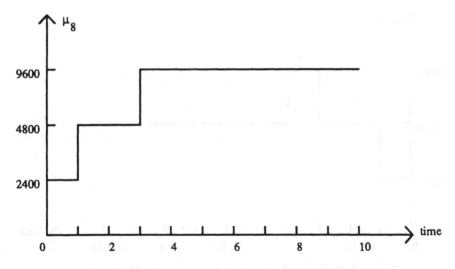

Figure 11. Evolution of processing rate μ_8 for 10 time steps

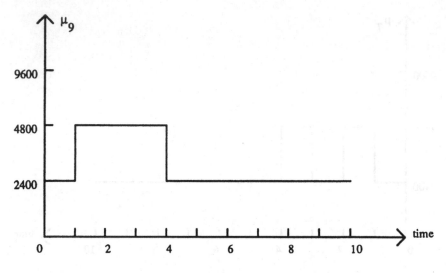

Figure 12. Evolution of processing rate μ_9 for 10 time steps

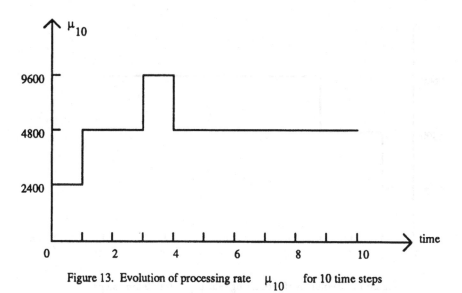

Figure 13. Evolution of processing rate μ_{10} for 10 time steps

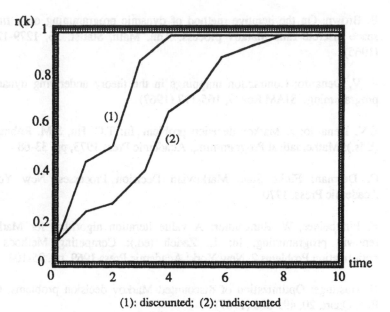

(1): discounted; (2): undiscounted

Figure 14. Evolution of learning measure r(k) during 10 time steps

References

1. J. Bather: Optimal decision procedures for finite Markov chains. Adv. in Appl. Prob. 5, Part I, 328-339 (1973)

2. R. Belman: Applied Dynamic Programming, Princeton University Press, Princeton, NJ, 1962

3. D.P. Bertsekas: Constrained Optimization and Lagrange Multiplier Methods. New York: Academic Press 1982

4. D.P. Bertesekas and J.N.Tsitsiklis: Parallel and Distributed Computation: Numerical Methods. Englewood Cliffs, NJ: Prentice-Hall 1989

5. D.P. Bertsekas and R. Gallager: Data Networks. Englewood Cliffs, NJ: Prentice-Hall 1992

6. B. Brown: On the iterative method of dynamic programming on a finite space discrete time Markov process. Ann. Math. Statist. 36, 1279-1285 (1965)

7. E. V. Denardo: Contraction mappings in the theory underlying dynamic programming. SIAM Rev. 9, 165-177 (1967)

8. E.V. Denardo: A Markov decision problem. In: T.C. Hu, S.M. Robinson (Eds.): Mathematical Programming. Academic Press 1973, pp. 33-68

9. C. Derman: Finite State Markovian Decision Processes. New York: Academic Press, 1970

10. B. Finkbeiner, W. Rungaldier: A value iteration algorithm for Markov renewal programming. In: L. Zadeh (ed.): Computing Methods in optimization Problems 2. New York: Academic Press 1969, pp. 95-104

11. N. Hastings: Optimization of discounted Markov decision problems. Op. Res. Quart. 20, 499-500 (1969)

12. N. Hastings: Bounds on the gain of a Markov decision process. Op. Res. 19, 240-244 (1971)

13. N. Hastings, J. Mello: Tests for suboptimal actions in discounted Markov programming. Man. Sci. 19, 1019-1022 (1973)

14. J. Hertz, A. Krogh, P.G. Palmer: Introduction to the Theory of Neural Computation. Redwood City: Addison-Wesley Publishing Company 1991

15. R. Howard: Dynamic Programming and Markov Processes, John Wiley, New York, 1960

16. H. Kushner, A. Kleinman: Accelerated procedures for the solution of discrete Markov control problems. IEEE Trans. Automatic Control AC-16, 147-152 (1971)

17. V. Lakshmikantham, S. Leela, A.A. Marttynyuk: Stability Analysis of Nonlinear Systems. New York: Marcel Dekker, Inc. 1989

18. J. MacQueen: A test for suboptimal actions in Markovian decision problems. Op. Res. 15, 559-561 (1967)

19. T. Morton, W. Wecker: Discounting, ergodicity and convergence for Markov decision processes. Man. Sci. 23, 890-900 (1977)

20. A. Odoni: On finding the maximal gain for Markov decision processes. Op. Res. 17, 857-860 (1969)

21. E. Porteus: Bounds and transformations for discounted finite Markov decision chains. Op. Res. 23, 761-784 (1975)

22. E. Porteus, S.J. Totten: Accelerated computation of the expected discounted return in a Markov chain. Op. Res. 26, 350-358 (1978)

23. D. Reetz: Solution of a Markovian decision problem by successive overrelaxation. Op. Res. 21, 29-32 (1973)

24. S.M. Ross: Stochastic Processes. New York: John Wiley & Sons 1983

25. M. Schwartz: Telecommunication Networks: Protocols, Modeling and Analysis. Reading, MA: Addison-Wesley Publishing Company 1987

26. P.J. Schweitzer: A turnpike theorem for undiscounted Markovian decision processes. ORSA/TIMS National Meeting, May 1968

27. P.J. Schweitzer: Iterative solution of the functional equations for undiscounted Markov renewal programming. J.M.A.A. 34, 495-501 (1971)

28. J. Shapiro: Turnpike planning horizons for a Markovian decision model. Man. Sci. 14, 292-300 (1968)

29. J. Van Nunen: A set of successive approximation methods for discounted Markovian decision problems. Op. Res. 20, 203-209 (1976)

30. D. White: Dynamic programming, Markov chains, and the method of successive approximations. J.M.A.A. 6, 373-376 (1963)

31. W. Zangwill: Nonlinear Programming. A Unified approach. Englewood Cliffs, NJ: Prentice Hall 1969

"Blob" analysis of biomedical image sequences: A model-based and an inductive approach

Sören Molander
Department of Applied Electronics
Chalmers University of Technology
412 96 Göteborg

Abstract

New directions in robotics and computer vision indicate that useful behaviours of artificial systems can be achieved with simple reactive control strategies and with a task dependent representation of incoming data.

This philosophy has been used in the design and implementation of an offline system for automatic object detection and delineation in biomedical image sequences, where the task is to estimate object area vs. time. The image is represented with feature vectors on a coarse resolution and scale, and the image is processed using a scene model and a procedural model. The scene model expresses relations between objects in the scene, the objects being represented by regions, or "blobs". The objects are labeled with a relaxation labeling algorithm, and a constraint satisfaction algorithm. The procedural model - a finite state automaton - expresses the different processing paths due to possible model-data mismatches, including processing on a higher resolution and the use of alternative scene models. The automaton comprises a top level supervisor as well as a lower level reactive control mechanism. The control mechanism and the scene model are exchangeable between different applications. Images in the sequence are re-processed if the objects of interest in consecutive images do not overlap.

As a comparison, unsupervised learning with Kohonen's self-organizing feature maps were used to train the system to perform segmentation and delineation. The feature map was trained with a low dimensional feature vector randomly sampled from a population of representative images. During the association phase, for each feature vector, the map nodes are searched for the best matching node, and the nodes that correspond to the desired object are grouped into larger regions. The object is delineated from the largest region retrieved from the map.

Two domains have been analyzed by the system, ultrasound image sequences of the heart and gamma camera sequences of the heart.

1 Introduction

1.1 Computer vision in perspective

Since the late eighties there has been a great deal of interest in what has become known as active and purposive vision systems. This can be understood in the light of the lack of real progress in designing truly autonomous task driven vision systems. Many systems in the seventies and eighties were rule-based[59,34,52,46], schema-based[13,42] or blackboard systems[47], the two latter being generalizations of production rules where more complex behaviour can be built in. Pattern recognition and statistical methods[29] have been more successful, and examples range from document processing[9], zip code analysis[35] to road tracking for vehicle guidance[7,51], usually in combination with some symbolic control mechanism. It is fairly safe to state that to this date, however, there are no examples of task driven systems with visual input that are capable of performing well in a noisy and partly unknown environment. Researchers in Artificial Intelligence have gradually realized the limitations of the pure symbolist approach, and are looking for other metaphors such as autonomous agents in robotics[8], distributed systems[11], and artificial neural networks[23,49,56]. Similarly, the computer vision field is undergoing a shift of paradigm[28], and scientists are realizing that vision must be put in a more well-defined context[2,3,10,15]. With the advent of new hardware architectures, there is an even more pressing need to come up with efficient design principles for computer vision systems. Although the retinocortical pathways in animate vision systems have been mapped in great detail[26,62,63,65,17,55] little is known as to what information is used, how it is used, how it is coded, and how it is related to higher brain functions such as planning and specific recognition tasks[14,12,43]. This leaves researchers in the computer vision and related fields much in the dark as how to mimic the most effective vision systems known.

1.2 The segmentation-model loop

The task at hand is to estimate the area of a region of interest - an object - over time in image sequences, a problem of interest in for instance biomedical domains. To this end, one of the most important problems that needs to be addressed - and in most computer vision applications - is the segmentation-model loop. This is related to the so called frame problem or the focus of attention problem[4], which can be formulated as the catch-22 of image understanding: how does one segment the image efficiently without first knowing what the image depicts, and conversely, how does one know what the image depicts without an initial segmentation? One way to solve this problem is to bootstrap the system with a model of the scene, use an initial segmentation as a basis for object classification and labeling and check if the results are consistent with the model. If not, alternative actions can be taken, such as a new higher resolution segmentation phase, or the use of a new scene model. In addition to the model-based approach, a continuity constraint is used to re-process images with objects that do not have a sufficient degree of overlap within the sequence. One problem with early systems was that the condition-action control mechanisms were too cumbersome and sophisticated to be practical. This has been avoided by making the current system weakly hierarchical; at the highest abstraction level, a supervisor initiates task dependent actions chunked together in execution and evaluation states. In doing so, the reactive control level at the next level needs only to address pertinent problems at a particular time in the processing chain. The term weakly hierarchical means that most processing time is spent in the processing states and not in the communication between the different levels. The condition-actions pairs in the control mechanism are derived from the experience of a human operator, and depend on previous actions and the current processing state. A "good" domain model is one that captures the relevant information for the majority of images, and that allows for the use of alternative strategies only when it is absolutely necessary. It should be stressed that "bad" images in the context of biomedical imagery do not indicate abnormal medical conditions, it only refers to images that do not fit the default model. In fact, automated medical diagnosis in image domains is a very difficult task in itself, and one that has been successful only in few cases[60].

1.3 The meaning of a "model"

A word on nomenclature is appropriate here: the word "model" in this article is used in the engineering sense, and not in the more formal computer science sense. To put it simply, in what follows, a model is a simplified description of the physical world. No stringent line is drawn between formal or informal descriptions - the language - and the meaning that emerges from this description through the use of various algorithms and procedures, which constitute the semantic aspects - the model. In fact, any system, both formal systems (axiomatic) and informal systems (such as this system and biological systems), must always be described in terms of these two entities, and they cannot be reduced into one or the other, a principle often called linguistic complementarity[39,40].

2 The image representation

At the heart of successful engineering applications involving signal-and image understanding, lies a good data representation. For the specific task and class of images discussed here - delineation of large scale objects - a convenient representation of the image takes the form of a resolution limited vector representation organized in a pyramid. This entails tessellating the original image into subimages, where the size of each subimage depends on the image size. The maximum resolution in the pyramid is 32 x 32 subimages, the next lowest resolution 16 x 16, etc. Thus if the image size is 256 x 256 pixels, the size of a maximum resolution subimage will be 8 x 8 pixels, which means that the image will be represented both on a coarse scale and on a coarse resolution. Each sub-image is associated with a feature vector. The elements in the feature vector are calculated from statistical averages taken over the subimage, or from from the global properties of the whole pyramid image. The most important features are grayscale, edges and regions. The two former are local features and the latter is a global feature. The subimage grayscale is derived from a straightforward average over the pixels in the subimage, and the edge feature is derived from a least mean square plane fit over the grayscale surface. An "edge" is defined as the cosine of the angle between the normal vector of the LMS fitted plane and the normal of the x-y plane; small angles indicate no edges and the cosine is close to unity. Both the direction and the strength of the edge are preserved and stored in the feature vector.

The region feature describes invariant properties in the image in terms of grayscale or texture, and has a physiological counterpart in the grouping phenomenon, well studied in Gestalt psychology[38]. Here we are mostly interested in grouping grayscale properties, for which there exists a myriad of methods. One example is the scale-space representation[64,37] which essentially entails smearing the image with a kernel that satisfies a discrete diffusion equation[37], followed by a "blob" detection operation. Blob detection can most effectively be visualized by imagining a bucket of water being poured on top of a grayscale landscape; the puddles left in the landscape are the blobs. The blob information is coded in the feature vector as (say) "subimage no 3 belongs to blob 5".

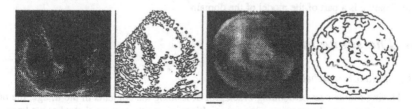

Figure 2-1

An ultrasound image and a gamma camera image of the heart, (1:st and 3:d image) and their zero-crossings (2:nd and 4:th image). The images were smoothed with a solution to the discrete diffusion equation (very similar to a Gaussian). A a simple Laplacian operator ($1-\frac{4}{1}$ 1) was applied to extract the zero-crossings, and the results displayed were thresholded.

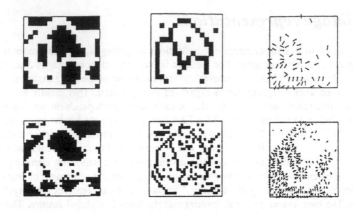

Figure 2-2

These images show some examples of elements in the feature vector. The top row shows the elements at the 16x16 resolution level, and the bottom row the same features at the 32x32 level. Going from left to right, the leftmost images depicts blobs, the middle images ridges, and the rightmost images depict edges. The latter is defined as the projection of the normal vector of a grayscale surface fitted LMS plane onto the x-y plane of the subimage. The ridge measure in the middle was calculated from a simple local decision rule involving the number of directions that have a maximum value at each point.

The images depicting the zero-crossings in figure 2-1 clearly show that is there is an enormous information (not in the Shannon sense, but in a semantic sense) redundancy in the image. All we want is to delineate the left ventricle, and for that purpose the representation on the coarse scale of resolution in figure 2-2 is more appropriate. It turns out that for different applications, other measures of ridges and edges than those described here are more suitable, and this information is a part of the model of the domain.

3 The scene model

3.1 Relaxation labeling

Having found a set of building blocks for describing the contents in the image, the next step is to find a way of characterizing the scene. Since we tacitly assume that what we know what is in the image, a natural way of describing the scene is in terms of objects and relations, where the objects are the blobs. A convenient framework for this purpose is the relaxation labeling formalism. This is a well established method in image processing, closely related to probabilistic optimization[61] and Bayesian inference[50]. For each object $i = 1,2,...M$, assign a object-label assignment vector $p_i(l) = p_i(l_1, l_2, ...l_L)$ normalized such that $\Sigma_{l=1} p_i(l) = 1$ (to simplify the notation, the labels themselves are sometimes treated as indices).

Geometrically, the object p is a simplex with L corners on the coordinates axes (a generalized tetraeder), and the tip of the vector p moves on the manifold spanned by the simplex. The goal is to achieve an *unambiguous* labeling, so that $l = (0,0,..1,..0,0)$, for some label index j. The next step is to define a neighbour relation over objects and labels. A natural way to encode this relation is to use a constraint matrix $r_{ij}(l,m)$ with meaning as "*object i with label l is related to object j with label m with strength r*". The constraints can in principle assume any real value, and here we adopted the convention that no relation is coded as zero, otherwise with the numbers 0.5 or 1.0, with the interpretation "weak" and "strong".

Using the constraints and the object-label assignment vector p, the goal is to adjust the labels iteratively to achieve an unambiguous labeling. Two of the most commonly used updating formulas are Rosenfeld's[53] formula and the Hummel- Zucker algorithm[27]. The latter algorithm solves the problem rigorously, and is more complex in that it takes into account what happens with the p vector at edges of the simplex. In our application - though not so in general - it turned out that the faster Rosenfeld update formula worked equally well, and so it was used in all applications.

3.2 Model examples

The objects in an "ideal" ultrasound four chamber heart image are the right and left ventricles (LV and RV) and the left and right atria (LA and RA), and the relations between them are particularly simple to define: LV is to the right of RV, RA is to the right of LA etc. The direction relations were measured from the center of mass positions of the blobs, and quantized into 4 directions, N,S,E,W. North is defined in the sector from 45 to 135 degrees, and similarly for the other directions, see figure 3-1. Similar relations and objects are defined for the gamma camera domain. The ideal result of the relaxation labeling procedure is an object-label assignment vector with an unambiguous labeling, i.e. with all labels zero except one label, which is unity. In practice most values will be different from zero, so a decision has to be taken which one to choose. There are several ways of doing this, and the simplest strategy is to choose the label with the maximum value, i.e. $\{l_j: \max_j p_i(l_j)\}$, for a certain object i, and all label indices j.

Figure 3-1

The left figure shows the scene model for the ultrasound domain; it takes a total of 16 numbers to describe the scene. The model in the middle is for the gamma camera domain, and here it takes a total of 10 numbers. In the ultrasound domain, the objects were represented with black blobs - extracted from local minima - and in the gamma camera domain they were bright blobs. Some of the constraints numbers $r_{i,j}(l,m)$ are shown in the figures. The relations are fuzzy ("LV is relatively east of (r=0.5) RA" and "LV is most certainly east of (r=1.0) RV"), which is sensible since most natural relations are not crisp. The quantized direction relations are indicated to the right.

3.3 Constraint satisfaction re-labeling

The model described above only contains information on the relative position of blobs, and is blind to other image cues. Innumerable studies have shown that edge features are highly relevant in both animate and artificial systems, and this must also incorporated in a good model. Here it is done in the following way: after the initial relaxation labeling, a number of blobs turn out with identical labels, and several of them do not usually belong to the same object. In the ultrasound domain, for example, if there are three blobs labeled LV, only two of them might actually be situated in the left ventricle in the image, while the other one might sit on top of the left atrium, LA. In this case, we want to use the fact that there are white ridges across the image, separating the blobs (fig. 2-1) . This would enable a re-labeling of the label LA to LV in an iterative manner, using the predicate *Ridge_Between*(LV,LA,strength_threshold). So, for all objects with duplicated labels, the existence of a strong enough ridge is checked, and if strong enough ridges are found, the labels are changed recursively. This procedure is sometime called

174

constraint satisfaction, since the goal is to achieve a certain goal given a set of constraints. The procedure works if there are no spurious ridges and edges in the image, which of course is a rare event; in some cases the algorithm plunges into an infinite loop, which must terminated at a sufficient recursive depth. The results are stored, and in the subsequent evaluation phase the system determines if auxiliary processing is needed (se below).

4 Control and supervision

4.1 Single frame mode

So far the declarative aspects of the system have been described, and no clues as to what to do with this information - the procedural aspects - have been given. Since the objective is to delineate images, at the top level some temporal ordering (or task abstraction) of processing events must be given. It is natural to picture these events - or states - as elements in a finite state automaton. The states are either *execution* or *evaluation* states. In the execution states, the image is processed and in the evaluation states, the results are tallied with the model of the domain. Each state represents a useful chunk of actions and reflects the results of experiments on a testbed of images. The pertaining procedural model information is:

- The number of expected objects in the image (all evaluation states)

- Their internal direction relations (the second evaluation state, FEVA)

- Geometrical predicates over the objects (the FEVA state) such as overlaps between objects in different directions

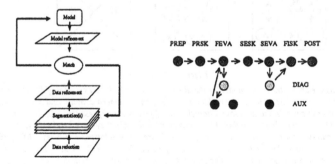

Figure 4-1

The left figure shows the general principle of a model-based vision system. The system is bootstrapped with a preconceived idea of what it is seeing in order to enter the segmentation-model loop. The figure to the right shows the system operation as a set of discrete set of states. For the domain described here, the automaton has the following meaning: PREP=pre processing (calculation of the pyramid representation), PRSK = primal sketch (initial labeling and constraint satisfaction label check), FEVA = the first evaluation state, SESK - the secondary state (region growing and merge of the labeled blobs), SEVA = second evaluation state (including geometrical predicates), FISK = final sketch state (region growing), and POST = post processing state (domain specific processing to make final adjustments of the labeled blobs).

The top row to the right in figure 4-1 shows the default processing chain, which the system uses when no model-data inconsistencies are found. In control theory, this would be called the supervision level of the system, since this level has no knowledge at all of what is being done in other parts if the system. If there are model-data inconsistencies, the action is determined in the DIAG state, which represents the control level. The actions depend on the image domain; for the ultrasound domain, more actions were necessary to obtain a robust delineation of the left ventricle than in the gamma camera domain. What action to execute and when, depends on the

175

type of model-data mismatch, the history of previous actions, and the priority of actions. For example, in the ultrasound domain, if the number of labeled objects are too few (less than 4, if the initial model consists of 4 objects), then the first action is to try a model with three objects instead. If this fails, the system tries a the original 4-chamber model on a higher resolution. For each system state and model-data mismatch condition, there is a either a unique action or no action. If there is no action available, the system simply proceeds with the default processing chain as if nothing has happened. In the post processing state, a final evaluation is done and if the system finds any remaining incompatibilities with the model, it produces an error log file.

4.2 Sequence mode

The analysis of sequences proceeds frame by frame with the system as described. After this analysis, an interpolation procedure is triggered if the area overlap between consecutive relevant labels is too small or too large. If the there is an overlap, the results from the previous image is superimposed on the current image, using the pyramid image from the current image in the region growing. The exact conditions are defined in Appendix II. In principle, one could try this for the whole sequence: use the model-based system for the first frame, and simply use the interpolation scheme for the rest of the sequence. This was in fact attempted, but failed because of a "blob wandering" effect: the blobs eventually skidded off the correct positions on top of the relevant objects in the image.

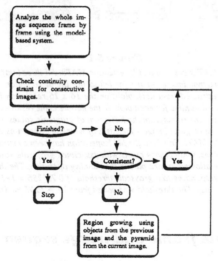

Figure 4-2

The continuity constraint algorithm applied to the sequence after the model-based frame-by-frame analysis.

5 Region growing

As we have seen, regions defined as "blobs" are convenient segmentation primitives for the types of images considered here. For a robust region based delineation procedure, however, a more refined method had to be devised. To this end, an algorithm based on a version of the so called medial axis transform (MAT) has been developed, and the extended algorithm is called MATS[44] (MAT with support) . The image is first transformed into a binary image, for example by a thresholding operation. A set of operators is subsequently applied to the entire binary image iteratively from different direction (N-S and E-W), such that in each point, corners or one-pixel binary protrusions are removed. The operation continues until only a "skeleton" remains, which is why the operation is also called thinning. The idea behind the MATS is to constrain the MAT operation to certain parts of the image, and to "carve" out holes in the binary image instead of creating a skeleton. The constraint initially imposed is that only the (say) darkest pixels in the original image are allowed to take part in the thinning process. The MAT operators proceed on the binary image *only* if the current binary pixel has a associated grayscale value below the current threshold. The thinning constraints are then relaxed so that brighter regions can take part in the thinning, after which the MAT operators attack the these enlarged areas of the binary image. The process is repeated until a suitable stop criterion is given; here we have chosen to stop when the thinning grayscale threshold is high enough for the desired features to appear in the binary image (figure 5-1). The result image is usually full of glitches, and these are removed with a local decision procedure[44].

Figure 5-1

An illustration of the MAT algorithm and its extension MATS. All three images are pyramid representations of the original image. At the far left is the initial condition of the matrix to be thinned. These were extracted from dark blobs on the 16 x 16 resolution level, and the white points represent the local minima for each blob. In the middle is the result - at the 32 x 32 level - of the MAT without the constraints from the histogram of grayscale values. Not surprisingly, the correspondence with the original image - the edges of which are shown as arrows - is poor. To the right is result of the MATS, in which the MAT operation has been constrained with the histogram of grayscale values. Only binary pixels with associated grayscale values below a certain variable threshold are allowed to take part in the thinning operation. The thinning was stopped at the 75 % level, corresponding to a grayscale threshold of 0.75*256 = 192 (the original image was quantized with 8 bits). The threshold was varied from 10 % to 75 %, in steps of 10, and in the last step 5.

6 Results: single frame and image sequences

6.1 A system trace for the ultrasound domain

Figures 6-1, 6-2, and 6-3 show a trace of an analysis an ultrasound image in which the "blob dynamics" is unveiled. Essentially, the system tries out different ways of fitting blobs into desired parts of the image, and to grow these into a desired shape. The system has several redundant methods of checks and cross checks to make sure that the outcome is consistent with the model. The example also indicates why it is so difficult to design and implement general purpose vision systems; no matter whether the scene model is cast in terms of semantic networks[34], multi relational grammars[58], or the relaxation framework above, there will always be a case when the model is either too clumsy or insufficient to deal with real world data. That is still why the most effective systems are tailor-made for specific applications, with no or little generality built in. The current system is a hybrid in this sense: it is tailored for the task of delineating blobby images, but general in its declarative and procedural model-based approach.

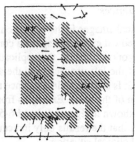 

Figure 6-1 a, b

figure 6-1 a shows the initial blob segmentation of a four-chamber heart image on the 16x16 resolution level. The foreground objects shown here have been separated from the background objects using the distance from the center. Figure 6-1 b shows the result after the constraint satisfaction algorithm. The large blob in the top left corner have been relabeled background (BG), using information on the existence of a ridge between the "inside" and the "outside" of the heart. After the segmentation, the system passes the First EVAluation (FEVA) state with no model-match errors, and proceeds to the SEcondary SKetch (SESK) state.

Figure 6-2 a, b

After the SESK state, the system enters the Second EVAluation state. The geometrical error tracker finds an overlap in the North-South direction between the objects labeled RA and LV. The system then re-iterates the analysis on the 32x32 resolution level (figure 6-2 a). The number of objects are reduced in a series of model-based pruning operations, including an intersection with the 16x16 segmented image (figure 6-2 b).

Figure 6-3 a, b

The results from the previous image are scrutinized in the SEVA state, and several direction-relation anomalies are found. The objects with the most frequent deviations from the relations in model are removed (figure 6-3 a). As one can see, the LA object spikes into the Left Ventricle (LV), which is not in accordance with the model. However, the control mechanism has found no more pertinent actions, and the system proceeds to the FInal SKetch state and the POST processing state. The latter state is domain dependent, and here the LA-LV anomaly is detected - the algorithm finds the junction between the four chambers - and remedied (figure 6-3 b).

178

6.2 Extracting useful information from image sequences

One of the motives behind automatic analyses of medical images is the explosion of image data, which now is a serious problem, and which forces medical experts to spend a considerable amount of time in front of interactive image processing workstations. In the applications here, the main interest lies in estimating the ejection fraction of the left ventricle, i.e. the amount of blood being pumped out in the aorta over one heart cycle. It is a crude estimate, since the data is two dimensional but it gives an indication of the condition of the heart. Examples from two ultrasound sequences and one gamma camera sequence are shown in figures 6-4 and 6-5, together with the associated area-time graphs. It should be stressed that the results only indicate the feasibility of the system and they have not been used in a real clinical situation.

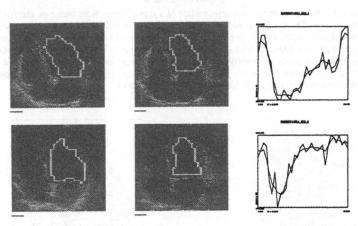

Figure 6-4

The images on the far left show the diastolic phase (the heart at its maximum volume), and the ones in the middle show the systolic phase (minimum volume) in a sequence of four chamber ultrasound heart images. The sequence on the top consisted of 29 images, and the bottom one 33 images. The contours were extracted at the 32 x 32 resolution level, and up-sampled to 64 x 64 and outlined on the images. The graphs on the right show area vs. time for the sequences. The boundaries are not well defined, so the curves are rather noisy. The dotted smooth graph in the diagram is a Fourier interpolation. (seven first harmonics, truncated with a triangular window).

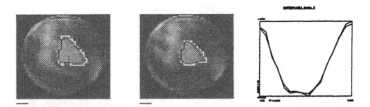

Figure 6-5

Diastolic (left) and systolic (middle) phases in a cycle of gamma camera images (16 images per sequence). Note that the ejection fraction graph to the right is much smoother than that from the ultrasound images. The reason is that the graph was derived from integrating the grayscale surface over the delineated region, and small changes in the boundary will not greatly affect the integration.

7 Segmentation with self organizing feature maps

7.1 Preliminaries

The method for segmentation and labeling described is active, in the sense that the system almost always has a second chance to detect and correct erroneously labeled and delineated images. This redundancy is vital if a system has to cope with difficult images. On the other hand, it is assumed that one knows the domain and how to model the contents in the scene in terms of objects and relations. As a contrast, if the system could learn to recognize a scene and learn the segmentation and recognition part, much would be gained. There are two main classification paradigms: supervised and unsupervised classification, and here the unsupervised paradigm is followed. Unsupervised classification is often called clustering, since the goal is to find patterns or structures in the data without beforehand knowledge of the right answer. There are many methods for unsupervised classification - e.g. K-means clustering and the LBG-algorithm - and we have chosen to use Kohonen's self-organizing feature maps. This is essentially a method for mapping a (usually) high dimensional space of feature vectors to a lower-dimensional classification space, $R^n \rightarrow R^k$. The main difference between Kohonen feature map method and the classic clustering algorithms is that adjacent feature vectors will also be adjacent in classification space; this is why the method is sometimes called "topological sorting". In what follows we will use a 2-dimensional classification space, with a square grid neighbourhood (simply called the Kohonen map, K^2), so that the mapping is from $R^n \rightarrow K^2$. The rationale is the convenience of regarding the learning process as a mapping from a 2-dimensional image (represented by n-dimensional feature vectors) to a 2-dimensional space - the Kohonen map, and the association phase as the reverse mapping. Starting with a set of randomized weights w_{ij}, where the index i is the map coordinate (which is one dimensional here to keep things simple), and j is the j:th element in the vector w_i, the goal is to search the space of feature vectors x and update a specific w_{ij}^* if it minimizes the distance between x and w_i. The training algorithm is as follows:

> Initialize w_{ij}.
> Transform (rotate) the space of feature vectors to equalize variance.
> Repeat
>   Choose m feature vectors randomly from the entire set.
>   For each of the m vectors, choose winner node i^* such that $|w_i^* - x| = min_i |w_i - x|$, for all nodes i.
> $w_{ij}^{n+1} := w_{ij}^n + k\,G(i,i^*)\,(x_i - w_{ij}^n)$ for all m vectors, if node i^* wins.
> $w_{ij}^{n+1} := w_{ij}^n$, for other nodes.
>   Adjust k and G according to suitable rules.
> Until n=n_stop.

7.2 The training phase

In the setup here, the neighbour credit assignment function G was a Gaussian. The learning rate k was decreased exponentially with each iteration n, starting from 0.3 and finishing with 0.05. Similarly, the standard deviation of the Gaussian was decreased exponentially from s=3 to s=0.1, and n_stop was set to 10000 iterations. The map size was 8x8, so initially the winning node affects many of its neighbours. An interesting feature with this algorithm, and one of the reasons it was chosen, is that it tends to under-sample high probability areas and over-sample low probability areas, i.e. the variance of the input probability function is overestimated. Analytically this has been shown in the case of densely sampled weight-spaces[18], and in fact the estimated probability density function is proportional to $f(x)^{d/(d+2)}$, where d is the dimension of the feature vector.

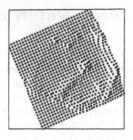 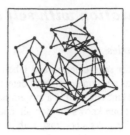

Figure 7-1

The picture on the left is a 2-D projection of the 3-D space consisting of feature vectors. The vector has 3 elements (x,y,f(x,y))., and the figures shows the (x,y) projection of the rotated feature vectors from one image, where each dot represents a vector. When projecting in the (x,y) -direction one sees the rotated outline of the four chamber ultrasound image. The transformation was derived from the eigenvectors of the diagonalized 3x3 covariance matrix C (this is sometimes called the Hotelling transform). The covariance matrix is defined as $C_x=E[(x-m)(x-m)^T]$, and the transformed diagonalized matrix $C_y = AC_xA^T = AC_xA^{-1}$. The data that goes into the Kohonen algorithm is $y=L^{-1}Ax$, where L is a diagonal matrix of the inverse of the eigenvalues. This means that the new covariance matrix will have unity variance in each direction, ensuring a non-biased sampling of the feature space. To the right are the Kohonen weights shown in the same projection, with the neighbour relations indicated by the lines between vertices.

7.3 The image representation.

The same pyramid representation of the image was used here as in the model-based system. Several experiments were undertaken with different combinations of elements and dimensions of the feature vector, ranging from grayscale, variance, "local divergence" derived from the normal vector of the LMS fitted plane, neighbouring grayscale values, and the image coordinates. The best results were achieved with a three-dimensional vector, consisting of the image coordinate pair and the pyramid grayscale value, i.e. *(x,y,f(x,y))*, each element being normalized to the range [0.0..1.0]. This means that the data is actually sorted before it is clustered, since the coordinate pairs constitute a natural ordering, and where the grayscale is a "jitter" component. In the learning phase, m feature vectors are chosen randomly (uniformly) from the entire set of feature vectors. The map will thus encode an average of the properties of the images in the training set. The main reason why this coding turned out most successful, is that the sequence of data is quasi-stationary; the sequence does not vary drastically from image to image. This is very different from speech signals, where the data can be considered stationary only for short periods. For truly non-stationary image sequences, a better representation would be to let one feature vector represent one image. The set of feature vectors would then represent different images in the sequence, and during training the map would be more likely to encode the time varying properties of the sequence. The problem then remains to find a suitable feature vector to represent the entire image, most likely a difficult task in realistic domains.

7.4 The association phase

During the association phase, for each feature vector, the map is searched for the best matching node. If the best node ends up in an LV node, the result image is set to unity, otherwise to zero. Thus, the association phase embodies a classical search problem, and the search strategy here is basically the classical "British Museum" approach. There exists more efficient algorithms for searching Kohonen maps - an active research field in communication theory[18] - but since the images here are heavily down-sampled, the search times are not prohibitive.

7.5 The ultrasound domain

The total data set consisted of 10 sequences, and the map was trained with data from 10 ultrasound images from 5 sequences - two images per sequence taken at the diastolic and systolic points - at the highest resolution level, which amounts 10x32x32=10024 data vectors. With 8x8=64 classes, this amounts to an average of 16 points per class per image. During each iteration, 5 samples were selected randomly from the entire set after which the weights were adjusted. Thus, the learning is a combination of off-line "batch" learning and incremental on-line learning; the former is useful if the data set contains redundant samples or if the data set is small, and the latter allows for a more efficient search of weight space. The relevant map nodes encoding the location of the LV - a total of 5 in this domain - were derived by visual inspection using 2 sequences, and the nodes were used as a scene model in the association phase.

7.6 Results

The Kohonen map was trained on half of the sequences, and evaluated for all 10 sequences. Some "cheating" was done to this end; frequently, the output from the map consisted of several spurious regions, of which only the largest one was retained. As in the model-based case, a local filter operation was applied to remove glitches, after which the resulting contour was upsampled to 64 x 64 and superposed onto the image. In 6 sequences of the grand total of 10, the results were acceptable (by visual inspection at least), although the general impression was that the segmented areas were too small.

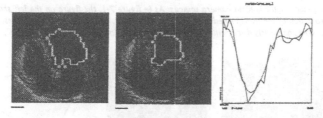

Figure 7-2

A sequence of ultrasound images, segmented with Kohonen feature maps, which is the same sequence as in figure 6-4. This sequence was actually used in the training, but the map had seen only 2 images from the sequence. Note that the area-time graph is smoother, an effect of the combined averaging of spatial and temporal properties of the data.

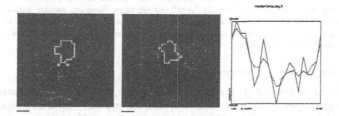

Figure 7-3

Another Kohonen segmented sequence, in which the LV is full of white echoes and only partially delineated. Even the model-based method has problems with this sequence (not shown here), but nevertheless yields a more consistent delineation, where the whole chamber is covered.

7.7 Conclusions

As seen in figure 7-2 the method works if the LV is well defined and not "polluted" with white echoes, as in figure 7-3 where the method clearly fails. The remedy in this situation would be to use a different set of map coordinates for this class of images, in the same spirit as the method of using allophones in speech recognition[33,31], where each allophone refers to a different instantiation of a phoneme. It is also quite possible that one can achieve better results in the current domains with a different feature vector representation, and with a larger training set. A major difference between the feature vector representation of data in image understanding and in speech understanding, is that one is interested in both the internal structure *within* the image, as well as the difference in structure between images in the sequence. A tentative and more biologically plausible solution would be to cascade two maps, one which encodes spatial structure and one which encodes temporal structure.

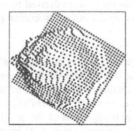
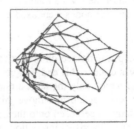

Figure 7-4

Kohonen map for the gamma camera images. As in figure 7-1, the figure on the left shows the (x,y) projection of the transformed space of feature vectors, and to the right the corresponding weights. One can clearly discern the different objects in the image.

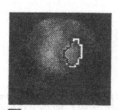

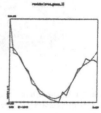

Figure 7-5

A Kohonen segmented gamma camera sequence. The sequence seen here was actually in the training set, so the map had already "seen" this sequence. However, another sequence (not shown) which also was a part of the training set, was not satisfactorily segmented. As in the ultrasound domain, this demonstrates the weakness of the method; some image sequences worked well, others did not work at all.

7.8 The gamma camera domain

The gamma camera data set consisted of 6 sequences, the sizes of which ranged from 16 to 33 images per sequence. As in the ultrasound case, the map was trained on images from half of this set, which amounts to 3x2x1024 = 6144 training vectors. The initial conditions otherwise were identical to that of the ultrasound domain, and altogether three map coordinates were used to model the scene in this domain. An example of a segmented sequence is seen in figure 7-5, which visually is an acceptable result. As in the ultrasound, domain, however, the method could not handle too diverse data sets, and only 3 sequences of 6 were properly segmented.

8 Conclusions and final remarks

In this paper we have tried to do two things: first, we have described a task oriented system for automatic analysis of images, with a simple but efficient control structure. The major contributions comprises:

- A vector representation of image features.

- An exchangeable relational model of the scene, implemented with relaxation labeling and constraint satisfaction.

- A hierarchical representation of the procedural aspects of the domain, comprising an exchangeable reactive control mechanism and a fixed top level supervisor.

- A robust region growing method for object delineation, based on the Medial Axis Transform.

- A continuity constraint check for images in the sequence, based on overlapping objects.

- A rudimentary self-knowledge of the domain, used to assess the quality of the result.

- A completely automatic detection and delineation of sequences 4-chamber heart images and gamma camera images of the heart.

Second, we have contrasted the model-based approach with the alternative method of segmentation by an unsupervised learning procedure. The salient features here are:

- A scene "model" induced by examples from image sequences, encoding both temporal and structural properties of images.

- Region extraction by searching the Kohonen map.

The main advantage with the model-based method lies in its flexibility and its ability to adapt itself to unexpected situations. The drawback is the need for knowledge engineering and that the whole method is more complex. The main advantage with the unsupervised learning is that the method is conceptually easy to grasp, straightforward to use, and obliterates the need for an analytical scene model. The disadvantage lies in its inflexible mapping; Once the map has been learned, there is no way of achieving on-line "re-learning". On-line learning methods have indeed been investigated, e.g. with genetic algorithms[6,20,25] and supervised feedforward neural networks[57]. The eternal problem with on-line learning, however, is that of designing a useful credit assignment procedure: how much should the new data affect the old result, and how to discover when new data is not readily modeled with a stationary probability density function? In the statistical world, this can partly remedied by extracting piecewise stationary signal segments, and use a multitude of classes to represent different flavours of one class. One can never hope to cover all situations, however, so eventually one must resort to some sort of on-line re-categorization.

There is an informal law in the science of human organizations that "structure comes before strategy". If transferred into the present context, this could be interpreted as "If the system is not robust enough on all levels, it doesn't matter how much sophisticated planning you do at the top level". This is one of the most important lessons that we can learn from biological systems; there is almost always a backup system or different pathways to take care of unexpected situations. It has taken evolution millions of years to evolve humans to our present state. In designing truly adaptive autonomous systems, researchers have to mimic both the result of the evolutional blueprints as well as the capabilities of learning on part of the individual. It is highly improbable that this will be a successful undertaking unless we fully grasp the mechanisms behind reasoning and perception. This will most likely be achieved someday, but as of yet we are a long way from that goal.

9 Appendices

I Implementational details

The system runs in a VMS environment, and was developed on VAX workstations. The major bulk of code was written in VAX PASCAL, a dialect of Wirth PASCAL, with additional system programming in FORTRAN. The supervisor was written in OPS-5, an easy to use rule based language. Each state in the automaton in figure 4-1 was implemented as a VMS command file - the equivalent of a UNIX script - and consists of other command files or executable programs. Each program communicates with a public blackboard and image data files, and the results are also stored as files. While this slows down the system during runtime, it is very convenient during the development phase, since intermediate results are easily obtained and new procedures are can readily be added. Another convenient detail was the use of a mail box facility, which enables swift communication between the supervisor and the scripts. It furthermore makes it possible to spawn scripts from programs and vice versa up to a pre-defined nesting level. The control software was written is PASCAL and not in OPS-5, the main reason being that the priority order of the condition-action matching in the OPS-5 inference engine was rather artificial and of no use here.

II Logical conditions in image sequence interpolation

The overlap conditions are most easily phrased as follows. Define the overlap (intersection) between the current and next image

$$\overline{B}(n) \equiv B(n) \cap B(n+1)$$

Define an interval within which the geometrical overlap between two consecutive has to roam $I=[I_{low}, I_{high}]$, where the lower and upper bounds are related to the quotient between two images. Three cases can be discerned: no overlap between consecutive images; the images overlap but the next one is too large; the images overlap, but the next one is too small. The first condition takes the following logical form

$$\left(\frac{Area(\overline{B}(n))}{Area(B(n))}\right) \leq I_{low} \wedge \left(\frac{Area(\overline{B}(n))}{Area(B(n+1))}\right) \leq I_{low} \rightarrow B(n) \Rightarrow B(n+1)$$

The RHS expression has the following meaning: calculate the pyramid of the next image $B(n+1)$, but instead of using the result of the region growing from $B(n+1)$, use the result from the previous image $B(n)$. Similarly for the other cases: the images overlap but the next image is too large is translated as

$$\left(\frac{Area(\overline{B}(n))}{Area(B(n))}\right) \in I \wedge \left(\frac{Area(\overline{B}(n))}{Area(B(n+1))}\right) \leq I_{low} \rightarrow B(n) \Rightarrow B(n+1)$$

and the last condition is obtained by interchanging $B(n)$ and $B(n+1)$. In the applications here, $I_{low} = 0.5$ and $I_{high} = 1.5$.

10 Acknowledgments

Special thanks to Holger Broman who has provided constant support and ideas throughout the project. Tomas Gustavsson at the Medical Image Processing Laboratory MEDNET at the

University of Gothenburg supplied the ultrasound image sequences, and Mikael Ekman at the department of Radiophysics at Sahlgrenska Hospital supplied the gamma camera image sequences. Petter Knagenhjelm at the Department of Information Theory at Chalmers University of Technology gave insightful information on data representation in Kohonen maps. Thanks also to the organizers and participants at the summer school in Huddinge, for many stimulating talks and discussions. This research was supported by the Technical Research Council, research grant no. 93-237.

11 References

[1] Aloimonos J.Y., and Rosenfeld A., 1991: Computer Vision. Science, Vol 253, 13 Sept., pp.1249-1254.

[2] Aloimonos J.Y.,1990: Purposive and Qualitative Active Vision. DARPA IU Workshop. Pittsburgh PA, pp. 816-828.

[3] Ballard D.H., 1989: Animate Vision. Proceedings of the International Joint Conference on Artificial Intelligence (IJCAI), Detriot, (Morgan Kaufmann, Los Altos, CA, 1989) pp. 1635-1641.

[4] Barrow G.H & Tenenbaum J.M., 1981: Computational Vision. Invited paper in the proc. of IEEE, Vol. 69, No. 5, May 1981, pp. 572-595.

[5] Bilbro G.L., White M., and Snyder W., 1989: Neural Computers, Springer Verlag, Eckmiller R., and Malsburg C. v.d. (eds), pp 72-79.

[6] Booker L.B., Goldberg D.E., and Holland J.H., 1989: Classifier Systems and Genetic Algorithms. Artificial Intelligence, Elsevier (North Holland), No. 40, pp. 235-282.

[7] Del Bimbo A., Landi L., and Santini S., 1993: Determination of road directions using feedback neural nets. Signal processing, Elsevier, special issue on intelligent systems for signal and image understanding, Vol. 32, Nos.1-2, pp. 147-160.

[8] Brooks, R.A., ,1991: New Approaches to Robotics. Science, Vol. 253, 13 Sept., pp. 1227-1232.

[9] Chauvet P., Lopez-Krahe J., Taflin E., and Maître H., 1993: System for and intelligent office document analysis, recognition and description. Signal Processing, Elsevier, special issue on intelligent systems for signal and image understanding, Vol. 32, Nos. 1-2, May, pp. 161-190.

[10] Crowley, J.L, (1991): Towards Continously Operating Integrated Vision Systems for Robotics Applications. Proceedings of the Seventh Scandinavian Conference on Image Analysis (7SCIA), Aalborg, Denmark, AUG 13-16.

[11] Demazeau Y. & Müller J. P. (eds.), 1990: Decentralized AI. North-Holland/Elsevier.

[12] Dennet D., 1991: Consciousness Explained.Allen Lane, The Penguin Press.

[13] Draper B.A., Collins R.T., Brolio J., Hanson A.R., and Riesman E.M., 1988: Issues in the Development of a Blackboard-Based Schema System for Image Understanding. Blackboard systems (Eds. Engelmore R. and Morgan T.) Addison-Wesley, pp. 189-218.

[14] Edelman G., 1988:The Remembered Present: A Biological Theory of Consciousness.Basic Books, Inc., New York

[15] Eklundh J-O., (ed) 1991: The 1991 Stockholm Workshop on Computational Vision. Report from the Computational Vision and Active Perception Laboratory at KTH, Stockholm, TRITA-NA-P9210, CVAP 102.

[16] Engelmore R. S., and Morgan T.,1988: Blackboard systems. AddisonWesley.

[17] Van Essen D.C., Andersson C.H., and D.J.Fellerman, 1992: Information processing in the primate visual system: an intergrated systems perspective. Science, Vol. 255, Jan. 24, pp. 419-423.

[18] Gersho A and Gray. R.M., 1992: Vector Quantization and Signal Compression. Kluwer Academic Publishers.

[19] Gibson, J.J., 1979: The Ecological Approach to Visual Perception. Boston, Houghton Mifflin.

[20] Goldberg D.E, 1987: Computer-Aided Pipeline Operation Using Genetic Algorithms and Rule LEarning. PART I: Genetic Algorithms in Pipeline Optimization. Engineering with Computers (Springer Verlag), No. 3., pp 35-44.

[21] Vision Systems, Academic Press, New York, pp. 303-33.

[22] Hebb, D. O.,1949: The organization of behavior. New York: Wiley.

[23] Hertz J., Krogh., and Palmer R., 1991: Introduction to the theory of neural computation. Lecture Notes Volume I, Santa Fe Institute Studies in the Sciences of Complexity (ISBN 0-201-51560-1).

[24] Hanson A. R., and Riesman E. M., 1987: The VISIONS image understanding system - 1986. In: Advances in computer vision, Brown C. (ed), L. Erlbaum.

[25] Holland J.H., 1975: Adaption in Natural and Artificial Systems. The Unversity of Michigan Press, Ann Arbor.

[26] Hubel D.H. and Wiesel T.N., 1962: Receptive fields, binocular interaction and functional architecture in the cat's visual cortex. Journal of physiology (London) 166, 106-154.

[27] Hummel R. and Zucker S.W., 1983: On the Foundations of Relaxation Labeling Processes. IEEE transactions on pattern analysis and machine intelligence, Vol. PAMI-5, NO.3, May pp. 267-287.

[28] Jain R.C., and Binford T.O., 1991: Ignorance, Muopia, and Naiveté in Computer Vision Systems. In the DIALOGUE section of CVGIP Image Understanding, Vol. 53, No. 1, January, pp. 112-117.

[29] Jain R.C. and Dubes A.K., 1988: Algorithms for Clustering Data. Prentice-Hall, Englewood Cliffs, New Jersey.

[30] Julesz B., 1975: Experiments in the visual perception of texture. Scientific American 232, 34-43.

[31] Knagenhjelm P., 1993: Competetive Learning in Robust Communication. Ph. D. thesis, Department of Information Theory, Chalmers University of Technology, Technical Report No. 242, ISBN 91-7032-827-7.

[32] Kohonen T., 1989: Self-Organization and Associative Memory. Berlin, Springer Verlag (3:rd ed).

[33] Kohonen T., Mäkisara K., and Saramäki T., 1984: Phonotopic Maps - Insightful Representation of Phonological Structures for Speech Recognition. IEEE proc. of 7:th International Conference on Pattern Recognition, Montreal, Canada.

[34] Kummert F., Niemann H., Prechtel R., and Sagerer G., 1993: Control and explanation in a signal understanding environment. Signal processing, Elsevier, special issue on intelligent systems for signal and image understanding, Vol. 32, Nos.1-2, May, pp. 111-145.

[35] Le Cun Y., Boser B., Denker J.S., Henderson D., Howard R.E., Hubbard W., and Jackel L.D. (1989): Backpropagation Applied to Handwritten Zip Code Recognition. Neural Compuation 1., 541-551.

[36] Lindeberg T., 1990: Scale-Space for Discrete Signals. IEEE Transactions on Pattern Analysis and Machine Intelligence, Mar. Vol. 12, NO. 3, pp. 234-253.

[37] Lindeberg T., 1991: Discrete Scale-Space Theory and the Scale-Space Primary Sketch. Ph.D. thesis, Department of Numerical Analysis and Computing Science, Royal Institute of Technology, Stockholm. TRITA-NA-P9108, ISSN 1101-2250, CVAP84

[38] Lowe D.G., 1985: Perceptual Organization and Visual Recognition. Kluwer Academic Publishers, Boston.

[39] Löfgren L., 1987: Complexity of Systems. In Systems & Control Encyclopedia, Cambridge, Pergamon Press, pp. 704-709.

[40] _____,1987: Autology, ibid., pp. 326-333.

[41] Marr D.,1982: Vision: A computational Investigation into the Human Representation and Processing of Visual Information. Freeman & Co.

[42] Minsky, M.,, 1975: A Framework for Representing Knowledge. The Psychology of Computer Vision, New York: McGraw-Hill, 1975.

[43] _____,1986: The Society of Mind. Simon and Schuster, New York.

[44] Molander S., Broman H., Gustavsson T., 1990: A Controllable Medial Axis Transform. Proceedings of the SSAB symposium on Image Analysis in Linköping, March 6-7, pp. 240-243.

[45] Molander S., and Broman H., 1993: Knowledge-based segmentation and state-based control in image analysis: Two examples from the biomedical domain. Signal processing, (Elsevier), special issue on intelligent systems for signal and image understanding, Vol. 32, Nos. 1-2, pp. 201-215.

[46] Mulder J. A., Mackworth A. K., and Havens W. S., 1988: Knowledge Structuring and Constraint Satisfaction: The Mapsee Approach. IEEE transactions on Pattern Analysis and Machine Intelligence, Vol 10, No. 6, pp. 866-878.

[47] Nagao M., Matsuyama T., and Mori H., 1979: Structured Analysis of Complex Aerial Photographs. Proceedings of the Sixth International Joint Conference on Artificial Intelligence (IJCAI-79), pp. 610616.

[48] Nicolin B., and Gabler R., 1987: A knowledge-Based System for the Analysis of Aerial Images. IEEE Transactions on Geoscience and Remote Sensing, May, vol. GE-25, no. 3, pp. 317-329.

[49] Papert S., and Minsky M., 1988: Perceptrons: An Introduction to Computational Geometry (expanded ed.). MIT press.

[50] Pearl J., 1988: Probabilistic Reasoning in Intelligent Systems. Morgan kauffman.

[51] Pomerleau, D.A, 1992: Progress in Neural Network Vision for Automomous Robot Driving. Proc. of Intelligent Vehicles, I. Masaki (ed), Detroit, Michigan.

[52] Reiter R. and Mackworth A. K., 1989: A logical Framework for Depiction and Image Interpretation. Artificial Intelligence. Vol 41, No. 2.

[53] Rosenfeld A., Hummel R. A., and Zucker S. W., 1976: Scene labeling by relaxation operations. IEEE Trans. Syst. Man, Cybern. 6, pp. 420-433.

[54] Rosenfeld, A., 1993: SURVEY: Image Analysis and Computer Vision: 1992. CVGIP, Image Understanding, Vol. 58, No. 1.pp. 85-135

[55] Roska T., et al., 1993: The use of CNN Models in the Subcortical Visual Pathway. IEEE Transactions on Circuits and Systems-1:Fundamental Theory and Applications, Vol. 40, No. 3, March, pp. 182-195.

[56] Rumelhart, D. E. et al.,1987: Parallel Distributed Processing: Explorations in the Microstructures of Cognition, Vol I & II. A Bradford Book, MIT Press.

[57] Sztipanovits J., 1992: Dynamic Backpropagation Algorithm for Neural Network Controlled Resonator-Bank Architecture. IEEE Trans. on Circuits and Systems, Vol. 39., No.2, February, pp. 99-108.

[58] Truvé S., 1992: Symbolic Image Interpretation by Parsing, Interpretation, and Pruning. Ph.D. thesis, Department of Computer Science, Chalmers University of Technology, Sweden. ISBN 91-7032-712-2

[59] Tsostos J.K., 1985: Knowledge organization and its role in representation and interpreation for time-varying data: the ALVEN system. Comput. Intell. 1, pp. 16-32.

[60] Uckun S., 1992: Model-based reasoning in biomedicine. In: (Bourne J., ed.) Critical Reviews in Biomedical Engineering, Vol. 19, No. 4, pp. 261-292.

[61] Wedelin D., 1993: Efficient Algorithms for Probabilistic Inference, Combinatorial Optimization and the Discovery of Causal Structure from Data. Ph. D. thesis, Department of Computer Science, Chalmers University of Technology, Göteborg, Sweden .(ISBN 91-7032-796-3).

[62] Werblin F.S., 1992: A Functional Analysis of the Tiger Salamander Retina. Univ. of California, Berkeley.

[63] _____, 1991: Synaptic Connections, Receptive Fields and Patterns of Activity in the Tiger Salamander Retina. Investigative Ophtalomology & Visual Science, Vol. 32, March.

[64] Witkin A. P., 1983: Scale-Space filtering. Proceedings of the Seventh International Joint Conference on Artificial Intelligence (IJCAI), pp. 1019-1022.

[65] Zeki S.M., & Shipp S., 1988: The functional logic of cortical connections. Nature 335, pp. 311-317.

Simplicity criteria for dynamical systems

Petr Kůrka

Faculty of Mathematics and Physics, Charles University in Prague
Malostranské náměstí 25, 118 00 Praha 1, Czechia

Abstract

We formulate two simplicity criteria for dynamical systems based on the concepts of finite automata and regular languages. Finite automata are regarded as dynamical systems on discontinuum and their factors yield the first simplicity class. A finite cover of a topological space is almost disjoint, if it consists of closed sets which have the same dimension as the space, and meet in sets whose dimension is smaller. A dynamical system is regular, if it yields a regular language when observed through any finite almost disjoint cover. Next we formulate two topological simplicity criteria. A dynamical system has finite attractors, if the ω-limit of its state space is finite. A dynamical system has chaotic limits, if every point is included in a set whose ω-limit is either a finite orbit or a chaotic subsystem. We show the relations between these criteria and classify according to them several classes of zero-dimensional and one-dimensional dynamical systems.

1   Introduction

Most deterministic dynamical systems encountered in science as models of physical, biological, sociological or economical processes display two types of behaviour. One type (ordered systems) is characterized by the prevalence of stable states or stable periodic trajectories in which the system eventually settles. These systems are well described by linear models which capture the essence of their behaviour in the vicinity of stable states. The other type is that of chaotic systems which move through their state space in apparently chaotic manner, recurrently visiting every its region. Chaotic systems are best characterized in probabilistic terms by a density distribution specifying the frequency of visits in particular regions of the state space.

There has recently been a surge of interest in dynamical systems "at the edge of chaos", whose behaviour is more complex than both the ordered and chaotic systems (see e.g. Langton [20], Kauffman [14], Mitchell et al. [22], or Crutchfield and Young [5], who speak about systems at the onset of chaos). These systems originated in the study of cellular automata. Wolfram introduces in

[27] four complexity classes of linear (i.e. one-dimensional) cellular automata. The first two classes comprise ordered systems: systems in class I reach the stable null state, and the systems in class II reach a periodic orbit. Systems in class III produce apparently random patterns and correspond to chaotic systems. Most interesting behaviour display systems in class IV which produce irregular patterns with great diversity. Wolfram conjectures that they have the power of universal computation, i.e. that they are able to simulate universal Turing machines. It seems that the power of universal computation is the key feature of the systems at the edge of chaos. In [3] von Neumann postulates it as a criterion of nontriviality for his self-reproducing automata. The power of universal computation is also possessed by the Game of Life (see Langton [20]) and by some linear cellular automata (see Smith [25]).

In view of these considerations it is desirable to characterize the complexity of a general dynamical system by its computing power. This is quite natural for dynamical systems whose state space is a discontinuum, i.e. a compact metrizable topological space which is perfect and zero-dimensional. Discontinuum is unique, as any two discontinua are homeomorphic. It is the state space of an infinite tape or more general infinite data structures like trees. The majority of automata studied in computer science as models of computing processes can be naturally regarded as dynamical systems on discontinuum. A finite automaton becomes a dynamical system on discontinuum, if it is attached to input tape, on which all its future inputs are written. A Turing machine becomes a dynamical system on discontinuum, if we allow for infinite content of its tape (see Moore [23]). Cellular automata are dynamical systems on discontinuum by definition, and they can be generalized to infinite automata networks consisting of different kinds of automata (see Koiran et al. [15]). In this way we obtain a hierarchy of dynamical systems on discontinuum with increasing computing power.

The connection to more general dynamical systems is provided by Alexandrov Theorem, which says that every compact metrizable topological space is a factor of discontinuum. A generalization proved in Balcar and Simon [1] reads that every dynamical system on a compact metrizable space is a factor of some dynamical system on discontinuum. Since factorization might only lead to simplification of the dynamics, the hierarchy of automata induces a hierarchy of dynamical systems on compact metrizable spaces via factorization. From this point of view, the simplest dynamical systems are the factors of finite automata.

An alternative approach to the study of complexity of dynamical systems uses the Chomsky complexity hierarchy of languages. This approach has been used for the classification of Z-subshifts (see e.g. Culik II et al. [6, 7]). Any Z-subshift determines a language of words occurring in its elements. The obtained language is central, and conversely any central language defines a unique Z-subshift. The complexity of Z-subshifts is identified with the position of the corresponding central language in the Chomsky hierarchy. This procedure can be generalized, as any finite cover of the state space of a dynamical system determines a language. This language corresponds to possible observations of the dynamical system, when we can observe only the set of the cover, in which the system presently finds itself.

To get a satisfactory theory, we restrict ourselves to covers which are sufficiently simple. Otherwise the complexity of the obtained language would reflect the complexity of the cover, rather than the complexity of the dynamics. In zero-dimensional spaces the appropriate covers are finite clopen partitions. In the case of a compact real interval we use interval covers consisting of closed intervals which meet at their endpoints. A general concept is almost disjoint cover, and it requires an additional assumption on the state space. We say that a topological space is homogenous, if it has at each point the same dimension. An almost disjoint cover of a homogenous space consists of homogenous closed sets of the same dimension as the state space, which meet in sets whose dimension is smaller. We say that a dynamical system is regular, if it generates a regular language on each almost disjoint cover.

Besides these two simplicity criteria based on the dual concepts of finite automata and regular languages, we formulate also a topological simplicity criterion. The distinction of the systems at the edge of chaos from both ordered and chaotic systems suggests that the two latter have something in common. Indeed in the light of a recent result of Brooks et al in [2], it is quite natural to view a periodic orbit as a special case of a chaotic dynamical system. These authors proved that any infinite, topologically transitive system with a dense set of periodic points is sensitive to initial conditions. Thus it is natural to drop the condition of sensitivity to initial condition from the standard definition of chaotic systems (see Devaney [8]). According to this relaxed definition, however, a finite system, which consist of a single periodic orbit, is chaotic too. Our topological simplicity criterion states that from every state some chaotic subsystem is reached: either periodic orbit, or infinite chaotic subsystem. We say that a dynamical system has chaotic limits, if every point is included in a set whose ω-limit is chaotic.

Thus we get three alternative simplicity criteria for dynamical systems: to be a factor of a finite automaton, to be regular, and to have chaotic limits. The present work is devoted to the study of these criteria both in general and in special classes of dynamical systems. We show that the simplest dynamical systems, which have only finite attractors and therefore chaotic limits, are both regular and factors of finite automata. Factors of finite automata have chaotic limits. This can be used for proving that some systems are not factors of finite automata. We exhibit examples showing that no other inclusions among these classes of dynamical systems occur.

On zero-dimensional spaces, any factor of a finite automaton is regular, but not vice versa. However, a subshift is a factor of a finite automaton iff it is regular. This is a reformulation of a theorem on sofic systems, which are regular Z-subshifts. The theorem states that a Z-subshift is regular iff it is a factor of a subshift of finite type (see Culik II and Yu [6]). This means that a sofic system can be generated by paths in a finite labelled graph. This graph, in turn, can be viewed as a finite automaton, which accepts the corresponding regular language. The topological properties of the subshift can be deduced from the structure of its graph. This approach can be generalized to nonregular subshifts. Any subshift may be generated by a labelled graph, which is infinite, if the subshift is not regular (see Fiebig and Fiebig [9]). We adopt this approach to derive the topological properties of the subshift from the structure of its graph.

Next we study one-dimensional dynamical systems which are one-to-one. On real interval every one-to-one dynamical system is a regular factor of a finite automaton. On the circle, a one-to-one dynamical system is necessarily a homeomorphism. If its rotation number is rational, then it is again a regular factor of a finite automaton. If not, then it is not a factor of a finite automaton and may be even nonrecursive. In particular no irrational rotation of the circle is recursive, so there is nothing between regularity and recursivity.

Finally we investigate unimodal systems on the real interval. These systems display surprisingly large variety of behaviour and their structure is fairly well understood (see Devaney [8], Collet and Eckmann [4], or Guckenheimer and Holmes [13]). The basic topological invariant of a unimodal system is its kneading sequence, which is determined by the trajectory of the turning point. If the kneading sequence is finite or periodic, then the system has a periodic orbit which attracts a dense set of points. If the kneading sequence is preperiodic, then the system has chaotic attractor which again attracts a dense set of points. In both these cases we get a system with chaotic limits. Some unimodal systems with aperiodic kneading sequences have chaotic limits too, but not all.

We use these results to classify unimodal systems according to our criteria. A unimodal system is regular iff it has a finite number of periodic points. This coincides with systems whose topological entropy is zero (except at the boundary: the first system with infinite number of periodic points has topological entropy zero too - see Milnor and Thuston [21]). Factors of finite automata form a larger class. They include all systems with finite, periodic and preperiodic kneading sequences. There is also a large class of unimodal systems which have not chaotic limits, and therefore are not factors of finite automata. They do occur in vicinity of both stable and chaotic systems, so they are truly at the edge of chaos. We leave open the question whether there exist aperiodic unimodal systems, which are factors of finite automata.

The present work is relatively self-contained except for some hard results on S-unimodal systems reproduced from Collet and Eckmann [4]. Some results about regular systems and factors of finite automata are in Kůrka [18, 19]. The generalized Alexandrov Theorem in Section 3 has been proved in Balcar and Simon [1]. The correspondence between subshifts and right central languages in Section 4 is adapted from Culik II and Yu [6] for the case of N-subshifts. Our treatment of the unimodal systems in Section 9 is a variant of the standard approach in Devaney [8] or Collet and Eckmann [4]. However, by differentiating between lower and upper itineraries, we get stronger results. In particular we are able to characterize the language of any interval cover without any differentiability assumptions.

2 Preliminaries

We work in the context of topological dynamics, where a dynamical system is conceived as a semigroup action on a metrizable topological space. We concentrate on discrete time, noninvertible dynamical systems, where the semigroup is the set of nonnegative integers with addition. In this case, the dynamical system can be defined as a self-map of the state space. We assume that the state space is compact and homogenous with respect to dimension. An assumption of this kind seems to be necessary for investigating languages generated by covers of the state space.

Recall that the dimension $dim(X)$ of a space X, and dimension $dim_x(X)$ of a space X at its point $x \in X$ are defined by simultaneous induction as follows:

$$dim(X) = -1 \qquad \text{if } X = 0,$$
$$dim(X) = \sup\{dim_x(X); x \in X\} \qquad \text{if } X \neq 0$$

$$dim_x(X) \leq n+1 \quad \text{if } (\forall U \text{ neighb. of } x)(\exists V \subseteq U \text{ neighb. of } x)$$
$$(dim(Fr(V)) \leq n)$$
$$dim_x(X) = n+1 \quad \text{if } dim_x(X) \leq n+1 \text{ and } dim_x(X) \not\leq n,$$
$$dim_x(X) = \infty, \qquad \text{if } (\forall n)(dim_x(X) \not\leq n)$$

Definition 1 *A space X is homogenous, if $(\forall x \in X)(dim_x(X) = dim(X))$.*

Every zero-dimensional space is homogenous, and a real interval or circle are homogenous one-dimensional spaces. A space is perfect, if each its point is a limit point. A discontinuum is a compact metrizable space which is perfect and zero-dimensional. Any two discontinua are homeomorphic, and any countable product of discrete finite spaces, which have at least two elements, is a discontinuum (see Hocking [10]).

A compact metrizable space admits the unique uniformity compatible with the topology. The uniformity is formed by all neighbourhoods of the diagonal

$$\mathcal{U}(X) = \{U \subseteq X \times X; (\forall x \in X)(\exists V \text{ neighb. of } x)(V \times V \subseteq U\}$$

A cover of space X is a system $\alpha = \{V_a; a \in A\}$ of its subsets whose union is X. A cover α is finer than $\beta = \{V_b; b \in B\}$, if $(\forall a \in A)(\exists b \in B)(V_a \subseteq V_b)$ (we write $\alpha \geq \beta$). The join of α and β is $\alpha \vee \beta = \{V_a \cap V_b; a \in A, b \in B\}$. If $f : X \to Y$ is a map and $\alpha = \{V_a; a \in A\}$ a cover of Y, then $f^{-1}(\alpha) = \{f^{-1}(V_a); a \in A\}$.

Definition 2 *A cover $\alpha = \{V_a; a \in A\}$ of X is almost disjoint if A is finite, each V_a is a closed homogenous set with $dim(V_a) = dim(X)$, and $dim(V_a \cap V_{a'}) < dim(X)$ for $a \neq a'$.*

A cover of a zero-dimensional compact space is almost disjoint iff it is a finite clopen partition, i.e. if it consists of disjoint clopen (closed and open) sets. A cover $\alpha = \{V_a; a \in A\}$ of a compact real interval is almost disjoint, iff each V_a is a finite union of closed intervals, and each $V_a \cap V_b$ $(a \neq b)$ consists of endpoints of these intervals. We say that α is an interval cover, if each V_a is a closed interval. The endpoints of these intervals are referred to as cutpoints of α.

Let A be a finite alphabet. We frequently use alphabet $2 = \{0, 1\}$. Denote $N = \{0, 1, 2, ...\}$ the set of non-negative integers, and $A^N = \{u = u_0 u_1 ...; u_i \in A\}$ the space of infinite sequences of letters of A with product topology. Thus A^N is a discontinuum. For $n \in N$ denote A^n the set of sequences of length n, $A^* = \cup_{n \in N} A^n$ the set of finite sequences, $A^+ = \cup_{n>0} A^n$, and $\overline{A^*} = A^* \cup A^N$. Denote $|u|$ the length of a sequence $u \in \overline{A^*}$, $(0 \leq |u| \leq \infty)$, and λ the word of zero length, so $u_i = \lambda$ for $i > |u|$. Write $u \trianglelefteq v$ if u is a substring of v, and $u \sqsubseteq v$ if u is an initial substring of v. Denote $u_{|i} = u_0 ... u_{i-1}$ the initial substring of u of length i. If $u, v \in A^*$, denote uv their concatenation, u^k the concatenation of u k times, and $\overline{u} \in A^N$ the infinite concatenation of u, defined by $\overline{u}_{k|u|+i} = u_i$. For $u \in 2^+$ denote $\hat{u}$ the string obtained form u by changing the last bit. The set A^* with concatenation operation and unity λ is a monoid. For $u \in \overline{A^*}$ define $\sigma(u) \in \overline{A^*}$ by $\sigma(u)_i = u_{i+1}$. Thus $\sigma(\lambda) = \lambda$. The restriction of σ to A^N is a continuous mapping. For $u \in A^n$ denote $[u] = \{v \in A^N; v_{|n} = u\} \subseteq A^N$ the cylinder determined by u. It is a clopen set. Then $\alpha_n = \{[u]; u \in A^n\}$ is a clopen partition of A^N.

A language over an alphabet A is any subset $L \subseteq A^*$. If A, B are alphabets, then $h : A^* \to B^*$ is a monoid homomorphism, if $h(\lambda) = \lambda$ and $h(uv) = h(u)h(v)$. Any map $h : A \to B^*$ defines a homomorphism by extension. A homomorphism h is λ-free, if $h^{-1}(\lambda) = \lambda$. Regular and recursive languages are closed with respect to λ-free homomorphism, i.e. if $h : A^* \to B^*$ is a λ-free homomorphism and $L \subseteq A^*$ is a regular (recursive) language, then so is $h(L)$.

3 Dynamical systems

Definition 3 *A dynamical system (X, f) is a continuous map $f : X \to X$ of a compact, homogenous, metrizable topological space X to itself.*

The n-th iteration of a dynamical system (X, f) is defined by $f^0(x) = x$, $f^{n+1}(x) = f(f^n(x))$. The orbit of a point $x \in X$ is $\mathcal{O}(x) = \{f^i(x); i \geq 0\}$. A point $x \in X$ is periodic with period $n > 0$, if $f^n(x) = x$ and $f^i(x) \neq x$ for $0 < i < n$. A point x is eventually periodic, if there exists $i \geq 0$, such that $f^i(x)$ is periodic. It is preperiodic, if it is eventually periodic but not periodic. It is aperiodic, if it is not eventually periodic. A subset $Y \subseteq X$ is invariant, if $f(Y) \subseteq Y$, and strongly invariant if $f(Y) = Y$. The ω-limit set of a set $Y \subseteq X$

is defined by

$$\omega(Y) = \bigcap_n \overline{\bigcup_{m>n} f^m(Y)}$$

If Y is nonempty, then $\omega(Y)$ is a nonempty, closed, strongly invariant set.

Definition 4 *A set $Y \subseteq X$ is an attractor of a dynamical system (X, f), if it has a closed neighbourhood V such that $f(V) \subseteq int(V)$, and $Y = \omega(V)$.*

Lemma 1 *Any attractor is stable, i.e. any its neighbourhood contains an invariant neighbourhood.*

Proof: Let ϱ be a metric compatible with the topology. For $A \subseteq X$ denote $B_\varepsilon(A) = \{x \in X; \varrho(y, A) < \varepsilon\}$. Let $V \subseteq X$ be closed neighbourhood of Y, $f(V) \subseteq int(V)$, and $Y = \omega(V)$. Let $U \subseteq V$ be an open neighbourhood of Y. Suppose that U does not contain any invariant neighbourhood of Y. For $k > 0$ denote $W_k = \cup_{n \in N} f^n(B_{1/k}(Y))$. Then W_k is an invariant neighbourhood of Y, so $W_k \not\subseteq U$. Thus there exists some $x_k \in B_{1/k}(Y)$, and an integer n_k with $f^{n_k}(x_k) \in X - U$. Since $X - U$ is compact, there exists a subsequence k_i, with $f^{n_{k_i}}(x_{k_i}) \to x \in X - U$. For all sufficiently large k, $x_k \in V$, so $x \in \omega(V) = Y$, which is a contradiction. $\square$

Definition 5 *A dynamical system (X, f) is topologically transitive, if for any nonempty open sets $U, V \subseteq X$ there exists $k > 0$ with $f^k(U) \cap V \neq 0$. It is sensitive to initial conditions, if*

$$(\exists U \in \mathcal{U}(X))(\forall x \in X)(\forall V \text{ neighb. of } x)(\exists y \in V)(\exists n)((f^n(x), f^n(y)) \notin U)$$

According to a standard definition (see Devaney [8]), a chaotic system is a dynamical system which has a dense set of periodic points, is topologically transitive, and sensitive to initial conditions. However, Brooks et al. have proved in [2] that infinite dynamical systems, which are topologically transitive, and have dense set of periodic points, are also sensitive to initial conditions. In view of this result it is natural to relax the definition of chaotic systems to topological transitivity and density of periodic points. According to this relaxed definition, any finite system consisting of a single periodic orbit is chaotic too. This leads to a topological definition of simple dynamical systems: We say that a dynamical system has chaotic limits, if every point is included in a set, whose ω-limit is chaotic. We define also an auxiliary concept of systems with bounded limits.

Definition 6 *Let (X, f) be a dynamical system.*
1. (X, f) is chaotic, if it is topologically transitive, and has a dense set of periodic points.
2. (X, f) has finite attractors, if $\omega(X)$ is finite.
3. (X, f) has chaotic limits, if every $x \in X$ is contained in some $Y \subseteq X$ such

that $(\omega(Y), f)$ is chaotic.

4. (X, f) has bounded limits, if there exists an integer n such that the set $\{x \in X; card(\omega(x)) \leq n\}$ is dense in X.

A finite system is chaotic iff it consists of a single periodic orbit. A system with finite attractors has both bounded limits and chaotic limits. Example 1 below and Example 2 from Section 7 show, that the last two properties are incomparable.

Example 1 *There exists a dynamical system with bounded limits, which has not chaotic limits.*

Proof: Let a be an irrational number, and define a dynamical system $(\overline{C}, f)$ on the complex sphere by $f(re^{2\pi i x}) = r^2 e^{2\pi i(x+a)}$, $f(\infty) = \infty$. If $|z| \neq 1$, then either $\omega(z) = \{0\}$ or $\omega(z) = \{\infty\}$, so $(\overline{C}, f)$ has bounded limits. If $|z| = 1$, and $z \in Y \subseteq \overline{C}$, then either $\omega(Y) = S_1 = \{z; |z| = 1\}$, and then it has no periodic point, or $\omega(Y) \supset S_1$, and then it is not topologically transitive. $\square$

Definition 7 *A homomorphism $\varphi : (X, f) \to (Y, g)$ of dynamical systems is a continuous mapping $\varphi : X \to Y$ such that $g\varphi = \varphi f$. We say that (Y, g) is a factor of (X, f), if φ is a factorization (i.e. a surjective map).*

$$
\begin{array}{ccc}
X & \xrightarrow{\ \ f\ \ } & X \\
\varphi \downarrow & & \downarrow \varphi \\
Y & \xrightarrow{\ \ g\ \ } & Y
\end{array}
$$

Lemma 2 *If $\varphi : (X, f) \to (Y, g)$ is a homomorphism, then $\varphi(\omega_f(A)) = \omega_g(\varphi(A))$.*

Proof: If $x \in \omega_f(A)$, then there exist $x_i \in A$ and n_i with $f^{n_i}(x_i) \to x$, so $\varphi(x_i) \in \varphi(A)$, $g^{n_i}(\varphi(x_i)) \to \varphi(x)$, and therefore $\varphi(x) \in \omega_g(\varphi(A))$. If $y \in \omega_g(\varphi(A))$, then there exist $x_k \in A$ and n_k with $g^{n_k}\varphi(x_k) \to y$. By compactness there exists a convergent subsequence $f^{n_{k_i}}(x_{k_i}) \to x \in \omega_f(A)$, so $\varphi f^{n_{k_i}}(x_{k_i}) \to \varphi(x)$. It follows $y = \varphi(x)$, and therefore $y \in \varphi(\omega_f(A))$. $\square$

Proposition 1
1. *A factor of a chaotic system is chaotic.*
2. *A factor of a system with finite attractors has finite attractors.*
3. *A factor of a system with chaotic limits has chaotic limits.*
4. *A factor of a system with bounded limits has bounded limits.*

Proof: Lemma 2

The following Theorem generalizes Alexandrov Theorem, which says that every compact metric space is a factor of discontinuum.

Theorem 1 *Any dynamical system is a factor of some dynamical system on discontinuum.*

Proof: Let (X, f) be a dynamical system. Let $U_i \in \mathcal{U}(X)$ be neighbourhoods of diagonal with $U_{i+1} \subseteq U_i$ and $\cap_i U_i = \Delta_X = \{(x, x); x \in X\}$. We first construct a sequence of finite open covers α_i of X, such that $(\forall V \in \alpha_i)(V \times V \subseteq U_i)$, and

$$(\forall V \in \alpha_i)(\exists V_1, V_2 \in \alpha_{i-1})(V \subseteq V_1, f(V) \subseteq V_2).$$

To construct α_0 just choose an open cover of X, whose sets satisfy $V \times V \subseteq U_1$, and select a finite subcover. Suppose that α_{i-1} has been constructed. For each $x \in X$ let V_x be an open neighbourhood of x such that $V_x \times V_x \subseteq U_i$, and there exist $V_1, V_2 \in \alpha_{i-1}$ with $V_x \subseteq V_1 \cap f^{-1}(V_2)$. Then we define α_i as a finite subcover of $\{V_x; x \in X\}$. We construct now a sequence of positive integers $k_0, k_1, \ldots$ and for each finite sequence $u_0, \ldots, u_{n-1}$ of binary words with lengths $|u_i| = k_i$ a nonempty closed set $W_{u_0 \ldots u_{n-1}} \subseteq X$. Let $2^{k_0} \geq card(\alpha_0)$, and define $\{W_{u_0}; u_0 \in 2^{k_0}\} = \{\overline{V}; V \in \alpha_0\}$ as the system of closures of all elements of α_0 (with possible repetitions). Suppose we have defined $W_{u_0, \ldots u_{n-1}}$ for all $u_0, \ldots u_{n-1}$ with lengths $k_0, \ldots, k_{n-1}$. Let k_n be such that for all such $u_0, \ldots, u_{n-1}$ we have $2^{k_n} \geq card(\{V \in \alpha_n; \overline{V} \cap W_{u_0 \ldots u_{n-1}} \neq 0\})$. For given $u_0, \ldots, u_{n-1}$ define

$$\{W_{u_0 \ldots u_n}; u_n \in 2^{k_n}\} = \{W_{u_0 \ldots u_{n-1}} \cap \overline{V}; V \in \alpha_n, W_{u_0 \ldots u_{n-1}} \cap \overline{V} \neq 0\}.$$

Then for each sequence $u_0, \ldots, u_n$, $W_{u_0 \ldots u_n} \subseteq W_{u_0 \ldots u_{n-1}}$, and there is a sequence $v_0, \ldots, v_{n-1}$ such that $f(W_{u_0 \ldots u_n}) \subseteq W_{v_0 \ldots v_{n-1}}$. Denote $g(u_1 \ldots u_n) = v_1 \ldots v_{n-1}$. For $u \in 2^N$ define $W_u = \cap \{W_{u_{|n}}; n > 0\}$ and $g(u) = \lim\{g(u_{|n}); n > 0\}$. Then W_u is a closed set, which contains unique point $\varphi(u) \in W_u$. It is easy to see that $\varphi : 2^N \to X$ and $g : 2^N \to 2^N$ are continuous and $f\varphi = \varphi g$. $\square$

4 Subshifts and languages

Subshifts are special dynamical systems on zero-dimensional spaces. They can be characterized by languages. If this language is regular, then there exists a finite accepting automaton, which recognizes it. This automaton can be viewed as a graph, whose vertices are the states, and whose labelled edges correspond to transitions. The graph properties of the accepting automaton say a great deal about the dynamical properties of the subshifts. This graph approach can be generalized to nonregular subshifts, which are recognized by infinite accepting automata.

Definition 8 *A subshift over an alphabet A is any nonempty subset of A^N, which is closed and σ-invariant. If $\Sigma \subseteq A^N$ is a subshift, then (Σ, σ) is a dynamical system. A nonempty language $L \subseteq A^*$ is right central, if*
1. $(\forall u \in L)(\forall v \in A^*)(v \trianglelefteq u \Rightarrow v \in L)$, and
2. $(\forall u \in L)(\exists a \in A)(ua \in L)$.

Definition 9 *The adherence of $L \subseteq A^*$ and the language of $\Sigma \subseteq A^N$ are defined by*

$$\begin{aligned}
\mathcal{A}(L) &= \{u \in A^N ; (\forall v \in A^*)(v \trianglelefteq u \Rightarrow v \in L)\} \\
\mathcal{L}(\Sigma) &= \{v \in A^* ; (\exists u \in \Sigma)(v \trianglelefteq u)\}
\end{aligned}$$

Proposition 2 *If $L \subseteq A^*$ is a right central language and $\Sigma \subseteq A^N$ a subshift, then $\mathcal{A}(L)$ is a subshift, $\mathcal{L}(\Sigma)$ is a right central language, $\mathcal{A}\mathcal{L}(\Sigma) = \Sigma$, and $\mathcal{L}\mathcal{A}(L) = L$.*

Proof: $\mathcal{A}(L)$ is clearly σ-invariant. If $u \notin \mathcal{A}(L)$, then there exists $wv \sqsubseteq u$ with $v \notin L$. It follows $[wv] \cap \mathcal{A}(L) = 0$, so $\mathcal{A}(L)$ is closed. If Σ is a subshift, then $\mathcal{L}(\Sigma)$ is a right central language by definition. If $v \in L$, then there exists a sequence $v^{(i)} \in L$, $v^{(0)} = v$, $v^{(i)} \sqsubseteq v^{(i+1)}$. Denote $u = \lim_{i \to \infty} v^{(i)}$. Then $u \in \mathcal{A}(L)$, so $v \sqsubseteq u$, and $v \in \mathcal{L}\mathcal{A}(L)$. Thus $L \subseteq \mathcal{L}\mathcal{A}(L)$. If $v \in \mathcal{L}\mathcal{A}(L)$, then there exists $u \in \mathcal{A}(L)$, with $v \trianglelefteq u$, so $v \in L$, and we get $\mathcal{L}\mathcal{A}(L) \subseteq L$. If $u \in \Sigma$, $v \in A^*$, and $v \trianglelefteq u$, then $v \in \mathcal{L}(\Sigma)$, so $u \in \mathcal{A}\mathcal{L}(\Sigma)$, thus $\Sigma \subseteq \mathcal{A}\mathcal{L}(\Sigma)$. If $u \notin \Sigma$, then for some integer n, $[u_{|n}] \cap \Sigma = 0$, since Σ is closed. Suppose $u \in \mathcal{A}\mathcal{L}(\Sigma)$. Then $u_{|n} \in \mathcal{L}(\Sigma)$, so there exists $w \in \Sigma$ with $u_{|n} \trianglelefteq w$, and therefore $\sigma^i(w) \in [u_{|n}]$ for some i. Since Σ is σ-invariant, $\sigma^i(w) \in \Sigma$, which is a contradiction with $[u_{|n}] \cap \Sigma = 0$. Thus $u \notin \mathcal{A}\mathcal{L}(\Sigma)$, and we get $\mathcal{A}\mathcal{L}(\Sigma) \subseteq \Sigma$. $\square$

Definition 10 *An accepting automaton over an alphabet A is a quadruple $Q = (Q', \delta, q_0, Q)$, where Q' is a set of states, $\delta : Q' \times A \to Q'$ is a transition function, $q_0 \in Q'$ is the initial state, and $Q \subseteq Q'$ is the set of accepting states. We assume $(\forall q \in Q')(\exists u \in A^*)(\delta^*(q_0, u) = q)$. The language recognized by Q is $L_Q = \{u \in A^* ; \delta^*(q_0, u) \in Q\}$*

Here $\delta^* : Q' \times A^* \to Q'$ is the extended transition function defined by $\delta^*(q, \lambda) = q$, $\delta^*(q, ua) = \delta(\delta^*(q, u), a)$. For any language $L \subseteq A^*$ there is a minimal accepting automaton Q_L: Define a relation $\sim$ on A^* by

$$u \sim v \quad \text{iff} \quad (\forall w \in A^*)(uw \in L \Leftrightarrow vw \in L)$$

Then $\sim$ is an equivalence, and a right congruence, i.e. $u \sim v$ implies $uw \sim vw$ for any $w \in A^*$. Moreover L is a union of some equivalence classes of $\sim$. Denote Q' the set of all equivalence classes of $\sim$, and define a transition function $\delta : Q' \times A \to Q'$ by

$$\delta(q, a) = p \text{ iff } (\exists u \in q)(ua \in p)$$

Let $q_0 \in Q'$ be the equivalence class containing the empty word λ, and $Q = \{q \in Q' ; q \subseteq L\}$. Then $Q_L = (Q', \delta, q_0, Q)$ is an accepting automaton for L. A language L is regular, iff it is recognized by a finite accepting automaton iff the set of equivalence classes of $\sim$ is finite (Nerode Theorem, see [11]).

Proposition 3 *An accepting automaton $Q = (Q', \delta, q_0, Q)$ recognizes a right central language iff following conditions are satisfied:*
1. *$q_0 \in Q$*
2. *$(\forall q \in Q)(\exists a \in A)(\delta(q, a) \in Q)$*
3. *$(\forall q \in Q' - Q)(\forall a \in A)(\delta(q, a) \in Q' - Q)$*
4. *$(\forall q \in Q)(\forall u \in A^*)(\delta^*(q, u) \in Q \Rightarrow \delta^*(q_0, u) \in Q)$*

Proof: Suppose that L_Q is right central. Since $\lambda \in L_Q$, $q_0 = \delta^*(q_0, \lambda) \in Q$. For any $q \in Q$, then there exists $u \in A^*$ with $\delta^*(q_0, u) = q$, and consequently $u \in L$. Since L is right central, there exists $a \in A$ with $ua \in L$, so $\delta(q, a) = \delta^*(q_0, ua) \in Q$. If $q \in Q' - Q$, then there is again $u \in A^*$ with $\delta^*(q_0, u) = q$. If $\delta(q, a) \in Q$ for some $a \in A$, then $ua \in L$, and $u \in L$, which is a contradiction. Thus $\delta(q, u) \in Q' - Q$. If $q \in Q$, and $\delta(q, u) \in Q$, then there exists $v \in A^*$ with $\delta^*(q_0, v) = q$, so $vu \in L$. It follows $u \in L$, so $\delta^*(q_0, u) \in Q$. Thus Q satisfies all four conditions of the Proposition. Conversely, if the conditions are satisfied, then $L_Q = \{u \in A^*; \delta^*(q_0, u) \in Q\}$ is a right central language. $\square$

Proposition 4 *If Q is an accepting automaton, and if L_Q is a right central language, then*

$$
\begin{aligned}
L_Q &= \{u \in A^*; (\exists q \in Q)(\delta^*(q, u) \in Q)\} \\
A(L_Q) &= \{u \in A^N; (\forall n)(\delta^*(q_0, u_{|n}) \in Q)\} \\
&= \{u \in A^N; (\exists q \in Q)(\forall n)(\delta^*(q, u_{|n}) \in Q)\} \\
&= \{u \in A^N; (\forall n)(\exists q \in Q)(\delta^*(q, u_{|n}) \in Q)\}
\end{aligned}
$$

Proof: Proposition 3

Definition 11 *Let $Q = (Q', \delta, q_0, Q)$ be an accepting automaton. We say that two states $p, q \in Q$ communicate $(p \approx q)$, if there exist $u, v \in A^*$ with $\delta^*(p, u) = q$, and $\delta^*(q, v) = p$. A state in Q is ergodic, if it communicates with itself, otherwise it is transient. A set $P \subseteq Q$ is a communication set, if*
1. *$(\forall p, q \in P)(p \approx q)$, and*
2. *$(\forall p \in P)(\forall q \in Q)(p \approx q \Rightarrow q \in P)$.*
A set $P \subseteq Q$ is final, if $(\forall p \in P)(\forall u \in A^*)(\delta^*(p, u) \in Q \Rightarrow \delta^*(p, u) \in P)$.

The communication relation is an equivalence on the set of ergodic states. A communication set is its equivalence class.

Proposition 5 *Let $\Sigma \subseteq A^N$, be a subshift, and $Q = (Q', \delta, q_0, Q)$ an accepting automaton for $\mathcal{L}(\Sigma)$. If $P \subseteq Q$ is a communication set, then*

$$
\Sigma_P = \{u \in \Sigma; (\forall n)(\exists p, q \in P)(\delta^*(p, u_{|n}) = q)\}
$$

is a chaotic subshift. In particular, if Q is a communication set, then (Σ, σ) is chaotic. If P is a final set, whose every state is ergodic, then Σ_P is a subshift, which is an attractor.

Proof: Suppose that P is a communication set. If $u \in \Sigma_P$, then for all n there exist $p, q \in P$ with $\delta^*(p, u_{|(n+1)}) = q$. Since P is a communication set, $\delta(p, u_0) = p' \in Q$, and $\delta^*(p', \sigma(u)_{|n}) = q$, so $\sigma(u) \in \Sigma_P$. Clearly Σ_P is closed, so it is a subshift. If $U, U' \subseteq \Sigma_P$ are nonempty open sets, then there exists an integer n and sequences $u \in [u_{|n}] \cap \Sigma_P \subseteq U$, $u' \in [u'_{|n}] \cap \Sigma_P \subseteq U'$. It follows there exist $p, q, p', q' \in P$ with $\delta^*(p, u_{|n}) = q$, $\delta^*(p', u'_{|n}) = q'$. There exists $v \in A^*$ with $\delta^*(q, v) = p'$, so $u_{|n} v u' \in \Sigma$. If $|v| = m$, then $u' \in U' \cap \sigma^{m+n}(U)$. Thus Σ_P is topologically transitive. Similarly, there exists $w \in A^*$ with $\delta^*(q, w) = p$, so $\overline{u_{|n} w} \in U$ is a periodic point. Thus Σ_P is chaotic. If P is a final set, then similar arguments show that Σ_P is again a subshift. Moreover, Σ_P is open in Σ, and $\sigma(\Sigma_P) = \Sigma_P = int(\Sigma_P)$, since every state in Σ_P is ergodic. Thus $\omega(\Sigma_P) = \Sigma_P$, and Σ_P is an attractor. $\square$

Proposition 6 *Let $\Sigma \subseteq A^N$, be a subshift, and $Q = (Q', \delta, q_0, Q)$ an accepting automaton which recognizes $\mathcal{L}(\Sigma)$.*
1. If there is a finite set $E \subseteq Q$ of ergodic states with $(\forall q \in Q)(\exists u \in A^*)(\delta^*(q, u) \in E)$, then (Σ, σ) has bounded limits.
2. If there exists a finite system $\{P_c; c \in C\}$ of communication sets such that $(\forall w \in \Sigma)(\exists c \in C)(\exists n)(\forall m \geq n)(\delta^*(q_0, w_{|m}) \in P_c)$, then (Σ, σ) has chaotic limits.

Proof. 1. For every $p \in E$ there exists $w_{(p)} \in A^*$ with $|w_{(p)}| = n_p$ and $\delta^*(p, w_{(p)}) = p$. Let $n = \max\{|w_{(p)}|; p \in E\}$. If $U \subseteq \Sigma$ is a nonempty open set and $u \in [u_{|m}] \cap \Sigma \subseteq U$, then there exists v with $\delta^*(q_0, u_{|m} v) = p \in E$. Then $u_{|m} v \overline{w_{(p)}} \in U$ is an eventually periodic point with period at most n. 2. If w satisfies the condition with communication class P_c, denote $n = min\{i; \delta^*(q_0, w_{|i}) \in P_c\}$. Let $W = \{u \in \Sigma; (\forall i < n)(u_i = w_i)$ & $((\forall i \geq n)(\delta^*(q_0, u_{;i}) \in P_c)\}$. Then $\omega(W) = \Sigma_{P_c}$ is chaotic by Proposition 5. $\square$

5 Regular systems

Definition 12 *Let (X, f) be a dynamical system, and $\alpha = \{V_a; a \in A\}$ a finite cover of X. The language and subshift generated by (X, f) on α are*

$$\mathcal{L}(X, f, \alpha) = \{u \in A^*; V_u \neq 0\}, \quad \mathcal{S}(X, f, \alpha) = \{u \in A^N; V_u \neq 0\},$$

where $V_u = \{x \in X; (\forall i < |u|)(f^i(x) \in V_{u_i})\}$ is the set of points, which visit sets of α according to the sequence u. We say that (X, f) is regular (recursive), if $\mathcal{L}(X, f, \alpha)$ is regular (recursive) for every finite almost disjoint cover α of X.

Proposition 7 *Let (X, f) be a dynamical system and $\alpha = \{V_a; a \in A\}$ a closed cover of X. Then $\mathcal{L}(X, f, \alpha)$ is a right central language, $\mathcal{S}(X, f, \alpha)$ is a subshift, $\mathcal{L}(\mathcal{S}(X, f, \alpha)) = \mathcal{L}(X, f, \alpha)$, and $\mathcal{A}(\mathcal{L}(X, f, \alpha)) = \mathcal{S}(X, f, \alpha)$.*

Proof: Clearly $\mathcal{L}(X, f, \alpha)$ is a right central language, and $\mathcal{S}(X, f, \alpha)$ is σ-invariant. If $u \in A^N$, then $V_u = \cap\{V_{u|_n}; n \geq 0\}$, and if all $V_{u|_n}$ are nonempty, then V_u is nonempty by compactness. If $u \notin \Sigma$, then $V_{u|_n} = 0$ for some n, and $[u|_n] \cap \Sigma = 0$, so $\mathcal{S}(X, f, \alpha)$ is closed. The rest of the proof is obvious. $\square$

If $\Sigma \subseteq A^N$ is a subshift, then $\mathcal{L}(\Sigma) = \mathcal{L}(\Sigma, \sigma, \{[a]; a \in A\})$, so the language of a subshift is a special case of the language of a dynamical system. We say that a cover α is a generator for (X, f), if for every $u \in A^N$, V_u contains at most one point.

Proposition 8 *If (X, f) has a generator, then it is a factor of a subshift.*

Proof: If α is a generator, define $\psi : (\mathcal{S}(X, f, \alpha), \sigma) \to (X, f)$ by $\psi(u) \in V_u$. Then ψ is a factorization. $\square$

Proposition 9 *Let (X, f) be a dynamical system and $\alpha \geq \beta$ almost disjoint covers of X. If $\mathcal{L}(X, f, \alpha)$ is regular (recursive), then so is $\mathcal{L}(X, f, \beta)$.*

Proof: Suppose that there exist $a \in A$, and $b \neq b' \in B$ with $V_a \subseteq V_b \cap V_{b'}$. Then $dim(V_a) \leq dim(V_b \cap V_{b'}) < dim(X)$, which is a contradiction. Thus for each $a \in A$ there exists a unique $h(a) \in B$ with $V_a \subseteq V_{h(a)}$. We prove $V_b = \cup\{V_a; h(a) = b\}$. If not, denote $Y = V_b - \cup\{V_a; h(a) = b\}$. For each $y \in Y$ there exist a', b', with $y \in V_{a'} \subseteq V_{b'}$, thus $Y \subseteq \cup\{V_b \cap V_{b'}; b' \neq b\}$, and $dim(Y) < dim(V_b)$. Since Y is open in V_b, V_b is not homogenous, which is a contradiction. The map h can be extended to a monoid homomorphism $h : A^* \to B^*$, which is λ-free, i.e. $h^{-1}(\lambda) = \lambda$. We get $\mathcal{L}(X, f, \beta) = h(\mathcal{L}(X, f, \alpha))$, and since both regular and recursive languages are closed to λ-free homomorphisms (see [11]), $\mathcal{L}(X, f, \beta)$ is regular (recursive). $\square$

Proposition 10 *A dynamical system (X, f) on a zero-dimensional space X is regular (recursive) iff there exists a sequence $\alpha_n = \{V_a; a \in A_n\}$ of clopen partitions of X such that $(\forall U \in \mathcal{U}(X))(\exists n)(\forall a \in A_n)(V_a \times V_a \subseteq U)$, (this means $diam(\alpha_n) \to 0$ for a metric compatible with the topology), $\alpha_{n+1} \geq \alpha_n$, and $\mathcal{L}(X, f, \alpha_n)$ is regular (recursive).*

Proof: Let β be a finite clopen partition. For each $x \in X$ there exists n_x and $W_x \in \alpha_{n_x}$ with $x \in W_x$, such that $\{W_x; x \in X\} \geq \beta$. Let $\{W_x; x \in A\}$ be a finite subcover of $\{W_x; x \in X\}$, and $n = \max\{n_x; x \in A\}$. Then $\alpha_n \geq \beta$, and we get the result by Proposition 9. $\square$

Proposition 11 *A dynamical system (I, g) on a compact real interval is regular (recursive) iff $\mathcal{L}(I, g, \alpha)$ is regular (recursive) for each interval cover α.*

Proof: If α is an almost disjoint cover of I, then each V_a is a finite union of closed nontrivial intervals, so there exists an interval cover, which refines it, and we apply Proposition 9. $\square$

Theorem 2 *Any dynamical system with finite attractors is regular.*

Proof: Let (X, f) be a dynamical system with finite $\omega(X)$. Since $\omega(X)$ is strictly invariant, it consists of the periodic points of X. Let $\alpha = \{V_a; a \in A\}$ be an almost disjoint cover of X. By Lemma 1 there exists a system $\{U_p; p \in \omega(X)\}$, such that U_p is a neighbourhood of p, if $p \notin V_a$, then $U_p \cap V_a = 0$, and $U = \cup_{p \in \omega(X)} U_p$ is invariant. Denote $L = \{u \in A^*; (\exists p \in \omega(X))(\forall i < |u|)(f^i(p) \in V_{u_i})\}$. Since $\omega(X)$ is finite, L is a regular language. For each $x \in X$ there exists n_x with $f^{n_x}(x) \in U$, and open neighbourhood W_x of x with $f^{n_x}(W_x) \subseteq U$. Since X is compact, there exists a finite subcover $\{W_x; x \in K\}$. Let $n = \max\{n_x; x \in K\}$. Then for any $u \in \mathcal{L}(X, f, \alpha)$, $\sigma^n(u) \in L$. It follows that $\mathcal{L}(X, f, \alpha)$ is a regular language. $\square$

6 Finite automata

A multitape Turing machine is a finite control device connected to several potentially infinite tapes. We can assume that the state of the control device itself is written on an additional tape, which does not move. The resulting machine is conceptually simpler. It consists of a finite number of doubly infinite tapes, which interact at fields with addresses zero. The contents of the zeroth fields of all tapes determine the move of the machine, which consists in rewriting the zero fields and shifting the tapes left or right. In contrast to computer science we assume that the tapes are actually infinite and have also infinite content. Thus we obtain a dynamical system on discontinuum (cf. Moore [23]). In a similar manner we conceive a multitape finite automaton. It is a finite control attached to input tapes, on which all future inputs are written. Moreover there are output tapes, to which the results are written. Thus the tapes move in one direction only.

Let A be a finite alphabet and denote $Z = \{\dots -1, 0, 1, \dots\}$ the set of integers. A tape is a data structure containing the letters of A indexed by Z. The state of a tape is a map $u : Z \to A$, thus its state space is A^Z, which is a discontinuum. The tape state may be updated by rewriting the zeroth position and shifting the tape left or right. This is accomplished by (continuous) updating functions $\sigma_a^i : A^Z \to A^Z$ defined by $\sigma_a^i(u)_j = a$ if $i + j = 0$, and $\sigma_a^i(u)_j = u_{i+j}$ otherwise. Note that $\sigma^i(u) = \sigma_{u_0}^i(u)$ coincides with the previous use:

$$\begin{aligned}
\sigma_a^{-1}(\dots u_{-2}u_{-1}u_0u_1u_2\dots) &= \dots u_{-3}u_{-2}u_{-1}au_1\dots \\
\sigma_a^0(\dots u_{-2}u_{-1}u_0u_1u_2\dots) &= \dots u_{-2}u_{-1}au_1u_2\dots \\
\sigma_a^1(\dots u_{-2}u_{-1}u_0u_1u_2\dots) &= \dots u_{-1}au_1u_2u_3\dots
\end{aligned}$$

Definition 13 *A Turing automaton is a finite system of finite alphabets $(A_t)_{t \in T}$ together with transition functions $\delta_t : Q \to A_t$, $\eta_t : Q \to \{-1, 0, 1\}$, where $Q = \prod_{t \in T} A_t$ is the set of inner states. A Turing automaton determines a* dynamical system (X, f), where $X = \prod_{t \in T} A_t^Z$, $\pi : X \to Q$ is the projection $\pi(u)_t = u_{t0}$, and

$$f(u)_t = \sigma_a^i(u_t), \quad \text{where } a = \delta_t \pi(u), \ i = \eta_t \pi(u), \quad u \in X, \ t \in T.$$

The advance of a tape $t \in T$ on $u \in X$ in n steps is $d_t(u, n) = \sum_{i=0}^{n-1} \eta_t \pi f^i(u)$. A finite automaton is a Turing automaton, whose all tapes move to the left, i.e. $(\forall t \in T)(\forall q \in Q)(\eta_t(q) \geq 0)$.

For $u \in X$ denote $u_{|n} = (u_{ti})_{t \in T, |i| < n} \in \prod_{t \in T} A_t^{2n-1}$. For $v \in \prod_{t \in T} A_t^{2n-1}$ denote $[v] = \{u \in X; u_{|n} = v\}$ the cylinder around v.

Proposition 12 If (Y, g^n) is a factor of a finite automaton, then so is (Y, g).

Proof: Let $\varphi : (X, f) \to (Y, g^n)$ be a factorization. We construct a dynamical system (X', f') with $X' = X \times n$, $f'(x, i) = (x, i + 1)$ for $0 \leq i < n - 1$, and $f'(x, n - 1) = (f(x), 0)$. Define $\psi : (X', f') \to (Y, g)$ by $\psi(x, i) = g^i \varphi(x)$. Then ψ is a factorization. The system (X', f') is in turn a factor of a finite automaton obtained from (X, f) by adding a new stationary tape with alphabet n. $\square$

Theorem 3 Any dynamical system with finite attractors is a factor of a finite automaton.

Proof: Let (Y, g) be a system with finite attractors. Then $\omega(Y)$ consists of periodic points. Denote t_p the period of $p \in \omega(Y)$. By Lemma 1 there exists a g^{t_p}-invariant neighbourhood U of $p \in \omega(Y)$, which does not meet $\omega(Y) - \{p\}$. It follows that $Y_p = \{y \in Y; g^{kt_p}(y) \to p\}$ is an open set, and $\{Y_p; p \in \omega(Y)\}$ is a clopen partition of Y. Construct a two-tape finite automaton (X, f) with alphabets $A_1 = A_2 = \omega(Y)$. Define transition functions by $\eta_1(p, q) = 0$, $\eta_2(p, q) = 1$, $\delta_1(p, q) = \delta_2(p, q) = g(p)$. Thus the first tape is stationary and only one of its fields is used. Its content changes according to g, and the results are written to the second (output) tape. As the time proceeds, the second tape contains ever longer periodic segments. For $p \in \omega(Y)$, $k \geq 0$ define $X_{pk} = \{(u, v) \in X; u_0 = p \ \& \ \min\{j > 0; g^j(v_{-j}) \neq u_0\} = k + 1\}$, $X_{p\infty} = \{u \in X; (\forall j > 0)(g(v_{-j}) = u_0 = p)\}$. Then $f(X_{pk}) = X_{g(p), k+1}$. For $k \geq 0$ define $\pi_k : X \to 2^N$ by $\pi_k(u, v) = v_{-k-1}v_{-k-2}\cdots$. Then $\pi_{k+1}f(u, v) = \pi_k(u, v)$ for $(u, v) \in X_{pk}$. Let $Y_{p0} = Y_p \cap \overline{Y - g(Y)}$. For $p \in \omega(Y)$ there exists a mapping $\psi_p : 2^N \to Y$ such that $\psi_p(2^N) = Y_{p0}$ if $Y_{p0} \neq 0$, and $\psi_p(2^N) = \{p\}$ if $Y_{p0} = 0$. For $(u, v) \in X_{pk}$ define $\varphi(u, v) = g^k \psi_{g-k(p)}\pi_k(u, v)$. For $(u, v) \in X_{p\infty}$ define $\varphi(u, v) = p$. For any $y \in Y - \omega(Y)$ there exists only a finite chain of preimages, so φ is surjective and it is clearly continuous. If $(u, v) \in X_{pk}$, then $\varphi f(u, v) = g^{k+1}\psi_{g-k-1(g(p))}\pi_{k+1}f(u, v) = g^{k+1}\psi_{g-k(p)}\pi_k(u, v) = g\varphi(u, v)$. If $(u, v) \in X_{p\infty}$, then $\varphi f(u, v) = g(p) = g\varphi(u, v)$. Thus $\varphi : (X, f) \to (Y, g)$ is a factorization. $\square$

Theorem 4 Every finite automaton has bounded limits.

Proof: Let (X, f) be a finite automaton and denote $n = card(Q)$. Let $U \subseteq X$ be an open set and $u \in [u_{|p}] \subseteq U$. Denote $T_u = \{t \in T; \lim_{k \to \infty} d_t(u, k) = \infty\}$ (T_u may be empty). There exist m, and $k \leq n$, such that $(\forall t \in T - T_u)(\forall i)(d_t(f^m(u), i) = 0)$, $(\forall t \in T_u)(d_t(u, m) \geq p)$, $\pi f^m(u) = \pi f^{m+k}(u)$. Denote $d_{t0} = d_t(u, m)$, $d_{t1} = d_t(u, m + k)$ and define $y \in X$ by

$y_{ti} = u_{ti}$ if $t \in T - T_u$

$y_{ti} = u_{ti}$ if $t \in T_u$, $i \leq d_{t1}$,

$y_{ti} = u_{tj}$ if $t \in T_u$, $i > d_{t1}$, $d_{t0} < j \leq d_{t1}$, and $d_{t1} - d_{t0}$ divides $i - j$.

Then $y \in [u_{|p}]$, and $g^i(y)$ converges to a periodic point with period $k \leq n$. $\square$

Theorem 5 *Any finite automaton has chaotic limits.*

Proof: Suppose (X, f) is a finite automaton and let $w \in X$. Denote

$$Q_0 = \{q \in Q; (\forall m)(\exists n \geq m)(\pi f^n(w) = q)\},$$

$$T_0 = \{t \in T; (\forall q \in Q_0)(\eta_t(q) = 0)\}, \quad T_1 = T - T_0.$$

Let j be the first integer, such that $(\forall n \geq j)(\pi f^n(w) \in Q_0)$ and denote

$Y = \{u \in X; (\forall i)(\pi f^i(u) \in Q_0) \ \& \ (\forall t \in T_0)(\forall i)(u_{ti} = f^j(w)_{ti})\}$.

Then Y is a closed invariant set. Let p be the least integer, for which there exists $z \in Y$ with $\{\pi f^i(z); i < p\} = Q_0$ and $\pi f^p(z) = \pi(z)$. Denote

$$W = \{u \in Y; (\forall q \in Q_0)(\forall k > 0)(card(\{i < kp; \pi f^i(u) = q\}) \geq k/2\})\}$$

We shall prove in the following lemmas that $\omega(W \cup \{w\})$ is chaotic.

Lemma 3 *Let $u \in Y$ and let for all k there exist n, m, and $v \in Y$, such that $f^n(v)_{|k} = u_{|k}$, for all $t \in T_1$ $d_t(v, n) \geq k$, and $d_t(f^n(v), m) \geq k$. Then $u \in \omega(W)$.*

Proof: Let $z \in Y$ be such that $\{\pi f^i(z); i < p\} = Q_0$, $\pi(z) = \pi(v)$, and $(\forall i)(\pi f^{p+i}(z) = \pi f^i(z))$. Let $r > n + m + p$ be an integer divisible by p. There exists $s \in Y$ satisfying $\pi(s) = \pi(v)$, and for all $t \in T_1$

$$
\begin{aligned}
s_{t,i} &= z_{t,i} & &\text{for} & i &\leq a = d_t(z, r) \\
s_{t,a+i} &= v_{t,i} & &\text{for} & 1 &\leq i \leq b = d_t(v, n + m) \\
s_{t,a+b+i} &= z_{t,\ell+i} & &\text{for} & 1 &\leq i, \text{ where } \pi f^{n+m}(v) = \pi f^\ell(z)
\end{aligned}
$$

Then $s \in W$, $f^r(s)_{ti} = v_{ti}$ for all $i \leq d_t(v, n + m)$, $t \in T_1$, and $f^{r+n}(s)_{|k} = f^n(v)_{|k}$, so $f^{r+n}(s)_{|k} = u_{|k}$, and therefore $u \in \omega(W)$. $\square$

Lemma 4 $\omega(W \cup \{w\}) = \omega(W)$.

Proof: It suffices to show $\omega(w) \subseteq \omega(W)$. Let $u \in \omega(w)$ and $k > 0$. We have $f^j(w) \in Y$ and there exist n, m, such that $d_t(f^j(w), n) \geq k$ for all $t \in T_1$, $f^{j+n}(w)_{|k} = u_{|k}$, and $d_t(f^{j+n}(w), m) \geq k$. By Lemma 3, $u \in \omega(W)$. $\square$

Lemma 5 *Periodic points are dense in $\omega(W)$.*

Proof: Let $U \subseteq X$ be an open set, and $U \cap \omega(W) \neq 0$. There exists $u \in [u_{|k}] \cap \omega(W) \subseteq U \cap \omega(W)$. There exist m and $v \in W$ such that $f^m(v)_{|k} = u_{|k}$, and $|d_t(v, m)| \geq k$ for all $t \in T_1$. There exists $r > m + k$ and $s \in Y$ such that $\pi(s) = \pi(v)$, and for all $t \in T_1$

$$
\begin{aligned}
s_{t,i} &= v_{t,i} && \text{for } 1 \leq i \leq d_t(v,m) + k \\
\{\pi f^i(s); i < r\} &= Q_0 \\
\pi f^r(s) &= \pi(v), \\
s_{t,a+i} &= s_{t,i} && \text{for } 1 \leq i, \ a = d_t(s,r) \\
s_{t,i} &= f^r(s)_{t,i} && \text{for } i < 0
\end{aligned}
$$

Then $s \in Y$, $f^r(s) = s$, $d_t(s,r) \geq k$ for $t \in T_1$, and $f^m(s)_{|k} = u_{|k}$. By Lemma 3, $s \in \omega(W)$. $\square$

Lemma 6 $\omega(W)$ *is topologically transitive.*

Proof: Let $U, U' \subseteq X$ be open sets, $u \in [u_{|k}] \cap \omega(W) \subseteq U \cap \omega(W)$, $u' \in [u'_{|k}] \cap \omega(W) \subseteq U' \cap \omega(W)$. There exist m, m' and $v, v' \in W$ with $f^m(v)_{|k} = u_{|k}$, $f^{m'}(v')_{|k} = u'_{|k}$, $d_t(v,m) \geq k$, $d_t(v',m') \geq k$ for all $t \in T_1$. There exist $r > m+k$, $r' > r + m' + k$ and $s \in Y$ such that for all $t \in T_1$

$$
\begin{aligned}
s_{t,i} &= v_{t,i} && \text{for } 1 \leq i \leq d_t(v,m) + k \\
\pi f^r(s) &= \pi(v'), \\
s_{t,a+i} &= v'_{t,i} && \text{for } 1 \leq i \leq d_t(v',m') + k, \ a = d_t(s,r) \\
\{\pi f^i(s); i < r'\} &= Q_0 \\
\pi f^{r'}(s) &= \pi(v), \\
s_{t,b+i} &= s_{t,i} && \text{for } 1 \leq i, \ b = d_t(s,r') \\
s_{t,i} &= f^{r'}(s)_{t,i} && \text{for } i < 0
\end{aligned}
$$

Then $f^m(s) \in U$, $f^{r+m'}(s) \in V$, and by Lemma 3, $s \in \omega(W)$. $\square$

7 Zero-dimensional systems

Proposition 13 *Any zero-dimensional factor of a regular system is regular.*

Proof: If $\varphi : (X, f) \rightarrow (Y, g)$ is a factorization, and α is a clopen partition of Y, then $\varphi^{-1}(\alpha) = \{\varphi^{-1}(A); A \in \alpha\}$ is a clopen partition of X and $\mathcal{L}(Y, g, \alpha) = \mathcal{L}(X, f, \varphi^{-1}(\alpha))$. $\square$

Theorem 6 *A subshift is regular iff it is a factor of a finite automaton.*

Proof: Let (X, f) be a finite automaton and let $\alpha_n = \{V_v; v \in A\}$ be the clopen partition of n-cylinders $V_v = [v]$ (here $A = \prod_{t \in T} A_t^{2n-1}$). Let us prove that if $V_{u_0 \ldots u_{m-1}} \neq 0$, and $V_{u_{m-1} u_m} \neq 0$, then $V_{u_0 \ldots u_m} \neq 0$. Choose $x \in V_{u_0 \ldots u_{m-1}}$,

$y \in V_{u_{m-1}u_m}$, and define

$z_{ti} = x_{ti}$ if $i < d_t(x, m) + n$,

$z_{ti} = y_{tj}$ if $i = d_t(x, m) + n$, $j = d_t(y, 1) + n - 1$.

Then $z \in V_{u_0...u_n}$. Thus $\mathcal{L}(X, f, \alpha_n)$ is regular (it is a finite complement language) and (X, f) is regular by Proposition 10. By Proposition 13 any subshift which is a factor of a finite automaton is regular. On the other hand if $\Sigma \subseteq A^N$ is a regular subshift, then its language is recognized by a finite accepting automaton $Q = (Q', \delta, q_0, Q)$, so $u \in \mathcal{L}(\Sigma)$ iff $\delta^*(q_0, u) \in Q$. Since $\mathcal{L}(\Sigma)$ is right central, for any $q \in Q$ there exists some $h(q) \in A$ with $\delta(q, h(q)) \in Q$. Define a two-tape automaton with alphabets $A_1 = Q$, $A_2 = A$, $\eta_1(q) = 0$, $\eta_2(q) = 1$ for any $q \in Q_1$. The transitions are defined by

$$\delta(q, a) \in Q \quad \Rightarrow \quad \delta_1(q, a) = \delta(q, a), \qquad \delta_2(q, a) = a,$$
$$\delta(q, a) \notin Q \quad \Rightarrow \quad \delta_1(q, a) = \delta(q, h(q)), \quad \delta_2(q, a) = h(q)$$

Define $\varphi : Q^Z \times A^Z \to A$ by $\varphi(u, v) = \delta_2(u_0, v_0)$. then $\varphi : (X, f) \to (\Sigma, \sigma)$ is a factorization. □

Example 2 *There exists a regular dynamical system with chaotic limits, which has not bounded limits.*

Proof: Take $X = 2^N$ and for $n > 0$ denote $U_n = [0^{n-1}1]$. For each $n > 0$ let $\{U_{nm}; 0 \le m < n\}$ be some clopen partition of U_n. Define $f : X \to X$ so that $f(\overline{0}) \doteq \overline{0}$, if $x \in U_{nm}$, then $f(x) \in U_{np}$, where $p = m + 1 \bmod n$, and $f^n(x) = x$. Thus each $x \in U_n$ is a periodic point with period n. Then (X, f) has chaotic limits, since every point is periodic, but not bounded limits, since their periods are arbitrarily large. To prove it is regular, consider for an integer k a clopen partition $\beta_k = \{U_{nm}; 0 \le m < n \le k\} \cup \{[0^k]\}$, and $\alpha_k = \{[u]; u \in 2^k\}$ the clopen partition of k-cylinders. Then $\mathcal{L}(X, f, \alpha_k \vee \beta_k)$ is regular, so (X, f) is regular by Proposition 10.

Example 3 *There exists a dynamical system with chaotic limits, which is neither regular nor has bounded limits.*

Proof: Consider the alphabet $3 = \{0, 1, 2\}$ with projection $\nu : 3 \to 2$ defined by $\nu(0) = 0$, $\nu(1) = \nu(2) = 1$. Denote
$X = \{u \in 3^N; (\forall k > 0)(\forall w \in 3^*)(w0^k 2 \sqsubseteq u \Rightarrow w0^k 2^k \sqsubseteq u)\}$.
Then X is a closed set, and $2^N \subseteq X$. We extend $f : 2^N \to 2^N$ from Example 2 to X as follows:

$$f(u) = f\nu(u) \quad \text{if} \quad 0^k 1 \sqsubseteq u, \ k \ge 0$$
$$f(u) = \nu\sigma(u) \quad \text{if} \quad 20 \sqsubseteq u \ \text{or} \ 21 \sqsubseteq u$$
$$f(u) = \sigma(u) \quad \text{if} \quad 22 \sqsubseteq u \ \text{or} \ 0^k 2 \sqsubseteq u, \ k > 0$$
$$f(\overline{2}) = \overline{2}$$

Then for any $u \in X$ with $u \neq \bar{2}$ there exists n with $f^n(u) \in 2^N$, so every point is either periodic or preperiodic. It follows that (X, f) has chaotic limits but not bounded limits. Let $\alpha_1 = \{[a]; a \in 3\}$ be the clopen partition of 1-cylinders. Then $\mathcal{L}(X, f, \alpha_1) = \{0^j 2^k u; 0 \leq j \leq k, u \in 2^*\}$ is not regular. $\square$

Example 4 *There exists a regular dynamical system which has no periodic point and therefore neither chaotic limits nor bounded limits.*

Proof: Define a length preserving map $f : 2^* \to 2^*$ by induction as follows: $f(\lambda) = \lambda$, $f(0) = 1$, $f(1) = 0$, $f(1^{n+1}) = 0^{n+1}$, $f(1^n 0) = 0^n 1$, $f(ua) = f(u)a$ for $u \in 2^n$, $u \neq 1^n$, $a \in 2$, $n > 0$. Then each $u \in 2^*$ is a periodic point with period $2^{|u|}$. (For example $00 \to 10 \to 01 \to 11 \to 00$.) Since $f(u) \sqsubseteq f(v)$ for $u \sqsubseteq v$, we can extend f to 2^N by $f(u) = \lim_{n \to \infty} f(u_{|n})$. Then $(2^N, f)$ is a dynamical system, which has no periodic point. If $\alpha_n = \{[u]; u \in 2^n\}$ is the partition of n-cylinders, then $v \in \mathcal{L}(2^N, f, \alpha_n)$ iff $v_{i+1} = f(v_i)$ for all $i < |v|$. Thus $\mathcal{L}(2^N, f, \alpha_n)$ is regular, and $(2^N, f)$ is regular by Proposition 10. $\square$

8 One-dimensional homeomorphisms

Theorem 7 *Any one-to-one dynamical system on real interval is regular.*

Proof: Let (I, g) be one-to-one and consider first a point-interval partition $\alpha = \{V_a; a \in A\}$ of I consisting of points and open intervals between them. Define order on A by $a < b$ iff $(\forall x \in V_a)(\forall y \in V_b)(x < y)$. Suppose first that g is increasing. Then each sequence $u \in \mathcal{L}(I, g, \alpha)$ is monotone. Moreover if V_a contains no fixed point, then there exists n_a such that if $a^k \in \mathcal{L}(I, g, \alpha)$, then $k < n_k$. Thus if $u \in \mathcal{L}(I, g, \alpha)$, then $u = a_0^{k_0} a_1^{k_1} ... a_j^{k_j} a_{j+1}^{k_{n+1}}$ with $k_i < n_i$ for $1 \leq i \leq j$. Thus $\mathcal{L}(I, g, \alpha)$ is regular. If g is decreasing, then g^2 is increasing, and $\alpha \vee g^{-1} \alpha = \{V_a \cap g^{-1}(V_b); a, b \in A\}$ is again a point-interval partition. Thus $\mathcal{L}(I, g, \alpha) = \mathcal{L}(I, g^2, \alpha \vee g^{-1} \alpha)$ is regular. If $\alpha = \{V_a; a \in A\}$ is an interval cover, and $\beta = \{V_b; b \in B\}$ a point-interval cover which refines it, then there exists a λ-free homomorphism $\nu : B^* \to A^*$ with $\mathcal{L}(I, g, \alpha) = \nu(\mathcal{L}(I, g, \beta))$, so (X, f, α) is regular. $\square$

Lemma 7 *Let $g : I \to I$ be an increasing continuous function on a compact real interval I, whose only fixed point is one of the endpoints of I. Then (I, g) is a factor of $(2^Z, \sigma_0^1)$.*

Proof: Let $I = [a, b]$, $g(a) > a$, $g(b) = b$ (the dual case $g(a) = a$ is similar). Denote $I_k = [g^k(a), g^{k+1}(a)]$, $X_k = \{u \in 2^Z; \max\{i < 0; u_i = 1\} = -k - 1\}$, where $k \geq 0$. Then $\sigma_0^1(X_k) = X_{k+1}$. For $u \in X_0$ define $\varphi_0(u) = a + (g(a) - a) \sum_{i=2}^{\infty} u_{-i} 2^{-i+1}$. For $u \in X_k$, $k \geq 0$ define $\varphi(u) = g^k \varphi_0 \sigma^{-k}(u)$, $\varphi(\bar{0}) = b$. Then for $u \in X_k$, $\sigma_0^1(u) \in X_{k+1}$, so $\varphi \sigma_0^1(u) = g^{k+1} \varphi_0 \sigma^{-k-1} \sigma_0^1(u) = g\varphi(u)$. Thus $\varphi : (2^N, \sigma_0^{-1}) \to (I, g)$ is a factorization. $\square$

Lemma 8 *Let (X, f) be a two-tape finite automaton with $A_0 = A_1 = 2$, $\eta_0(p, q) = 0$, $\eta_1(p, q) = 1$, $\delta_0(p, q) = max(p, q)$, $\delta_1(p, q) = p$. Let $g : I \to I$ be an increasing continuous function on a compact real interval I, whose only fixed points are the two endpoints of I. Then (I, g) is a factor of (X, f).*

Proof: Let $I = [a, b]$ and pick some $c \in (a, b)$. Suppose $f(c) > c$ (the dual case $f(c) < c$ is similar). For $k \in Z$ denote $I_k = [g^k(c), g^{k+1}(c)]$, and

$$
\begin{aligned}
X_{-\infty} &= \{(u, v) \in X; u_0 = 0 \ \& \ (\forall i \geq 0)(v_i = 0)\} \\
X_k &= \{(u, v) \in X; u_0 = 0 \ \& \ \min\{i \geq 0; v_i = 1\} = -k\}, \quad k \leq 0 \\
X_k &= \{(u, v) \in X; u_0 = 1 \ \& \ \max\{i < 0; v_i = 0\} = -k\}, \quad k > 0 \\
X_{\infty} &= \{(u, v) \in X; u_0 = 1 \ \& \ (\forall i < 0)(v_i = 1)\}
\end{aligned}
$$

Then $f(X_k) = X_{k+1}$. For $(u, v) \in X$ let $\varphi_0(u, v) = c + (g(c) - c) \sum_{i=1}^{\infty} u_i 2^{-i}$. Define $\varphi(u, v) = g^k \varphi_0(u, v)$ for $(u, v) \in X_k$, $\varphi(u, v) = a$ for $(u, v) \in X_{-\infty}$, and $\varphi(u, v) = b$ for $(u, v) \in X_{\infty}$. Then $\varphi : (X, f) \to (I, g)$ is a factorization. $\square$

Theorem 8 *Any one-to-one dynamical system on real interval is a factor of a finite automaton.*

Proof: Suppose first that (I, g) is an increasing map. Define a finite automaton (X, h), where $X = 2^Z \times 2^Z \times 2^Z \times 2^Z$, and $h(y, u, v, w) = (\sigma_0^1(y), f(u, v), w)$, where f is the finite automaton from Lemma 8. We shall prove that (I, g) is a factor of (X, h). Denote $F = \{x \in I; g(x) = x\}$ the set of fixed points. F is closed, therefore $I - F = \cup_{k \in K} A_k$ is a finite or countable union of open intervals $A_k = (a_k, b_k)$. Define $\theta : 2^Z \to I$ by $\theta(w) = a + (b - a) \sum_{i=0}^{\infty} w_i 2^{-i-1}$, where $I = [a, b]$. Each $\theta^{-1}(A_k)$ contains a cylinder $C_k = \{u \in 2^Z; u_0...u_{n-1} = v\}$ for some $v \in 2^n$. For $u, v \in 2^Z$ let $u \prec v$ if $u_0...u_{k-1} = v_0...v_{k-1}$, and $u_k < v_k$ for some $k \geq 0$. Denote $K_0 = \{k \in K; g(a_k) = a_k, \ g(b_k) = b_k\}$. If $k \notin K_0$, let $\psi_k : (2^Z, \sigma_0^1) \to (\overline{A_k}, g)$ be the factorization from Lemma 7. If $k \in K_0$, let $\psi_k : (2^Z \times 2^Z, f) \to (\overline{A_k}, g)$ be the factorization from Lemma 8. Define $\varphi : (X, h) \to (I, g)$ by

$$
\begin{aligned}
\varphi(y, u, v, w) &= \theta(w) & &\text{if} & &\theta(w) \in F \\
\varphi(y, u, v, w) &= a_k & &\text{if} & &\theta(w) \in A_k, \ \text{and} \ (\forall x \in C_k)(w \prec x) \\
\varphi(y, u, v, w) &= b_k & &\text{if} & &\theta(w) \in A_k, \ \text{and} \ (\forall x \in C_k)(x \prec w) \\
\varphi(y, u, v, w) &= \psi_k(y) & &\text{if} & &w \in C_k, \ \text{and} \ k \notin K_0 \\
\varphi(y, u, v, w) &= \psi_k(u, v) & &\text{if} & &w \in C_k, \ \text{and} \ k \in K_0
\end{aligned}
$$

Then φ is a factorization. If g is decreasing, then g^2 is increasing and (I, g) is a factor of a finite automaton by Proposition 12. $\square$

We turn now to dynamical systems on the circle, which we regard as the unit circle in the complex plane $S_1 = \{z \in C; |z| = 1\}$. There is a projection $\theta : R_1 \to S_1$ defined by $\theta(x) = \exp(2\pi i x)$. Occasionaly we identify S_1 with $[0, 1)$ with topology induced by θ. If (S_1, f) is a one-to-one dynamical system, then it is necessarily a homeomorphism which is either orientation preserving or orientation reversing. An increasing map $G : R_1 \rightarrow R_1$ is a lift of an orientation preserving homeomorphism (S_1, g), if $\theta : (R_1, G) \rightarrow (S_1, g)$ is a factorization. The rotation number $\varrho(g)$ of (S_1, g) is defined as the fractional part of $\varrho_0(G) = \lim_{n \rightarrow \infty} |G^n(y)|/n$, where G is any lift of g, and $y \in R_1$ ($\varrho(g)$ does not depend on the choice of either G or y). Then an orientation preserving homeomorphism of the circle (S_1, g) has a periodic point iff $\varrho(g)$ is rational (see Devaney [8] for a proof). If $\varrho(g) = p/q$, then $\varrho(g^q) = 0$. If (S_1, g) is an orientation reversing homeomorphism of the circle, then it has exactly two fixed points and g^2 is an orientation preserving homeomorphism. We define its rotation number as $\varrho(g) = \varrho(g^2)/2$. If $a \in [0, 1)$, denote $r_a : S_1 \rightarrow S_1$ the rotation of the circle defined by $r_a(z) = e^{2\pi i a} z$. Then $\varrho(r_a) = a$.

Theorem 9 *Any homeomorphism of the circle with rational rotation number is both regular and a factor of a finite automaton.*

Proof: If $\varrho(g)$ is rational, then $\varrho(g^n) = 0$ for some n, and g^n has a fixed point. We can suppose that 1 is a fixed point of g, and choose a lift G, with fixed points 0 and 1. Then (S_1, g^n) is a factor of $([0, 1], G)$, and this is a factor of a finite automaton by Theorem 8. By Proposition 12, (S_1, g) is a factor of a finite automaton too. If α is a point-interval partition of the circle containing point 1, then it is an inverse image, under θ, of some point interval cover β of $[0, 1]$. Thus $\mathcal{L}(S_1, g, \alpha) = \mathcal{L}([0, 1], G)$ is regular, so (I, g) is regular. $\square$

Theorem 10 *A homeomorphism of the circle with irrational rotation number is not a factor of a finite automaton.*

Proof: If (S_1, g) is a factor of a finite automaton, then it has bounded limits by Theorem 4, and therefore also periodic points. Thus its rotation number cannot be irrational.

Example 5 *The irrational rotation of the circle is not recursive.*

Proof: Suppose that $a \in (0, 1)$ is irrational, and $b \in (0, 1)$ is a number whose binary expansion is not recursive. Let α be an interval cover with cut points $0, \frac{1}{4}, \frac{1}{2}, \frac{3}{4}$, and b, and suppose that $\mathcal{L}(S_1, r_a, \alpha)$ is recursive. Then there is a recursive sequence u_n such that for each n, $u_n \in \{0, 1, 2, 3\}$, and there exists a real number x with $0 < x < \frac{1}{4}$, $\frac{u_n}{4} < x + na + m < \frac{u_n+1}{4}$ for some integer m. Thus $\frac{u_n-1}{4} < na - m < \frac{u_n+1}{4}$, and these inequalities yield a recursive procedure for the binary expansion of a. Similarly there exists a recursive sequence $v_n \in \{0, 1, 2, 3\}$ with $\frac{v_n}{4} < b + na - m < \frac{v_n+1}{4}$. These inequalities then yield a recursive procedure for the binary expansion of b, which is a contradiction. $\square$

9 Unimodal systems

Definition 14 *A dynamical system (I, g) defined on a real interval $I = [a, b]$ is unimodal, if $g(a) = g(b) = a$, and if there exists a turning point $c \in (a, b)$, such that g is increasing in $[a, c]$, and decreasing in $[c, b]$.*

If $g(x) < x$ for all $x > a$, then $\omega(X) = \{a\}$, so (I, g) is a system with finite attractors. If $g(c) \geq c$, then $\omega(X) = [a, g(c)]$. A powerful tool for investigating unimodal systems is the kneading theory, which is based on the concept of itineraries generated by the partition $\{[a, c), \{c\}, (c, b]\}$. We modify this theory using the cover $\{[a, c], [c, b]\}$, or more general interval covers instead. We lose the uniqueness of the itinerary, but the resulting theory is simpler. Moreover, we do not need any differentiability assumptions.

Throughout this section we assume that α is an interval cover, which contains among its cut points the turning point c. Thus let $a_0 < ... < a_n$ be an increasing sequence such that a_0, a_n are the endpoints of I, and $a_m = c$ is the turning point, where $0 < m < n$. Denote $V_j = [a_j, a_{j+1}]$, and $\alpha = \{V_j; j \in A\}$, where $A = \{0, ..., n-1\}$. For $j \in A$ denote $\tau(j) = 0$ if $j < m$, $\tau(j) = 1$ if $j \geq m$. For $u \in A^*$, $k \leq |u|$, denote $\tau_k(u) = \sum_{i=0}^{k-1} \tau(u_i) \bmod 2$. We say that $u \in A^*$ is even (odd), if $\eta_{|u|}(u)$ is zero (one).

Definition 15 *For $u, v \in \overline{A^*}$, define $u \prec v$ iff there exists k such that $u_{|k} = v_{|k}$, and either $\tau(u_{|k}) = 0$ and $u_k < v_k$, or $\tau(u_{|k}) = 1$ and $u_k > v_k$. Here $0 < ... < m - 1 < \lambda < m < ... < n - 1$. Define $u \preceq v$ iff either $u \prec v$, or $u \sqsubseteq v$ or $v \sqsubseteq u$.*

For example if $A = 2$, then we get $00 \prec 0 \prec 01 \prec \lambda \prec 11 \prec 1 \prec 10$. If $v \not\preceq u$ then $u \prec v$, and $u \preceq v$. If $v \not\prec u$, then $v \preceq u$. Thus $\prec$ is a strict ordering on $\overline{A^*}$, and $\preceq$ is an ordering on A^N.

Proposition 14 *If $u, v \in \overline{A^*}$, and $u \prec v$, then $(\forall x \in V_u)(\forall y \in V_v)(x \leq y)$.*

Proof: Let $u \prec v$, $k < \min(|u|, |v|)$, $u_{|k} = v_{|k}$, $u_k \neq v_k$, $x \in V_u$, $y \in V_v$. We prove $x \leq y$ by induction on k. If $k = 0$, then $u_0 < v_0$, and $x \leq y$. If $k > 0$, we apply the Proposition to $\sigma(u)$ and $\sigma(v)$. If $\tau(u_0) = \tau(v_0) = 0$, then $\sigma(u) \prec \sigma(v)$, so $g(x) \leq g(y)$, and $x \leq y$ since $x, y \leq c$. If $\tau(u_0) = \tau(v_0) = 1$, then $\sigma(u) \succ \sigma(v)$, so $g(x) \geq g(y)$, and $x \leq y$ since $x, y \geq c$. $\square$

Definition 16 *The upper and lower itineraries of $x \in I$ are defined by*

$$\mathcal{I}^\alpha(x) = \max\{u \in A^N; x \in V_u\}, \quad \mathcal{I}_\alpha(x) = \min\{u \in A^N; x \in V_u\}$$

(Here min and max are meant with respect to $\preceq$). For $w \in \overline{A^*}$, $a \in A$ denote

$$L^a(w) = \{u \in \overline{A^*}; (\forall i)(u_i = a \Rightarrow \sigma^{i+1}(u) \preceq w)\}$$
$$L_a(w) = \{u \in \overline{A^*}; (\forall i)(u_i = a \Rightarrow w \preceq \sigma^{i+1}(u))\}$$

The upper and lower itineraries might be obtained by inductive definition:

$$a_j < g^i(x) < a_{j+1} \quad \Rightarrow \quad \mathcal{I}^\alpha(x)_i = \mathcal{I}_\alpha(x)_i = j$$
$$g^i(x) = a_j, \ \tau(\mathcal{I}^\alpha(x)_{|i}) = 0 \quad \Rightarrow \quad \mathcal{I}^\alpha(x)_i = \min\{j, n-1\}$$
$$g^i(x) = a_j, \ \tau(\mathcal{I}^\alpha(x)_{|i}) = 1 \quad \Rightarrow \quad \mathcal{I}^\alpha(x)_i = \max\{j-1, 0\}$$
$$g^i(x) = a_j, \ \tau(\mathcal{I}_\alpha(x)_{|i}) = 0 \quad \Rightarrow \quad \mathcal{I}_\alpha(x)_i = \max\{j-1, 0\}$$
$$g^i(x) = a_j, \ \tau(\mathcal{I}_\alpha(x)_{|i}) = 1 \quad \Rightarrow \quad \mathcal{I}_\alpha(x)_i = \min\{j, n-1\}$$

Proposition 15 *If $x \in I$, then $x \in V_{\mathcal{I}^\alpha(x)}$, and $x \in V_{\mathcal{I}_\alpha(x)}$. If $x < y$, then $\mathcal{I}^\alpha(x) \preceq \mathcal{I}_\alpha(y)$.*

Proof: If $\mathcal{I}^\alpha(x) \succ \mathcal{I}_\alpha(y)$, then $x \geq y$ by Proposition 14.

Proposition 16 *If $u \in \overline{A^*}$, then $u \in \mathcal{L}(I, g, \alpha) \cup \mathcal{S}(I, g, \alpha)$ iff*

$$u \in L_j(\mathcal{I}_\alpha(g(a_j))) \cap L^j(\mathcal{I}^\alpha(g(a_{j+1}))) \quad \text{for each } j < m, \quad \text{and}$$
$$u \in L^j(\mathcal{I}^\alpha(g(a_j))) \cap L_j(\mathcal{I}_\alpha(g(a_{j+1}))) \quad \text{for each } j \geq m.$$

Proof: Let $u \in \mathcal{L}(I, g, \alpha) \cup \mathcal{S}(I, g, \alpha)$, and $u_i = j$. Pick up some $x \in V_u$. Then $g^i(x) \in V_{\sigma^i(u)} \subseteq V_j$. If $j < m$, then $g(a_j) \leq g^{i+1}(x) \leq g(a_{j+1})$. Suppose that $\mathcal{I}_\alpha(g(a_j)) \succ \sigma^{i+1}(u)$. Then $g(a_j) \geq g^{i+1}(x)$ by Proposition 14, so $g(a_j) = g^{i+1}(x) \in V_{\sigma^{i+1}(u)}$. This is however impossible, as $\mathcal{I}_\alpha(g(a_j))$ is the least sequence v, for which $g(a_j) \in V_v$. Thus $\mathcal{I}_\alpha(g(a_j)) \preceq \sigma^{i+1}(u)$, and $u \in L_j(\mathcal{I}_\alpha(g(a_j)))$. Similarly we prove the other inequalities, so the condition is necessary. To prove the sufficiency, we proceed by induction on the length of u. If $|u| \leq 1$, then $V_u \neq 0$. If $|u|$ is finite and the Proposition holds for $|u| - 1$, we apply it to $\sigma(u)$, which satisfies the condition of the Proposition. Thus there exists some $y \in V_{\sigma(u)}$. If $u_0 = j < m$, then $\mathcal{I}_\alpha(g(a_j)) \preceq \sigma(u) \preceq \mathcal{I}^\alpha(g(a_{j+1}))$. If $\mathcal{I}_\alpha(g(a_j)) \not\prec \sigma(u)$, then $\mathcal{I}_\alpha(g(a_j)) \sqsupseteq \sigma(u)$, and $a_j \in V_u \neq 0$. Similarly if $\sigma(u) \not\prec \mathcal{I}^\alpha(g(a_{j+1}))$, then $a_{j+1} \in V_u$. If $\mathcal{I}_\alpha(g(a_j)) \prec \sigma(u) \prec \mathcal{I}^\alpha(g(a_{j+1}))$, then $g(a_j) \leq y \leq g(a_{j+1})$ by Proposition 14, and there exists unique $x \in V_j$ with $g(x) = y$. Thus $x \in V_u \neq 0$. Similarly we proceed if $j \geq m$. If $|u|$ is infinite, $v \sqsubseteq u$ and $v \in 2^*$, then v satisfies the condition of the Proposition, so $V_v \neq 0$. Thus $V_u = \cap\{V_v; v \sqsubseteq u, v \in A^*\}$ is nonempty by compactness. $\square$

Proposition 17 *Let $w \in A^N$ be eventually periodic, and $j \in A$. If $w \in L^j(w)$, then $L^j(w) \cap A^*$ is regular. If $w \in L_j(w)$, then $L_j(w) \cap A^*$ is regular.*

Proof: For $u, v \in A^*$ define $u \sim v$ iff $(\forall s \in A^*)(us \in L^j(w) \Leftrightarrow vs \in L^j(w))$. By Nerode Theorem (see [11]) it suffices to prove, that $\sim$ has a finite number of equivalence classes. If $u_i = j$, and $u_k \neq j$ for $k < i$, then $u \sim \sigma^i(u)$. Thus it suffices to investigate words which begin with j. If $w = \overline{w_{|n}}$ is periodic even, then for all $u \in A^*$, $jw_{|2n}u \sim jw_{|n}u$, since for all i not divisible by n, $w_{i-1} - j$ implies $\sigma^i(w)_{|n} \prec w$, and $w_{|2n}u \preceq w$ iff $w_{|n}u \preceq w$. Thus there is at most $2n$ equivalence classes. If $w = \overline{w_{|n}}$ is periodic odd, then $jw_{|3n}u \approx jw_{|n}u$, so there is at most $3n$ equivalence classes. If w is preperiodic, then we can write $w = w_0...w_{m-1}\overline{w_m...w_{m+n-1}}$, where $w_m...w_{m+n-1}$ is even. Since the set $\{\sigma^i(w); i \in N\}$ is finite, there exists an integer p such that $\sigma^{i+1}(w)_{|p} \prec w_{|p}$ whenever $w_i = j$. Let k be an integer with $nk > p$ and put $r = m + nk$, $s = m + (n+1)k$. Then $jw_{|r}u \sim jw_{|s}u$, so there is at most s equivalence classes. $\square$

Theorem 11 $\mathcal{L}(I, g, \alpha)$ *is a regular (recursive) language iff all* $\mathcal{I}_\alpha(g(a_j))$ *and* $\mathcal{I}^\alpha(g(a_j))$ *are eventually periodic (recursive) sequences.*

Proof: If the condition is satisfied, then $L = \mathcal{L}(I, g, \alpha)$ is regular (recursive) since it is the intersection of a finite number of regular (recursive) languages (Propositions 16 and 17). To prove the converse, suppose that $w = \mathcal{I}^\alpha(g(a_j))$ is aperiodic and L is regular. Assume $j \geq m$. Since $a_j \in V_{jw}$, $jw_{|k} \in L$ for any $k > 0$. By Nerode Theorem (see [11]) there exists a right congruence $\sim$ of finite index on A^*, such that L is a union of its equivalence classes. There exists an infinite set $M \subseteq N$ such all $\{w_{|m}; m \in M\}$ have the same parity. Assume they are all even. Since $\sim$ is of finite index, there exist $m, n \in M$, $m \neq n$, such that $jw_{|m} \sim jw_{|n}$. Since $\sim$ is a right congruence, $jw_{|(m+k)} \sim jw_{|n}\sigma^m(w)_{|k} \in L$ for any k, so $w_{|n}\sigma^m(w)_{|k} \preceq w$, and $\sigma^m(w)_{|k} \preceq \sigma^n(w)_{|k}$, since $w_{|n}$ is even. Interchanging m and n in the above argument we get $\sigma^n(w)_{|k} \preceq \sigma^m(w)_{|k}$ for any k, and therefore $\sigma^n(w) = \sigma^m(w)$. This is, however, in contradiction with the aperiodicity of w. If all $\{w_{|m}; m \in M\}$ are odd, or if $j < m$, the proof is similar. For the case of recursivity suppose that $w = \mathcal{I}^\alpha(g(a_j))$ is nonrecursive and L is recursive. Again $jw_{|k} \in L$ for any k. If $j \geq m$, then $w_k = \max\{i; jw_{|k}i \in L\}$. If $j < m$, then $w_k = \min\{i; jw_{|k}i \in L\}$. Since L is recursive, we get a recursive procedure for the computation of w, which is a contradiction. $\square$

Definition 17 *Denote* $\alpha_g = \{[a, c], [c, b]\}$ *the standard cover of a unimodal system* (I, g) *with endpoints* a, b *and turning point* c. *The itinerary* $\mathcal{I}(x) \in \overline{2^*}$ *of* $x \in I$ *is defined by* $\mathcal{I}(x)_i = 0 \Leftrightarrow g^i(x) < c$, $\mathcal{I}(x)_i = 1 \Leftrightarrow g^i(x) > c$ *for* $0 \leq i < |\mathcal{I}(x)| = \inf\{n \geq 0; g^n(x) = c\}$. *The kneading sequence of* (I, g) *is* $\mathcal{K}(g) = \mathcal{I}(g(c))$. *Denote also* $\mathcal{I}_g(x) = \mathcal{I}_{\alpha_g}(x)$, $\mathcal{I}^g(x) = \mathcal{I}^{\alpha_g}(x)$, $\mathcal{K}_g = \mathcal{I}_g(g(c))$, $\mathcal{K}^g = \mathcal{I}^g(g(c))$.

Thus $\mathcal{I}(x)$ is finite iff the orbit of x contains c, e.g. $\mathcal{I}(c) = \lambda$ is the empty word. If $g(c) = c$, then $\mathcal{K}(g) = \lambda$. In general

$$
\begin{array}{lll}
\mathcal{K}_g = \mathcal{K}(g), & \mathcal{K}^g = \mathcal{K}(g) & \text{if } \mathcal{K}(g) \text{ is infinite,} \\
\mathcal{K}_g = \overline{\mathcal{K}(g)1}, & \mathcal{K}^g = \mathcal{K}(g)0\overline{\mathcal{K}(g)1} & \text{if } \mathcal{K}(g) \text{ is finite odd,} \\
\mathcal{K}_g = \overline{\mathcal{K}(g)0}, & \mathcal{K}^g = \mathcal{K}(g)1\overline{\mathcal{K}(g)0} & \text{if } \mathcal{K}(g) \text{ is finite even,} \\
\mathcal{I}_g(x) = \mathcal{I}(x), & \mathcal{I}^g(x) = \mathcal{I}(x) & \text{if } \mathcal{I}(x) \text{ is infinite,} \\
\mathcal{I}_g(x) = \mathcal{I}(x)1\mathcal{K}_g, & \mathcal{I}^g(x) = \mathcal{I}(x)0\mathcal{K}_g & \text{if } \mathcal{I}(x) \text{ is finite odd,} \\
\mathcal{I}_g(x) = \mathcal{I}(x)0\mathcal{K}_g, & \mathcal{I}^g(x) = \mathcal{I}(x)1\mathcal{K}_g & \text{if } \mathcal{I}(x) \text{ is finite even.}
\end{array}
$$

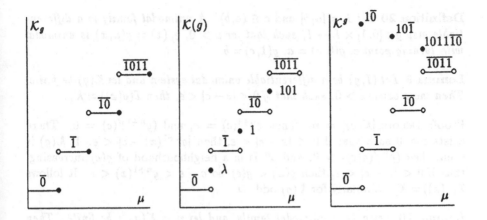

Fig. 1. Kneading sequences of $g_\mu(x) = \mu x(1-x)$

Definition 18 *For $w \in \overline{2^*}$ denote $\Sigma_w = \{u \in 2^N ; (\forall i > 0)(\sigma^i(u) \preceq w)\} = L^0(w) \cap L^1(w) \cap 2^N$. Σ_w is a subshift.*

Theorem 12 $\mathcal{L}(I, g, \alpha_g) = \mathcal{L}(\Sigma_{\mathcal{K}^g})$ *is regular iff $\mathcal{K}(g)$ is either finite or eventually periodic.*

Proof: Proposition 16 and Theorem 11.

Proposition 18 $\Sigma_{\mathcal{K}_g} = \{\mathcal{I}^g(x), \mathcal{I}_g(x); x \in I\}$.

Proof: If $i > 0$, then either $\sigma^i(\mathcal{I}^g(x)) = \mathcal{I}^g(g^i(x))$, or $\sigma^i(\mathcal{I}^g(x)) = \mathcal{I}_g(g^i(x))$. By Proposition 15, $\sigma^i(\mathcal{I}^g(x)) \preceq \mathcal{K}_g$ since $g^i(x) \leq g(c)$. Thus $\mathcal{I}^g(x) \in \Sigma_{\mathcal{K}_g}$. For the lower itinerary the proof is similar. If $u \in \Sigma_{\mathcal{K}_g}$, then $V_u \neq 0$ by Proposition 12. If there exists $x \in V_u$, whose orbit does not meet c, then $u = \mathcal{I}(x)$. If $g^i(x) = c$, then $\sigma^{i+1}(u) = \mathcal{K}_g$, so either $u = \mathcal{I}_g(x)$, or $u = \mathcal{I}^g(x)$. $\square$

10 Maximal sequences

Definition 19 *A sequence $w \in \overline{2^*}$ is maximal if $(\forall i > 0)(\sigma^i(w) \preceq w)$.*

Proposition 19 *If (I, g) is a unimodal system, then $\mathcal{K}(g)$, $\mathcal{K}_g$ and $\mathcal{K}^g$ are maximal.*

Proof: Propositions 16 and 18.

We prove now a converse of Proposition 19. This is possible for families of unimodal systems, which are differentiable (have continuous first derivatives) in both their state variable and parameter (cf. Collet and Eckmann [4]).

Definition 20 *Let $I = [a, b]$ and $c \in (a, b)$. A unimodal family is a differentiable map $g : [0, 1] \times I \rightarrow I$, such that for $\mu > 0$, $g_\mu(x) = g(\mu, x)$ is unimodal with turning point c, $g(0, c) = a$, $g(1, c) = b$.*

Lemma 9 *Let (I, g) be a differentiable unimodal system and let $\mathcal{K}(g)$ be finite. Then there exists $\varepsilon > 0$, such that if $0 < |x - c| < \varepsilon$, then $\mathcal{I}(g(x)) = \mathcal{K}_g$.*

Proof: Denote $|\mathcal{K}(g)| = n$. Then $g^{n+1}(c) = c$, and $(g^{n+1})'(c) = 0$. There exists $\varepsilon > 0$ such that if $0 < |x - c| < \varepsilon$, then $|g^{n+1}(x) - c| < \varepsilon$. If $\mathcal{K}(g)$ is even, then $(g^n)'(g(c)) > 0$, and g^n is in a neighbourhood of $g(c)$ increasing. thus if $0 < |x - c| < \varepsilon$, then $g(x) < g(c)$, so $c - \varepsilon < g^{n+1}(x) < c$. It follows $\mathcal{I}(g(x)) = \mathcal{K}_g$. Similarly for $\mathcal{K}(g)$ odd. $\square$

Lemma 10 *Let g be a unimodal family, and let $u = \mathcal{K}(g_{\mu_0})$ be finite. Then there is a neighbourhood U of μ_0, such that for all $\mu \in U$ either $\mathcal{K}(g_\mu) = u$, or $\mathcal{K}(g_\mu) = \overline{u0}$, or $\mathcal{K}(g_\mu) = \overline{u1}$.*

Proof: Denote $n = |u| + 1$. Then $g_{\mu_0}^n(c) = c$ and $(g_{\mu_0}^n)'(c) = 0$. There exists $\varepsilon > 0$ and a neighbourhood U of μ_0, such that for $\mu \in U$ and $|x - c| < \varepsilon$

$$u \sqsubseteq \mathcal{I}_\mu(x), \quad |(g_\mu^n)'(x)| < \frac{1}{2}, \quad |g_\mu^n(x) - c| < \frac{\varepsilon}{2}$$

We show that U is the required neighbourhood. Let $\mu \in U$ and denote $d = g_\mu^n(c) - c$. If $d = 0$, then $\mathcal{K}(g_\mu) = u$. Suppose $d > 0$. Then $2d < \varepsilon$ and for $x \in (c, c+2d)$ we have $|g^n(x) - g^n(c)| < |x-c|/2 < d$, so $g^n(c, c+2d) \subseteq (c, c+2d)$. It follows $\mathcal{K}(g_\mu)_i = 1$ for any i divisible by n, so $\mathcal{K}(g_\mu) = \overline{u1}$. Similarly for $d < 0$, $\mathcal{K}(g_\mu) = \overline{u0}$. $\square$

Lemma 11 *Let $w \in \overline{2^*}$ be maximal nonperiodic sequence, and let $v \in 2^*$ be finite. If $v \prec w$, then $\overline{v0} \prec w$, $\overline{v1} \prec w$. If $v \succ w$, then $\overline{v0} \succ w$, $\overline{v1} \succ w$.*

Proof: Let $v \prec w$ and denote $n = |v|$. If $v \prec w_{|n}$, then $\overline{v0} \prec w$ and $\overline{v1} \prec w$. Thus we assume $v = w_{|n}$. Since $v \prec w$, $|w| > n$. If v is even, then $w_n = 1$ and $\overline{v0} \prec w$. Since w is nonperiodic, $v1v1 \not\sqsubseteq w$ by Lemma 13. Let $w \prec \overline{v1}$. Since $v1$ is odd, $\sigma^{n+1}(w) \succ \sigma^{n+1}(\overline{v1}) = \overline{v1}$, so $\sigma^{n+1}(w) \succ w$ and this is a contradiction. Thus $\overline{v1} \prec w$. If v is odd, then $w_n = 0$, and $\overline{v1} \prec w$. If $w \prec \overline{v0}$, then $\sigma^{n+1}(w) \succ \sigma^{n+1}(\overline{v0}) = \overline{v0}$, and again $\sigma^{n+1}(w) \succ w$. Similarly if $w \prec v$. $\square$

Theorem 13 *Let g be a unimodal family, and w a maximal sequence. Then there exists μ with $\mathcal{K}(g_\mu) = w$.*

Proof: Assume $\overline{0} \neq w \neq 1\overline{0}$, since otherwise the Theorem is trivial. Suppose first that w is not periodic and denote $L_w = \{\mu; \mathcal{K}(g_\mu) \prec w\}$, $R_w = \{\mu; \mathcal{K}(g_\mu) \succ w\}$. Then both L_w and R_w are nonempty, since they contain $\overline{0}$ and $1\overline{0}$ respectively. We shall prove that they are open. Let $\mathcal{K}(g_{\mu_0}) = v \prec w$, and let n be the first integer with $v_n \neq w_n$. If $v_n \neq \lambda$, then the set $\{\mu; (\forall i \leq n)(\mathcal{K}(g_\mu)_i = v_i)\} \subseteq L_w$ is an open neighbourhood of μ_0. Let $v_n = \lambda$. By Lemma 10 there is a neighbourhood U of μ_0 such that for any $\mu \in U$, $\mathcal{K}(g_\mu) = v$, or $\mathcal{K}(g_\mu) = \overline{v0}$, or $\mathcal{K}(g_\mu) = \overline{v1}$. By Lemma 11, $\mathcal{K}(g_\mu) \prec w$, so $U \subseteq L_w$ is a neighbourhood of μ_0. Similarly we prove that R_w is open, so their complement is nonempty, and there exists μ with $\mathcal{K}(g_\mu) = w$. Suppose now that w is periodic, so either $w = \overline{v0}$ or $w = \overline{v1}$. Suppose $w = \overline{v0}$ and v is even. By the above proof there exists μ with $\mathcal{K}(g_\mu) = v$. Denote $\mu_0 = \inf\{\mu; \mathcal{K}(g_\mu) = v\}$. By Lemma 10 there exists ε such that for $\mu_0 - \varepsilon < \mu < \mu_0$ either $\mathcal{K}(g_\mu) = \overline{v0}$ or $\mathcal{K}(g_\mu) = \overline{v1}$. Suppose that the latter case occurs. Then for all $\mu < \mu_0$, $\mathcal{K}(g_\mu) \neq v$, and from continuity it follows that $\mathcal{K}(g_\mu) \succ v$. This is however in contradiction with $\mathcal{K}(g_0) = \overline{0}$. Thus for $\mu \in (\mu_0 - \varepsilon, \mu_0)$, $\mathcal{K}(g_\mu) = \overline{v0} = w$. In other cases the proof is similar. $\square$

Lemma 12 *A sequence $w \in 2^*$ is maximal, if both $\overline{w0}$ and $\overline{w1}$ are maximal. Moreover, if $x \in 2$ and wx is even, then $\Sigma_w = \Sigma_{\overline{wx}}$.*

Proof: Let w be maximal. Denote $n = |w|$ and $v = \overline{w0}$ or $v = \overline{w1}$. Suppose $\sigma^j(v) \succ v$, and let k be the first integer with $\sigma^j(v)_k \neq v_k$. We can assume $j \leq n$ and $k \leq n+1$. If $k < n-j$, then $\sigma^j(w) \succ w$, which is a contradiction. If $k \geq n-j$, then $w_j...w_{n-1} = w_0...w_{n-j-1}$. Since w is maximal, $w_j...w_{n-1} \prec w_0...w_{n-j}$, so $w_0...w_{n-j}$ is odd, and $w_j...w_n = w_0...w_{n-j}$. Since $w_j...w_n w_0...w_{j-1} \succ w_0...w_n$, we get $w_0...w_{j-1} \prec w_{n-j}...w_n$, and this is in contradiction with the maximality of w. Thus v is maximal. If $\overline{w0}$ and $\overline{w1}$ are maximal, and $w_j...w_{n-1} \succ w_0...w_{n-j}$, then $w_j...w_n \succ w_0...w_{n-j}$ where either $w_n = 0$, or $w_n = 1$, and this is a contradiction. If $u \in 2^N$ and $u \preceq w$, then $u \preceq wx$. Thus $\Sigma_w = \Sigma_{\overline{wx}}$. $\square$

Lemma 13 *Suppose that $w \in 2^*$ is nonperiodic odd, $\overline{w}$ is maximal, $v \in \Sigma_{\overline{w}}$, and $\sigma^i(v) = \overline{w}$. Then v is an isolated point in $\Sigma_{\overline{w}}$. Moreover if $n = |w|$, then $(\forall i < n)(\exists k < n)(\sigma^i(\overline{w})_{|k} \prec \overline{w}_{|k})$.*

Proof: If $wwu \in \Sigma_{\overline{w}}$ for some $u \in 2^N$, then $\sigma^{2n}(wwu) = u \preceq \overline{w}$ by definition, and $\sigma^n(wwu) = wu \preceq \overline{w}$, so $u \succeq \sigma^n(\overline{w}) = \overline{w}$, since w is odd. Thus $u = \overline{w}$, and $\overline{w}$ is an isolated point. If $\sigma^i(v) = \overline{w}$, then v is isolated since σ is continuous, and every point has only a finite number of preimages. $\square$

Definition 21 *For $w \in 2^+$ define $\hat{w}$ by changing the last bit, i.e. $\hat{w}_{|n} = w_{|n}$, $\hat{w}_n = 1 - w_n$, where $n = |w| - 1$. If $k > 0$, define $\mathcal{D}_k(w) = w(\hat{w})^k$. If k is a finite sequence of positive integers of length n, define $\mathcal{D}_k(w) = \mathcal{D}_{k_{n-1}}...\mathcal{D}_{k_0}(w)$. If k is an infinite sequence of positive integers, define $\mathcal{D}_k(w) = \lim_{n \to \infty} \mathcal{D}_{k_{|n}}(w)$.*

Thus for example

$$\mathcal{D}_{\overline{1}}(1) = 1011\ 1010\ 1011\ 1011\ 1011\ 1010\ 1011\ 1010\ ...$$
$$\mathcal{D}_{\overline{2}}(1) = 100101101\ 100101100\ 100101100...$$
$$\mathcal{D}_{\overline{21}}(1) = 100101\ 100100\ 100100\ 100101\ 100100\ 100101...$$

Lemma 14 *If $w \in 2^+$ is nonperiodic odd, and $\overline{w}$ is maximal, then $\hat{w} \prec w$, and $\overline{\hat{w}}$ is maximal.*

Proof: The first part is obvious. Suppose that $\sigma^j(\overline{\hat{w}}) \succ \overline{\hat{w}}$, $j < |w|$, and denote $k = \inf\{i; (\overline{\hat{w}})_{j+i} \neq (\overline{\hat{w}})_i\}$. We distinguish three cases: Let $k < n - j - 1$. Then $w_j...w_{k+j} \succ w_0...w_k$, so $\sigma^j(\overline{w}) \succ \overline{w}$, which is a contradiction. Let $k = n - j - 1$. Then $w_j...\hat{w}_{n-1} \succ w_0...w_{n-j-1}$, so $w_j...w_{n-1} = w_0...w_{n-j-1}$ is even. Since $w_j...w_{n-1}w_0...w_{j-1} \prec w_0...w_{n-j-1}w_{n-j}...w_{n-1}$, $w_0...w_{j-1} \prec w_{n-j}...w_{n-1}$, so $\overline{w} \prec \sigma^{n-j}(\overline{w})$ and this is again a contradiction. Let $k > n - j - 1$. Then $k < n$, and $w_j...\hat{w}_{n-1} = w_0...w_{n-j-1}$. Since $\sigma^j(\overline{w}) \prec \overline{w}$, we get $w_j...w_{n-1} \prec w_0...w_{n-j-1}$, so $w_0...w_{n-j-1}$ is odd. Since $w_j...\hat{w}_{n-1}w_0...w_{j-1} \succ w_0...w_{n-j-1}w_{n-j}...\hat{w}_{n-1}$, we get $w_0...w_{j-1} \prec w_{n-j}...\hat{w}_{n-1}$. Since $\sigma^{n-j}(\overline{w}) \prec \overline{w}$, we get $w_0...w_{j-1} = w_{n-j}...w_{n-1}$. It follows that $w_0...w_{j-1}w_j...\hat{w}_{n-1} = w_{n-j}...w_{n-1}w_0...w_{n-j-1}$. But the sequence on the left-hand side is even, while that on the right-hand side is odd, so we get a contradiction. $\square$

Proposition 20 *If k is a finite sequence of positive integers, $w \in 2^*$ is nonperiodic odd, and $\overline{w}$ is maximal, then $\mathcal{D}_k(w)$ is nonperiodic odd, $\overline{\mathcal{D}_k(w)}$ is maximal, and $\overline{w} \prec \overline{\mathcal{D}_k(w)}$. If k is an infinite sequence of positive integers, then $\mathcal{D}_k(w)$ is an aperiodic maximal sequence.*

Proof: Let $k > 0$ be an integer, $|w| = n$, and denote $v = \mathcal{D}_k(w)$. If $i < n$, then $\sigma^i(\overline{v}) = \sigma^i(w)\hat{w}^k... \prec \overline{v}$ by Lemma 13. If $jn \leq i < (j+1)n$, $0 < j \leq k+1$, then $\sigma^i(\overline{v}) = \sigma^{i-jn}(\hat{w})w^{k-j}... < \overline{v}$. Thus $\overline{\mathcal{D}_k(w)}$ is maximal, and $\mathcal{D}_k(w)$ is clearly nonperiodic odd. If k is a finite sequence, we get the result by induction. Let k be an infinite sequence and $\sigma^j(\mathcal{D}_k(w)) \succ \mathcal{D}_k(w)$. Let i be maximal integer for which $|\mathcal{D}_{k_{|i}}(w)| \geq j$. Then $\sigma^j(\mathcal{D}_{k_{|i}}(w)) \succ \mathcal{D}_{k_{|i}}(w)$, which is a contradiction. Thus $\mathcal{D}_k(w)$ is maximal. $\square$

Proposition 21 *If w is a maximal sequence, and $w \prec \mathcal{D}_{\overline{1}}(1) = 1011\,1010...$, then either $w = \overline{0}$ or $w = \overline{\mathcal{D}_{1^n}(1)}$ for some n. Moreover, Σ_w contains only eventually periodic points.*

Proof: Suppose $w \prec \mathcal{D}_{\overline{1}}(1)$, $w \neq \overline{0}$, and $w \neq \mathcal{D}_{1^n}(1)$ for any n. Then $1 = \mathcal{D}_\lambda(1) \sqsubseteq w$. Denote $v = \mathcal{D}_{1^n}(1)$ and suppose $v \sqsubseteq w$. If $vv \sqsubseteq w$, then $w = \overline{v}$, since v is odd, and w is maximal (Lemma 13. Thus there exists $u \sqsubseteq v$ with $vu \sqsubseteq w$. If $|u| < |v|$, then $\hat{u} \prec \hat{v}$, so $v\hat{u} \succ v\hat{v}$, and $w \succ \mathcal{D}_{\overline{1}}(1)$, which is a contradiction. Thus $|u| = |v|$, and $\mathcal{D}_{1^{n+1}}(1) \sqsubseteq w$. Thus $\mathcal{D}_{1^n}(1) \sqsubseteq w$ for any n, and therefore $\mathcal{D}_{\overline{1}}(1) = w$, which is a contradiction. If $u \in \Sigma_w$, then $u = 0^{n_0}1^{n_1}(10)^{n_2}...(\mathcal{D}_{1^{i-1}}(1))^{n_i}\overline{\mathcal{D}_{1^i}(1)}$ for some $i \leq n$ and $n_j \geq 0$. $\square$

Theorem 14 *A unimodal system (I, g) is regular if $K^g \prec \mathcal{D}_{\overline{1}}(1)$, and nonrecursive otherwise.*

Proof: If $\mathcal{K}^g \prec \mathcal{D}_{\bar{1}}(1)$, then the itinerary of any point $x \in I$ is either finite or eventually periodic, so $\mathcal{L}(I, g, \alpha)$ is regular by Theorem 11. If α is an arbitrary interval cover, then there exists a finer interval cover with cut point c, and we get the result by Proposition 9. On the other hand suppose that $\mathcal{K}^g \succeq \mathcal{D}_{\bar{1}}(1)$. Let n_i be any sequence of positive integers, and denote $w = 0^{n_0} 1^{n_1} (10)^{n_2} (1011)^{n_3} ... (\mathcal{D}_{1^i}(1))^{n_{i+1}} ...$. Then $w \in \mathcal{L}(I, g, \alpha_g)$, and there exists a point $x \in I$ with $\mathcal{I}(x) = w$. If the sequence n_i is nonrecursive, then so is w. $\square$

Thus regularity and recursivity distinguish unimodal systems similarly as topological entropy, since the topological entropy of (I, g) is zero iff $\mathcal{K}_g \preceq \mathcal{D}_{\bar{1}}(1)$ (see Milnor and Thurston [21]).

If $w \in 2^N$ is a maximal sequence, we deduce the properties of Σ_w from the graph structure of an accepting automaton of its language. If w is eventually periodic, then the accepting automaton is finite, so Σ_w is a factor of a finite automaton. By Proposition 6 it has bounded limits and chaotic limits. If $w \in 2^N$ is a maximal aperiodic sequence, denote $\mathcal{Q}_w = (Q', \delta, q_0, Q)$ the accepting automaton for Σ_w defined by $Q' = \{\wedge, -1, 0, 1, 2, ...\}$, $Q = \{-1, 0, 1, 2, ...\}$, $q_0 = -1$, and

$$
\begin{aligned}
\delta(\wedge, x) &= \wedge, \\
\delta(-1, x) &= 0, \\
\delta(q, w_q) &= q + 1 \\
\delta(q, \hat{w}_q) &= \wedge \quad \text{if } (\exists k \leq q+1)(\sigma^{q+1-k}(w_{|q} \hat{w}_q) \succ w_{|k}) \\
\delta(q, \hat{w}_q) &= \max\{k \leq q; \sigma^{q+1-k}(w_{|q} \hat{w}_q) = w_{|k}\} \text{ otherwise}
\end{aligned}
$$

If w is eventually periodic, we obtain a finite automaton by identifying some $r < s$ (see the proof of Proposition 17).

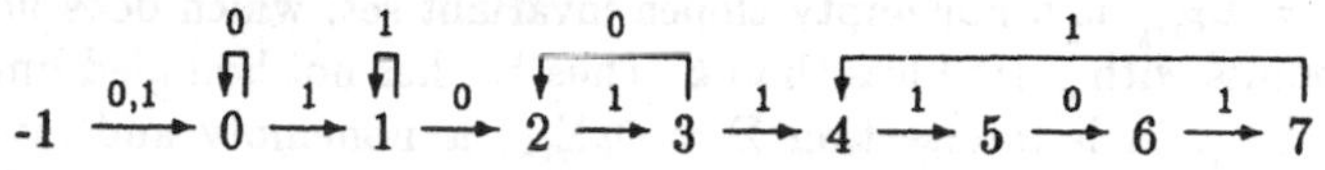

$$
-1 \xrightarrow{0,1} 0 \xrightarrow[1]{\overset{0}{\curvearrowleft}}{}^{1} 1 \xrightarrow[0]{\overset{1}{\curvearrowleft}} 2 \xrightarrow[1]{0} 3 \xrightarrow[1]{1} 4 \xrightarrow[1]{1} 5 \xrightarrow[0]{1} 6 \xrightarrow[1]{1} 7
$$

Fig.2. Accepting automaton for $\Sigma_{\overline{1011}}$

$$
-1 \xrightarrow{0,1} 0 \xrightarrow[1]{\overset{0}{\curvearrowleft}}{}^{1} 1 \xrightarrow[0]{\overset{1}{\curvearrowleft}} 2 \xrightarrow[1]{0} 3 \xrightarrow[1]{1} 4 \xrightarrow[1]{1} 5 \xrightarrow[0]{0,1} 6 \xrightarrow[1]{1} 7
$$

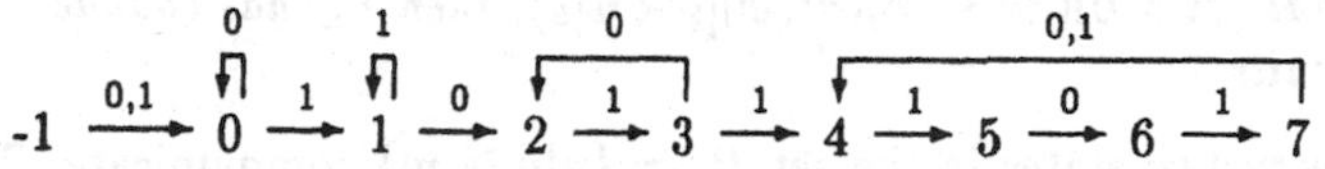

Fig. 3. Accepting automaton for $\Sigma_{1011\overline{10}}$

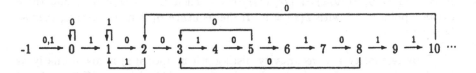

Fig. 4. Accepting automaton for $\Sigma_{D_{\bar{1}}(1)}$

$$
\begin{array}{ccccccccccc}
-1 & \xrightarrow{0,1} & 0 & \xrightarrow{1} & 1 & \xrightarrow{0} & 2 & \xrightarrow{0} & 3 & \xrightarrow{1} & 4 & \xrightarrow{0} & 5 & \xrightarrow{1} & 6 & \xrightarrow{1} & 7 & \xrightarrow{0} & 8 & \xrightarrow{1} & 9 & \xrightarrow{1} & 10 & \cdots
\end{array}
$$

Fig. 5. Accepting automaton for $\Sigma_{1001012013_0\ldots}$

Theorem 15 *Let $w \in 2^N$ be a maximal aperiodic sequence. Then (Σ_w, σ) has bounded limits iff it has chaotic limits iff Q_w has a final communication set.*

Proof: If Q_w has a final communication set, then it has chaotic limits and bounded limits by Proposition 6. If Q_w has no final communication set, then the set $M = \{m \in Q; (\forall u \in \mathcal{L}(\Sigma_w))(\delta(m, u) \geq m)\}$ is infinite. We shall prove that if $m \in M$, then the set $P_m = \{n; n \geq m\}$ does not contain cycles of length less than m. If so, then there exist $m \leq p < q$ with $\delta^*(q, \hat{w}_q) = p$, and $q - p + 1 < m$. It follows $w_{|q}\hat{w}_q \prec w_{|(q+1)}$, and $w_k \ldots \hat{w}_q = w_{|p}$, where $k = q - p + 1 < m$. Then $w_{|q} = w_{|k}w_0 \ldots w_{p-2}$, so $w_{|k}w_{|k} \sqsubseteq w$ since $k \leq p - 1$, and $w_{|k}$ is even by Lemma 13. Since $w_0 \ldots \hat{w}_q \prec w_0 \ldots w_q$, we get $w_k \ldots \hat{w}_q \prec w_k \ldots w_q$, so $w_k \ldots w_q$ is odd. Since $w_k \ldots w_q \preceq w_0 \ldots w_{p-1}$, and $w_k \ldots \hat{w}_q = w_0 \ldots w_{p-1}$, we get $w_k \ldots w_q \prec w_0 \ldots w_{p-1}$, so $w_k \ldots w_q$ is even, and this is a contradiction. Thus $P_m = \{n; n \geq m\}$ does not contain cycles of length less than m. If $\bar{u} \in \Sigma_{P_m}$, then $|u| \geq m$, and for some i, $w_{|m} \sqsubseteq \sigma^i(\bar{u})$. For a given integer k there exists j_k such that $w_{|j_k}$ is not initial substring of a periodic point with period k. Thus if $m_k \in M$ and $m_k > j$, then $\Sigma_{(k)} = \Sigma_{P_{m_k}}$ is a nonempty clopen invariant set, which does not contain periodic points with period less than k. Thus Σ_w has not bounded limits. Since $\Sigma_{(k+1)} \subseteq \Sigma_{(k)}$, their intersection $\Sigma = \cap_k \Sigma_{(k)}$ is nonempty and invariant. If $W \cap \Sigma \neq 0$, then either $\omega(W) \subseteq \Sigma$, so it has no periodic points, or $\omega(W) \not\subseteq \Sigma_{(k)}$ for some k, and then it is not transitive. Thus (Σ_w, σ) has not chaotic limits. $\square$

Proposition 22 *If $w \in 2^N$ is an infinite maximal sequence, and if there exists m such that $(\forall i > 0)(\exists k \leq m)(\sigma^i(w)_{|k} \prec w_{|k})$, then Σ_w has bounded limits and chaotic limits.*

Proof: Every two states in the set $P_m = \{n | n \geq m\}$ communicate. Thus P_q is a final communication set for some $q \leq m$. $\square$

Example 6 *For any sequence of positive integers n_i, $w = 1001^{n_1}01^{n_2}0\ldots$ is a maximal sequence, and Σ_w has chaotic limits and bounded limits.*

Theorem 16 *If $w = \mathcal{D}_k(v)$, where v is odd nonperiodic, $\bar{v}$ is maximal, and k is an infinite sequence of positive integers, then (Σ_w, σ) has neither bounded limits nor chaotic limits.*

Proof: Suppose first that k is an integer, let $u = \mathcal{D}_k(v)$, and $|v| = n$. If $v_{|m} \hat{v}_m \prec v_{|(m+1)}$, then either $m < n$, or $m = jn - 1$, $1 \le j \le k$. Indeed suppose $jn \le m < (j+1)n - 1$. Since $v_{|jn}$ is odd, we get $v_{|(m-jn)} \hat{v}_{m-jn} \succ v_{|(m-jn+1)}$, which is a contradiction. If k is a sequence of positive integers, denote $n_i = |\mathcal{D}_{k_{|i}}(v)|$. Then for $n_i \le m < n_i(k_i + 1) = n_{i+1}$ either $\delta^*(m, \hat{w}_j) = \wedge$, or $\delta^*(m, \hat{w}_j) \ge n_i$. Thus $\mathcal{Q}_w$ has no final communication set, so (Σ_w, σ) has not bounded limits by Theorem 15. $\square$

11 S-unimodal systems

Definition 22 *A unimodal system (I, g) is S-unimodal, if it has continuous third derivation, $g''(c) < 0$, and $Sg(x) < 0$ for all $x \ne c$, where $Sg(x) = \frac{g'''(x)}{g'(x)} - \frac{3}{2}(\frac{g''(x)}{g'(x)})^2$ is the Schwarzian derivative.*

The functions with negative Schwarzian derivative have nice geometrical properties, which are shared by all its iterates, as $Sg^n(x) < 0$ whenever $Sg(x) < 0$. We list several properties of S-unimodal systems, whose proofs can be found in Collet and Eckmann [4].

Definition 23 *A fixed point $x \in I$ of (I, g) is stable from the left (stable from the right), if there exists $\varepsilon > 0$, such that for all $y \in I \cap (x - \varepsilon, x)$ (for all $y \in I \cap (x, x + \varepsilon)$), $\lim_{i \to \infty} g^i(y) = x$. A fixed point y is stable, if it is either stable from the left or stable from the right (or both). A periodic point with period n is stable, if it is stable fixed point of (I, g^n).*

Lemma 15 *Let (I, g) be an S-unimodal system, and let $x \in I$ be a periodic point with period n. If $|(g^n)'(x)| < 1$, then x is stable from both sides. If $|(g^n)'(x)| = 1$, then x is stable from one side. If $|(g^n)'(x)| > 1$, then x is not stable.*

Proof: The Lemma holds in general for differentiable functions except in the case $|(g^n)'(x)| = 1$. Denote $h(x) = g^n(x)$ if $(g^n)'(x) = 1$, and $h(x) = g^{2n}(x)$, if $(g^n)'(x) = -1$. Then $h'(x) = 1$, and x is a stable periodic point of (I, g) iff it is a stable fixed point of (I, h). If $h''(x) > 0$, then x is stable from the left. If $h''(x) < 0$, then x is stable from the right. If $h''(x) = 0$, then $h'''(x) < 0$ by $Sh(x) < 0$, and x is stable from both sides. $\square$

Lemma 16 *If (I, g) is S-unimodal system then for each n there is only a finite number of periodic points with period n. If $x < y < z$ are consecutive fixed points of $h = g^n$, and if $h'(w) \ne 0$ for $w \in (x, z)$, then $h'(y) > 1$.*

Proposition 23 (Singer Theorem) *If (I, g) is an S-unimodal system, then it has at most one stable periodic orbit.*

Proposition 24 (II.6.2. and II.5.4. of [4]) *If (I, g) is S-unimodal, then it has a stable periodic point iff $\mathcal{K}(g)$ is either finite or periodic. If it has no stable periodic orbits, then different points in I have different itineraries.*

Proposition 25 (II.5.7. and II.5.8. of [4]) *Let (I, g) be an S-unimodal system, which has a stable periodic orbits and denote E_g the set of points, which is not attracted to this orbit. Then E_g has Lebesgue measure zero, and different points in E_g have different itineraries.*

We can reformulate these results as follows:

Theorem 17 *Let (I, g) be S-unimodal system, and let $\alpha = \{V_a; a \in 2\}$ be the standard cover. If $\mathcal{K}(g)$ is neither finite nor periodic, then V_u has empty interior for any $u \in 2^N$. If $\mathcal{K}(g)$ is either finite or periodic, and $u \in 2^N$, then V_u has nonempty interior iff $u \in \Sigma_{\mathcal{K}_g}$ and there exists $k \geq 0$ with $\mathcal{K}(g) \sqsubseteq \sigma^k(u)$.*

Proposition 26 *If (I, g) is S-unimodal, then there exists a homomorphism $\psi : (\Sigma_{\mathcal{K}_g}, \sigma) \to (I, g)$, with $(\forall u \in \Sigma_{\mathcal{K}_g})(\psi(u) \in V_u)$, which maps eventually periodic points to eventually periodic points with the same period.*

Proof: If $\mathcal{K}(g)$ is neither finite nor periodic, then by Theorem 17, V_u is a singleton for each $u \in \Sigma_{\mathcal{K}_g}$, so the condition $\psi(u) \in V_u$ defines ψ uniquely, and ψ is clearly surjective and continuous. Suppose now that $\mathcal{K}(g) = \overline{w}$ is periodic odd, and denote $n = |w|$. Then g^n is decreasing on $V_{\overline{w}}$ and has therefore a unique g^n-fixed point $\psi(\overline{w}) \in V_{\overline{w}}$. If $\sigma^i(u) = \overline{w}$, then there is a unique $\psi(u) \in V_u$ with $g^i \psi(u) = \psi(\overline{w})$. Thus ψ is first defined on the orbit of $\overline{w}$, and then on its preimages. By Lemma 13 each u for which $\sigma^i(u) = \overline{w}$, is an isolated point, so ψ is continuous. Finally we assume that $\mathcal{K}(g)$ is either finite or periodic even. Then $\mathcal{K}_g = \overline{w}$ is periodic even too. If $w = 0$, then $\Sigma_{\mathcal{K}_g} = \{\overline{0}, 1\overline{0}\}$, and the Proposition is trivial. If $w \neq 0$, then $10 \sqsubseteq w$, and $w^p 1 \in \Sigma_{\mathcal{K}_g}$ for any p. Thus $\overline{w}$ is a limit point and $V_{\overline{w}} \neq V_u$ whenever u is an initial substring of $\overline{w}$. Denote $W = \cap \{\overline{V_u - V_{\overline{w}}}; u \in 2^*, u \sqsubseteq \overline{w}\}$. Then W is a nonempty invariant set, and does not contain any inner point of $V_{\overline{w}}$. Suppose that W contains both endpoints of $V_{\overline{w}}$. Then they would be unstable fixed points for g^{2n}, and therefore there would exist a stable fixed point between them. This is, however, in contradiction with $Sg < 0$ (Lemma 16). Thus W consists of exactly one periodic point with period n and we define $\psi(\overline{w}) \in W$. The construction then proceeds as in the preceding case. It follows from the construction that ψ is continuous at $\overline{w}$, and also at all its preimages. $\square$

Theorem 18 *Let (I, g) be S-unimodal system such that $\mathcal{K}(g)$ is neither finite nor periodic. Then (I, g) is a factor of $(\Sigma_{\mathcal{K}_g}, \sigma)$.*

Proof: Theorem 17 and Proposition 26

Theorem 19 *If* (I, g) *is S-unimodal and* $\mathcal{K}(g)$ *is preperiodic, then* (I, g) *is a factor of a finite automaton.*

Proof: Theorems 12, 6 and 18.

Theorem 20 *If* (I, g) *is S-unimodal and* $\mathcal{K}(g)$ *is either finite or periodic, then* (I, g) *is a factor of a finite automaton.*

Proof: Denote $\mathcal{K}_g = \overline{w}$, $|w| = n$. Consider an alphabet $A = \{0, 1, 2, 3, 4, 5\}$, a projection $\nu : A \to \{0, 1\}$ defined by $\nu(a) = a \bmod 2$, and a subshift $\Sigma \subseteq A^N$ defined by $x \in \Sigma$ iff either $x \in \Sigma_{\overline{w}}$, or $x = uv$, where u, v satisfy following conditions: $u \in \{0, 1\}^*$, $v \in \{2, ..., 5\}^N$, $\nu(uv) \in \Sigma_{\overline{w}}$, $w_{n-1} w_0 ... w_{n-2}$ is not a final segment of u, $\nu\sigma(v) = \overline{w}$, and $v_i \in \{4, 5\}$ iff n divides i. Since $\Sigma_{\overline{w}}$ is regular, Σ is regular too, and there is a factorization $\varphi' : (X', f') \to (\Sigma, \sigma)$. By Proposition 26 there is a homomorphism $\psi : (\Sigma_{\overline{w}}, \sigma) \to (I, g)$. Thus we have a homomorphism $\psi\nu\varphi' : (X', f') \to (I, g)$, which is, however, not surjective. By Proposition 8 there is a factorization $\varphi'' : (X'', f'') \to (V_{\overline{w}}, g^n)$. We construct a finite automaton (X, f), where $X = X' \times X''$, and $f(u', u'') = (f'(u'), f''(u''))$ if $\varphi'(u')_0 \in \{4, 5\}$ (this depends only on u'_0 and u''_0), $f(u', u'') = (f'(u'), u'')$ otherwise. Then $\sigma\nu\varphi'(u') = \overline{w}$ provided $\varphi'(u')_0 \in \{4, 5\}$. Define $\varphi : X \to I$ by

$$\varphi(u', u'') = \psi\varphi'(u') \qquad \text{if} \quad \varphi'(u') \in 2^N$$
$$\varphi(u', u'') = g_{\nu\varphi'(u')_0} g^n \varphi''(u'') \qquad \text{if} \quad \varphi'(u')_0 \in \{4, 5\}$$
$$\varphi(u', u'') = g_{\nu\varphi'(u')_0} \varphi(f'(u'), u'') \qquad \text{if} \quad \varphi(f'(u'), u'') \text{ has been defined}$$

(Here $g_i : [a, g(c)] \to V_i$, are the two inverses of g.) Then $\varphi(u', u'') \in V_{\nu\varphi'(u')}$, and φ is a surjection. If $\varphi'(u')_0 \in \{4, 5\}$, then $\varphi f(u', u'') = \varphi(f'(u'), f''(u'')) = g_{\nu\varphi'f'(u')_0} \varphi((f')^2(u'), f''(u'')) = ... = g_{\nu\varphi'(u')_1} ... g_{\nu\varphi'(u')_{n-1}} \varphi((f')^n(u'), f''(u'')) = g_{\nu\varphi'(u')_1} ... g_{\nu\varphi'(u')_n} g^n \varphi'' f''(u'') = \varphi'' f''(u'') = g^n \varphi''(u'') = g\varphi(u', u'')$. If $\varphi'(u')$ is eventually periodic, and $\varphi'(u')_0 \notin \{4, 5\}$, then $\varphi f(u', u'') = \varphi(f'(u'), u'') = g\varphi(u', u'')$. Thus φ is a factorization. $\square$

Lemma 17 *Let* (I, g) *be an S-unimodal systems with aperiodic kneading sequence and let* $\psi : (\Sigma_{\mathcal{K}_g}, \sigma) \to (I, g)$ *be the homomorphism constructed in Proposition 26. Then* ψ *is one-to-one on eventually periodic points and if* $U \subseteq \Sigma_{\mathcal{K}_g}$ *is a nonempty open set, then* $\psi(U)$ *has a nonempty interior.*

Proof: By Proposition 18, ψ is a factorization and for all $x \in I$, $\psi^{-1}(x) = \{\mathcal{I}_g(x), \mathcal{I}^g(x)\}$. If x is eventually periodic, then $\mathcal{I}_g(x) = \mathcal{I}^g(x)$, so ψ is one-to-one on eventually periodic points. Let $U \subseteq \Sigma_{\mathcal{K}_g}$ be a nonempty open set, $u \in [u_{|n}] \cap \Sigma_{\mathcal{K}_g} \subseteq U$, and $x = \psi(u)$. If $u = \mathcal{I}_g(x)$, or $u = \mathcal{I}^g(x)$, then by Lemma 9 there exists y in a neighbourhood of x for which $\mathcal{I}(y)_{|n} = u_{|n}$, and therefore the interval between x and y is contained in $\psi(U)$. $\square$

Theorem 21 *Let (I, g) be an S-unimodal systems with aperiodic kneading sequence. Then (I, g) has bounded limits iff it has chaotic limits iff the accepting automaton $\mathcal{Q}_{\mathcal{K}(g)}$ has a final communication set.*

Proof: If $\mathcal{Q}_{\mathcal{K}_g}$ has a final communication set, then (I, g) has bounded limits and chaotic limits by Propositions 6 and Theorem 15. If $\mathcal{Q}_{\mathcal{K}_g}$ has no final communication set, then by the proof of Proposition 6 there exist open subshifts $\Sigma_{(k+1)} \subseteq \Sigma_{(k)} \subseteq \Sigma_{\mathcal{K}_g}$, which do not contain periodic points with periods less than k. Since $\psi(\Sigma_{(k)})$ have nonempty interior, (I, g) has not bounded limits by Proposition 17. If $y \in \cap_k \psi(\Sigma_{(k)})$, and $y \in W$, then $\omega(W)$ is not chaotic. $\square$

Corollary 1 *Let (I, g) be an S-unimodal system. If $\mathcal{K}(g)$ is either finite or eventually periodic, then (I, g) is a factor of a finite automaton. If $\mathcal{K}(g) = \mathcal{D}_k(v)$, where v is odd nonperiodic, $\bar{v}$ is maximal, and k is an infinite sequence of positive integers, then (I, g) is not a factor of a finite automaton.*

Proof: Theorems 19, 20, 21 and 16.

12 Conclusion

Corollary 2 *The relations between the simplicity classes of dynamical systems are given by the following diagram.*

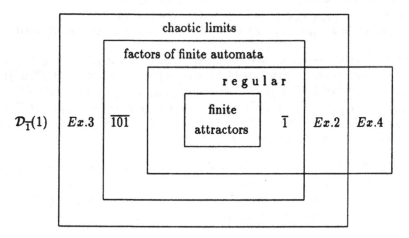

Fig. 6. Simplicity classes

Proof: By Theorem 2 any system with finite attractors is regular, and by Theorem 3 it is a factor of a finite automaton. By Theorem 5 any factor of a finite automaton has chaotic limits. An S-unimodal system with $\mathcal{K}(g) = \bar{1}$ is a regular factor of a finite automaton, which has not finite attractors. Example 2 shows that there exists a regular system with chaotic limits which is not a factor of a finite automaton. Example 4 shows that there exists a regular system which has not chaotic limits. An S-unimodal system with $\mathcal{K}(g) = \overline{101}$ is a factor of a finite automaton which is not regular. Example 3 shows that there exists a system with chaotic limits which is neither regular nor is a factor of a finite automaton. Finally S-unimodal system with $\mathcal{K}(g) = \mathcal{D}_{\overline{1}}(1)$, or any irrational rotation of the circle, are nonregular systems which have not chaotic limits. □

Thus our criteria express different aspects of simplicity of dynamical systems. Systems with finite attractors are simple by all three criteria. Regular systems are simple from the point of view of an observer: they generate simple time series. Factors of finite automata are simple from the engineering point of view: they have a simple construction. Finally the systems with chaotic limits are structurally simple. Their behaviour can be described in terms of simple dynamical structures.

References

[1] B.Balcar, P.Simon: Appendix on general topology. Handbook of Boolean Algebras (J.D.Monk and R.Bonnet, eds.), Elsevier Science Publishers B.V., 1241 (1989).

[2] J.Brooks, G.Cairns, G.Davis, P.Stacey: On Devaney's definition of chaos. The American Mathematical Monthly, 99,4, 332 (1992).

[3] A.W.Burks: Essays on Cellular automata. University of Illinois Press. Urbana 1970.

[4] P.Collet, J.P.Eckmann.: Iterated Maps on the Interval as Dynamical Systems. Birkhauser, Basel 1980.

[5] J.P.Crutchfield, K.Young: Computation at the onset of chaos. in: Complexity, Entropy and the Physics of Information, SFI Studies in the Sciences of Complexity, vol VIII (W.H.Zurek, ed.), Addison Wesley 223 (1990).

[6] K.Culik II, S.Yu: Cellular automata, $\omega\omega$-regular sets, and sofic systems. Discrete Applied Mathematics 32,85-101 (1991).

[7] K.Culik II, L.P.Hurd, S.Yu: Computation theoretic aspects of cellular automata. Physica D 45, 357-378 (1990).

[8] R.L.Devaney: An Introduction to Chaotic Dynamical Systems. Addison-Wesley, Redwood City 1989.

[9] D.Fiebig, U.R.Fiebig: Covers for coded systems. Symbolic Dynamics and its Applications (Peter Walters, ed.) American Mathematical Society, Providence 1992.

[10] J.G.Hocking: Topology. Addison-Wesley, Reading 1961.

[11] J.E.Hopcroft, J.D.Ullmann: Introduction to Automata Theory, Languages, and Computation. Addison-Wesley, Menlo Park 1990.

[12] J.Guckenheimer: Sensitive dependence to initial conditions for one dimensional maps. Commun.Math.Phys. 70, 133 (1979).

[13] J.Guckenheimer, P. Holmes: Nonlinear Oscillations, Dynamical Systems, and Bifurcations of Vector Fields. Applied Mathematical Sciences 42, Springer-Verlag, Berlin 1983.

[14] S.A.Kauffman, S.Johnson: Co-evolution at the edge of chaos: coupled fitness landscapes, poised states, and co-evolutionary avalanches. in: Artificial Life II. SFI Studies in the Sciences of Complexity, vol. X (Ch.G.Langton, Ch.Taylor, J.D.Farmer, S.Rasmunsen, eds.) Addison-Wesley, Redwood City 1992.

[15] P.Koiran,P.Cosnard,M.Garzon: Computability with low-dimensional dynamical systems. Rapport n 92-31, Ecole Normal Superieure de Lyon, 1992.

[16] C.Kuratowski: Topologie. Polskie Towarzystwo Matematyczne, Warszawa 1952.

[17] P.Kûrka: Ergodic languages. Theoretical Computer Science 21:351-355, (1982).

[18] P.Kûrka: One-dimensional dynamics and factors of finite automata. to appear in Acta Universitatis Carolinae, Mathematica et Physica.

[19] P.Kûrka: Regular unimodal systems and factors of finite automata. submitted to Theoretical Computer Science.

[20] Ch.G.Langton: Life at the Edge of Chaos. in: Artificial Life II. SFI Studies in the Sciences of Complexity, vol. X (Ch.G.Langton, Ch.Taylor, J.D.Farmer, S.Rasmunsen, eds.) Addison-Wesley, Redwood City 1992.

[21] J.Milnor, W.Thurston: On iterated maps of the interval. Dynamical Systems (J.C.Alexander, ed.) Lecture Notes in Mathematics 1342, pp. 465-563, Springer-Verlag, Berlin 1988.

[22] M.Mitchell, P.T.Hraber, J.P.Crutchfield: Revisiting the edge of chaos: evolving cellular automata to perform computations. Complex Systems, in press.

[23] Ch.Moore: Unpredictability and undecidability in dynamical systems. Physical Review Letters 64(20), 2354-2357, (1990).

[24] M.Shub: Global Stability of Dynamical Systems. Springer-Verlag, Berlin 1987.

[25] A.R.Smith: Simple computation-universal cellular spaces. J. ACM 18, 339-353 (1971).

[26] S.M.Ulam, J.von Neumann: On combinations of stochastic and deterministic processes. Bull.Amer.Math.Soc. 53, 1120, (1947).

[27] S.Wolfram: Theory and Applications of Cellular Automata. World Scientific, Singapore, 1986.

Analysis of Dynamical Systems using Predicate Transformers: Attraction and Composition

M. Sintzoff and F. Geurts *

University of Louvain, Unité d'Informatique,
Place Sainte Barbe 2, B-1348 Louvain-la-Neuve, Belgium
E-mail: {ms,gf}@info.ucl.ac.be

Abstract. We present a framework for the compositional analysis of
dynamical systems. This framework is based on set-valued functions, defined by predicate transformers. It integrates concepts from mathematics,
computing science, and neurosciences. We also introduce an additional
concept: the attraction between predicates. The main results of the paper are then presented. We propose composition rules which permit to
see a complex system as composed of simpler ones, to study these simple
systems using the concepts introduced before, and then to compose the
results for deriving the analysis of the initial complex system.

1 Introduction

Discrete autonomous dynamical systems [1, 2] can be analyzed using predicate
transformers for iterated guarded commands [3]. Initial results of such a study
have been reported [4]; they provide sufficient criteria on invariants of infinite iterations for detecting topological transitivity and sensitive dependence on initial
conditions.

The present paper tackles two further topics in the same framework, namely
the attraction of a predicate by another through a system and the composition
of systems.

The goals of this study of dynamical systems are the following ones. Firstly,
we would like to find technical tools for the analysis of simple systems (topological transitivity, sensitive dependence, attraction). Secondly, we would like to
establish an algebra of systems, i.e. composition laws and influence on the properties of the resulting (complex) systems. Thirdly, we would like to extend this
theory to systems with parameters and study the bifurcation phenomena. Finally, to complete this analysis approach, we could develop a synthesis approach
for complex dynamical systems.

The paper first reviews the framework used. Then, concepts and criteria for
attraction are developed. Basic cases of composition by sum and by product
are investigated. The usefulness of the proposed approach is discussed. Finally,
we compare our approach of analyzing dynamical systems with related ones in
mathematics, computing science, and neurosciences.

---

* Supported by the National Fund for Scientific Research (Belgium)

2 Dynamical Systems and Predicate Transformers

This section recalls previous results [4] and presents the main intuitions behind. Basically, dynamical systems are analyzed as maps between sets of states [5]. Sets of states are expressed as predicates, and thus dynamical systems are analyzed in terms of predicate transformers.

2.1 Discrete Dynamical Systems as Programs for Infinite Iterations

Discrete autonomous dynamical systems [1, 2] are defined by maps from states to states; these maps are iterated infinitely often. Such systems are readily expressed as programs on the form of simple guarded commands [3], to be repeated forever. In these simple commands, only assignments are used. Such programs allow to model sequential as well as asynchronous parallel computations in a clear and effective way.

For instance, the classical dyadic transformation, or shift map,

$$S(x) = \begin{cases} 2x & \text{if } 0 \leqslant x < \frac{1}{2} \\ 2x - 1 & \text{if } \frac{1}{2} \leqslant x \leqslant 1 \end{cases}$$

is expressed as follows:

$$S = [\quad 0 \leqslant x < \tfrac{1}{2} \rightarrow x := 2x$$
$$[] \ \tfrac{1}{2} \leqslant x \leqslant 1 \rightarrow x := 2x - 1 \ ]$$

This dyadic map is the kernel of the baker map [1, 2] and generates chaotic dynamics.

Dynamical systems in principle must be deterministic viz functional [2], but in general their inverses are not functional, and thus cannot serve as dynamical systems. That constraint of determinism is removed in the present framework: guarded-command programs in general are nondeterministic, their inverses are guarded-command programs again, and thus inverses of dynamical systems can now also be dynamical systems. This symmetry results from replacing a functional view by a relational one, and proves helpful in the analysis of the dynamics.

Unless stated otherwise, programs contain at least two guarded commands.

In programs, each guarded command must define a one-to-one function. Thus, the guarded commands $(0 \leqslant x < 1 \rightarrow x := 2x)$, $(x = 1 \rightarrow x := 2x)$, and $(x = 1 \rightarrow x := 2)$ are all acceptable, but $(0 \leqslant x < 1 \rightarrow x := 2)$ is not, because it is not one-to-one. This restriction does not dramatically reduce the applicability of the approach; moreover, it entails the continuity of predicate-transformers presented below and the invertibility of programs.

Suppose a dynamical system is defined by a function which is not invertible but which is composed of a finite number of invertible pieces, viz monotone and anti-monotone functions. Then, we can construct a program for this dynamical system simply by writing one guarded command for each invertible function piece. For instance, the system defined for $0 \leqslant x \leqslant 1$ by the logistic map

$$x := 4x(1 - x)$$

is expressed by the program

$$L = [\quad 0 \leqslant x \leqslant \tfrac{1}{2} \rightarrow x := 4x(1 - x)$$
$$\square \quad \tfrac{1}{2} \leqslant x \leqslant 1 \rightarrow x := 4x(1 - x)]$$

The first guarded command expresses an increasing function, and the second one a decreasing function.

Classically, the system (e.g. L) starts from an initial condition (e.g. 0.4) and iterates indefinitely from single values to single values (e.g. 0.4, 0.96, 0.1536, ...). It is also possible to start with an initial interval (e.g. $[0, \tfrac{1}{2}]$) and the successive iterations give a sequence of intervals (e.g. $[0, 1]$, $[0, 1]$, ...). In general, the initial condition can be specified by a predicate, or set of states; then the sets of states generated by the successive iterations are also expressed by predicates.

2.2 Predicate Transformers

The systems S considered here are analyzed using two predicate transformers, viz the **pre-image** S_- and the **post-image** S_+ [2]:

$$\text{if} \quad S \quad = [\square i \in \Sigma : S_i] = [\square i \in \Sigma : b_i \rightarrow x := e_i]$$

$$\text{then } S_+.P = \vee i \in \Sigma : S_{i+}.P$$
$$S_-.P = \vee i \in \Sigma : S_{i-}.P$$
$$S_{i+}.P = \exists u : P.u \wedge b_i.u \wedge (x = e_i.u)$$
$$S_{i-}.P = \exists v : b_i.x \wedge (v = e_i.x) \wedge P.v$$

The predicate $S_-.P$ defines the set of states *from which* a state in P can be reached by an application of S. The predicate $S_+.P$ gives the set of states *which* can be reached *from* a state in P by an application of S.

In [4], properties of predicate transformers have been introduced, among which and-continuity. Let F be a predicate transformer. F is **and-continuous** iff for every decreasing sequence of predicates $\{P_i\}$ such that $P_{n+1} \Rightarrow P_n$ for all n, $F(\bigwedge_i P_i) = \bigwedge_i F(P_i)$. This property is important because it will ensure the existence and unicity of a maximal fixed point of the invariant equations. It will also give us an iterative method for computing these invariants. We will come back to this in section 2.3.

Now we have to look for conditions ensuring and-continuity of the predicate transformer S_- obtained from a system S. Such a condition has been proposed in [3]: it is bounded non-determinism. This means that every state has a finite

---

[2] Let us recall the notations we use through this paper. If S is a program, b_i a guard of S or a predicate, and e_i the corresponding function, f another function: $e_i.x$ is the application of the function to the argument x, i.e. $e_i(x)$, $b_i.x$ is equivalent to a test of x matching the predicate b_i, $e_i.f.x$ is simply equivalent to $e_i(f(x))$ and $S_{i+}.b_i$ is equivalent to $\vee\{S_{i+}.x|b_i.x\}$. The notations b_i, e_i, and c_i keep x implicit.

number of direct successors. More formally, let $R_S \subseteq Q \times Q$ be the relation giving the behaviour of the system S; this relation R_S is **image-finite** (viz verifies the bounded non-determinism property) iff $\forall q \in Q : \exists k \in \mathbf{N} : \#\{r \in Q | (q, r) \in R_S\} < k$. As proven in [6], the condition of bounded non-determinism is necessary and sufficient to ensure and-continuity.

We also want that the dual predicate transformer S_+ be and-continuous. Therefore, we need the property of bounded non-determinism in both the future and the past. We need that the relation R_S describing the behaviour of S and its inverse S^{-1} be image-finite.

To obtain the bounded non-determinism in both the future and the past, it suffices to work with a finite number of guarded-commands describing one-to-one functions (i.e. functions for which inverses exist and whose inverses are one-to-one, too). Let us remark that the last property (one-to-one functions) entails the invertibility of programs.

2.3 Invariants as Fixed-Points

The **positive invariant** J_+ of S is the largest predicate which can be preserved by direct iterations of S, viz in the infinite future. It is the maximal solution of

$$J_+ \Rightarrow S_- . J_+$$

Indeed, J_+ must be a set of states from which J_+ can again be reached by an application of S. Dually, the **negative invariant** J_-, accessible from the infinite past, is the maximal solution of

$$J_- \Rightarrow S_+ . J_-$$

We can rewrite the above implications as equations:

$$J_\pm = J_\pm \wedge S_\mp . J_\pm$$

Thanks to Knaster-Tarski's fixed-point theorem, in the lattice of predicates, there exists only one maximal solution to each equation, if the predicate transformers used are monotonic. Moreover, the solutions can be computed iteratively, due to the and-continuity of the predicate transformers [7, 8, 6, 4]:

$$J_\pm = \lim_{n \to \infty} S_\mp^n . \mathbf{true}$$

$$J_\pm^{(n+1)} = S_\mp . J^{(n)} = S_\mp^{n+1} . \mathbf{true}$$

We have seen and-continuity is ensured by the assumption that each transition in a system must be a one-to-one function. And-continuity entails monotonicity, which lets $(S_\pm^n)_n$ be a monotone decreasing sequence of predicates, and thus entails the convergence of the limit.

The **invariant** J of S gives the largest set of states which are accessible in the past as well as in the future. It is the maximal solution of

$$J \Rightarrow S_+ . J \wedge S_- . J \qquad \text{viz} \qquad J \Rightarrow S_{+,-} . J$$

or

$$J = J \wedge S_{+,-}.J$$

where the combined predicate-transformer $S_{+,-}$ is given by

$$S_{+,-}.P = S_+.P \wedge S_-.P$$

In fact, the invariant can be decomposed as a conjunction of the negative and positive invariants [4]:

$$J = J_- \wedge J_+$$

It is the inner structure of the invariant J which determines the structure of the dynamics of the system S.

An invariant is *full* whenever, for any infinite sequence of transitions in the past and any infinite sequence of transitions in the future, it contains a state resulting from that past and beginning that future: all conceivable bi-infinite histories are thus realizable. If the invariant of a program S is full, then S generates a weak form of *topological transitivity* [2]: the set of traces, i.e. sequences of transitions, is uncountable, but not necessarily the set of orbits (sequences of state).

A full invariant is *atomic* (viz pulverized or atomized) if each bi-infinite history determines exactly one state. If the invariant is full as well as atomic, then the program is topologically transitive in the strong sense and it is *sensitively dependent on initial conditions*.

Fullness and atomicity are defined below in terms of trace-based invariants.

2.4 Trace-Based Invariants

A **one-way trace** σ is a finite or infinite word over the alphabet Σ of transition labels.

$$\sigma \in \Sigma^\omega, \quad \Sigma^\omega = \Sigma^* \cup \Sigma^\infty$$

where Σ^* and Σ^∞ are respectively the set of finite traces (i.e. words) over Σ and the set of infinite traces over Σ.

Two-way traces (σ, τ) are pairs of one-way traces, respectively representing the past and the future. It is easy to generalize S_{i+} and S_{i-} to $S_{\sigma+}$ and $S_{\tau-}$:

$$S_{\sigma i+}.P = S_{i+}.S_{\sigma+}.P$$
$$S_{i\tau-}.P = S_{i-}.S_{\tau-}.P$$

Then,

$$\begin{aligned}
J_+^{(n)} &= S_-^n.\textbf{true} \\
&= \vee \tau \in \Sigma^n : S_{\tau-}.\textbf{true} \\
&= \vee \tau \in \Sigma^n : J_{,\tau} \\
J_-^{(n)} &= \vee \sigma \in \Sigma^n : J_{\sigma,}
\end{aligned}$$

where

$$\begin{aligned}
J_{,\tau} &= S_{\tau-}.\textbf{true} \\
J_{\sigma,} &= S_{\sigma+}.\textbf{true}
\end{aligned}$$

viz

$$J_{,\tau} \xleftarrow{S_{\tau-}} \text{true}$$
$$\text{true} \xrightarrow{S_{\sigma+}} J_{\sigma,}$$

In other words, $J_{,\tau}$ (resp. $J_{\sigma,}$) gives the states which accept the future τ (resp. the past σ). Clearly,

$$J = \vee \sigma, \tau \in \Sigma^\infty : J_{\sigma,\tau}$$
$$J_{\sigma,\tau} = J_{\sigma,} \wedge J_{,\tau}$$
$$J_{\sigma i, j\tau} = S_{i+}.J_{\sigma,} \wedge S_{j-}.J_{,\tau}$$

2.5 Fullness

The invariant J of a system S is **full** iff, for each trace (σ, τ), $J_{\sigma,\tau}$ contains at least one state:

$$\forall \sigma, \tau \in \Sigma^\infty : J_{\sigma,\tau} \neq \textbf{false}$$

This means *each* trace (σ, τ) can be realized: there exists at least one state with past σ and future τ. If J is a full invariant of S, then S shows a *weak* form of *topological transitivity*:

$$\forall \sigma_1, \tau_1, \sigma_2, \tau_2 \in \Sigma^* : [S_{\tau_1 \sigma_2 +}.(J \wedge J_{\sigma_1,\tau_1}) \wedge (J \wedge J_{\sigma_2,\tau_2}) \neq \textbf{false}]$$
$$\wedge [(J \wedge J_{\sigma_1,\tau_1}) \wedge S_{\tau_1 \sigma_2 -}.(J \wedge J_{\sigma_2,\tau_2}) \neq \textbf{false}]$$

viz

$$J \wedge J_{\sigma_1,\tau_1} \neq \textbf{false} \underset{S_{\tau_1 \sigma_2 -}}{\overset{S_{\tau_1 \sigma_2 +}}{\rightleftharpoons}} J \wedge J_{\sigma_2,\tau_2} \neq \textbf{false}$$

For each pair of approximated components of the invariant, viz $J \wedge J_{\sigma_1,\tau_1}$ and $J \wedge J_{\sigma_2,\tau_2}$, there is a trace, viz $\tau_1 \sigma_2$, by which a state in the first component is transformed into a state in the second component.

Moreover, given a full invariant J,

$$\forall m, n \geqslant 0 : \forall \sigma, \tau \in \Sigma^m :$$
$$\exists p \geqslant 0 : \forall \sigma', \tau' \in \Sigma^n : \quad [S_+^p.(J \wedge J_{\sigma'.\tau'}) \wedge (J \wedge J_{\sigma,\tau}) \neq \textbf{false}]$$
$$\wedge [(J \wedge J_{\sigma,\tau}) \wedge S_-^p.(J \wedge J_{\sigma',\tau'}) \neq \textbf{false}]$$

Thus, **each** component $J \wedge J_{\sigma,\tau}$ contains **pre**-images, by **some** S_-^p, of states taken in **all** the components $J \wedge J_{\sigma',\tau'}$. Conversely, **each** $J \wedge J_{\sigma,\tau}$ contains **post**-images, by **some** S_+^p, of states taken in **all** $J \wedge J_{\sigma',\tau'}$.

Proposition 1. *The fullness of the invariant J of a system S can be proved by finding two sets Φ and Ψ of predicates verifying the following criteria:*

1. $\forall i \in \Sigma : (\exists A \in \Phi : A \Rightarrow J_{i,}) \wedge (\exists B \in \Psi : B \Rightarrow J_{,i})$

2. $\forall i \in \Sigma : (\forall A \in \Phi : S_{i+}.A \in \Phi) \wedge (\forall B \in \Psi : S_{i-}.B \in \Psi)$
3. $\forall A \in \Phi, B \in \Psi : A \wedge B \neq \mathbf{false}$

The predicates A, B in the sets Φ, Ψ approximate from below the components J_σ, and $J_{,\tau}$; if their conjunction $A \wedge B$ is never empty, then no $J_{\sigma,\tau} = J_{\sigma,} \wedge J_{,\tau}$ is empty either.

Proposition 2. *If the invariant is full and the system has at least two transitions, then the set of bi-infinite traces (σ, τ) contains*

1. *a countable infinity of periodic traces consisting of traces of all periods;*
2. *an uncountable infinity of nonperiodic traces;*
3. *a dense trace.*

Note fullness of the invariant implies fairness [9] of the system, but not conversely in general. For this, the definition of fullness should be generalized, by using sufficiently rich languages of feasible traces; here, we simply take the language containing *all* bi-infinite traces.

2.6 Atomicity

The invariant J of a system S is **atomic** iff each component $J_{\sigma,\tau}(\sigma, \tau \in \Sigma^\infty)$ *contains at most one state*: any past-plus-future history (σ, τ) is realized by one state at most. A state may well have different bi-infinite histories, but each feasible bi-infinite history must determine exactly one state.

Fullness plus atomicity entail *strict topological transitivity*: for $\sigma, \tau \in \Sigma^*$, the approximated components $J \wedge J_{\sigma,\tau}$, although never empty, can be taken as small as desired. Similarly, fullness plus atomicity entail *sensitive dependence on initial conditions* and *on final conditions*: each approximated component $J \wedge J_{\sigma,\tau}(\sigma, \tau \in \Sigma^m)$, however small, contains pre-images by some S_-^p, and post-images by some S_+^p, of states taken in all the components $J \wedge J_{\sigma',\tau'}(\sigma', \tau' \in \Sigma^n)$.

Proposition 3. *Atomicity can be detected by finding two sets Φ and Ψ of predicates, and a function μ from $\Phi \wedge \Psi$ to an ordered set M with minimum 0, which verify the following:*

1. $\forall i \in \Sigma : (\exists A \in \Phi : J_{i,} \Rightarrow A) \wedge (\exists B \in \Psi : J_{,i} \Rightarrow B)$
2. $\forall i \in \Sigma : (\forall A \in \Phi : S_{i+}.A \in \Phi) \wedge (\forall B \in \Psi : S_{i-}.B \in \Psi)$
3. $\forall A \in \Phi, B \in \Psi : \mu(A \wedge B) = 0 \Rightarrow \#(A \wedge B) \leqslant 1$
4. (a) $\exists k : 0 < k < 1 :$
   $\forall i, j \in \Sigma : \forall A \in \Phi, B \in \Psi : \mu(S_{i+}.A \wedge S_{j-}.B) \leqslant k \times \mu(A \wedge B)$, or
   (b) $\forall i, j \in \Sigma : \forall A \in \Phi, B \in \Psi :$
   $\mu(S_{i+}.A \wedge S_{j-}.B) = 0 \vee \mu(S_{i+}.A \wedge S_{j-}.B) < \mu(A \wedge B)$

One of (4.a) and (4.b) suffices. These conditions respectively correspond to Liapunov stability functions and to Floyd termination functions; they use the nonnegative reals and the naturals as set M.

We use the obvious notation

$$\Phi \wedge \Psi = \{A \wedge B | A \in \Phi, B \in \Psi\}$$

2.7 Examples

To help understanding, each transition S_i is given with a *pre-guard* b_i and a *post-guard* c_i:

$$S_i = [\, b_i \to x := e_i \to c_i \,]$$
$$c_i = S_{i+}.b_i$$

2.7.1 A Fixed Point Invariant. Consider

$$S = [0 \leqslant x \leqslant 1 \ \to \ x := 2x \ \to \ 0 \leqslant x \leqslant 2]$$

Clearly,

$$
\begin{aligned}
J_+^{(1)} &= S_-.\textbf{true} & &= 0 \leqslant x \leqslant 1 \\
J_+^{(2)} &= S_-.(0 \leqslant x \leqslant 1) &&= 0 \leqslant x \leqslant 2^{-1} \\
J_+^{(n+1)} &= 0 \leqslant x \leqslant 2^{-n} \\
J_+ &= (x = 0)
\end{aligned}
$$

$$
\begin{aligned}
J_-^{(1)} &= S_+.\textbf{true} & &= 0 \leqslant x \leqslant 2 \\
J_-^{(2)} &= S_+.(0 \leqslant x \leqslant 2) &&= 0 \leqslant x \leqslant 2 \\
&= J_-^{(1)} &&= J_- \\
\\
J &= J_- \wedge J_+ &&= (x = 0)
\end{aligned}
$$

There is only one infinite trace, viz $(1^\infty, 1^\infty)$, and one state in the invariant $J = (x = 0)$.

Let us add another expanding transition:

$$
\begin{aligned}
T = [\ & 0 \leqslant x \leqslant 1 \to x := 2x \to 0 \leqslant x \leqslant 2 \\
\parallel\ & 0 \leqslant x \leqslant 1 \to x := 3x \to 0 \leqslant x \leqslant 3\,]
\end{aligned}
$$

The invariant is still the singleton predicate $x = 0$, but the set of traces is now uncountable: there is one constant orbit but very many traces. The invariant is trivially full: all the bi-infinite traces determine the same, unique, fixed-point state $x = 0$.

Because of the duality between past and future, the same properties hold for the inverse program:

$$
\begin{aligned}
T^{-1} = [\ & 0 \leqslant x \leqslant 2 \to x := \tfrac{x}{2} \to 0 \leqslant x \leqslant 1 \\
\parallel\ & 0 \leqslant x \leqslant 3 \to x := \tfrac{x}{3} \to 0 \leqslant x \leqslant 1\,]
\end{aligned}
$$

2.7.2 The Dyadic Map. Let be

$$S = [\quad 0 \leqslant x \leqslant \tfrac{1}{2} \rightarrow x := 2x \qquad \rightarrow 0 \leqslant x \leqslant 1$$
$$\square \ \tfrac{1}{2} \leqslant x \leqslant 1 \rightarrow x := 2x - 1 \rightarrow 0 \leqslant x \leqslant 1]$$

Then,

$$J_+^{(1)} = S_-.\mathbf{true} \qquad = 0 \leqslant x \leqslant \tfrac{1}{2} \vee \tfrac{1}{2} \leqslant x \leqslant 1 = 0 \leqslant x \leqslant 1$$
$$J_+^{(2)} = S_-.(0 \leqslant x \leqslant 1) = 0 \leqslant x \leqslant 1$$
$$= J_+^{(1)} \qquad\qquad = J_+$$

$$J_-^{(1)} = S_+.\mathbf{true} \qquad = 0 \leqslant x \leqslant 1$$
$$J_-^{(2)} = S_+.(0 \leqslant x \leqslant 1) = 0 \leqslant x \leqslant 1$$
$$= J_-^{(1)} \qquad\qquad = J_-$$

$$J \quad = J_- \wedge J_+ \qquad = 0 \leqslant x \leqslant 1$$

2.7.2.1 Fullness of the Invariant. We choose the sets
Φ: the predicate $0 \leqslant x \leqslant 1$
Ψ: the predicates $p \leqslant x \leqslant q$ where $0 \leqslant p \leqslant q \leqslant 1$.

The condition (1) for fullness (§2.5) is immediate: for $i = 1, 2$,

$$J_{i,} = S_{i+}.\mathbf{true} = 0 \leqslant x \leqslant 1 \in \Phi$$
$$J_{,1} = S_{1-}.\mathbf{true} = 0 \leqslant x \leqslant \tfrac{1}{2} \in \Psi$$
$$J_{,2} = S_{2-}.\mathbf{true} = \tfrac{1}{2} \leqslant x \leqslant 1 \in \Psi$$

Condition (2) w.r.t. Φ is immediate too. W.r.t. Ψ, we have

$$S_{1-}.(p \leqslant x \leqslant q)$$
$$= \tfrac{p}{2} \leqslant x \leqslant \tfrac{q}{2}$$
$$\in \Psi \qquad (0 \leqslant \tfrac{p}{2} \leqslant \tfrac{q}{2} \leqslant 1)$$

$$S_{2-}.(p \leqslant x \leqslant q)$$
$$= \tfrac{p}{2} + \tfrac{1}{2} \leqslant x \leqslant \tfrac{q}{2} + \tfrac{1}{2}$$
$$\in \Psi \qquad (0 \leqslant \tfrac{p}{2} + \tfrac{1}{2} \leqslant \tfrac{q}{2} + \tfrac{1}{2} \leqslant 1)$$

The condition (3) holds since $(0 \leqslant x \leqslant 1) \wedge (p \leqslant x \leqslant q) \neq \mathbf{false}$, given $0 \leqslant p \leqslant q \leqslant 1$. Thus, each component $J_{\sigma,\tau}$ contains at least one state, and J is full. Moreover, since $J = 0 \leqslant x \leqslant 1$, *each* point in [0,1] is contained in some $J_{\sigma,\tau}$.

Note the exact expression of these components takes the following form, for $\sigma, \tau \in \Sigma^n$, and $0 \leqslant m < 2^n$:

$$J_{\sigma,\tau} = \tfrac{m}{2^n} \leqslant x \leqslant \tfrac{m+1}{2^n}$$

The use of the approximating predicates $p \leqslant x \leqslant q$ allows to prove fullness without having to compute these exact expressions explicitly; this explicit computation is still feasible here, but in general becomes too hard.

2.7.2.2 Atomicity. We prove it using the sufficient criteria in §2.6, with
  Φ: the predicate $0 \leqslant x \leqslant 1$
  Ψ: the predicates $p \leqslant x \leqslant q$ where $0 \leqslant p \leqslant q \leqslant 1$

  μ: $\Phi \wedge \Psi \rightarrow$ Nonnegative Reals
  $\mu(p \leqslant x \leqslant q) = q - p$

The conditions (1) and (2) are verified as above for fullness. Condition (3) holds since $q - p = 0$ entails $\#(\{x|p \leqslant x \leqslant q\}) = 1$. Condition (iv.a) is verified as follows, using $k = \frac{1}{2}$:

$$
\begin{aligned}
&\mu(S_{1+}.(0 \leqslant x \leqslant 1) \wedge S_{1-}.(p \leqslant x \leqslant q)) \\
&= \mu(\tfrac{p}{2} \leqslant x \leqslant \tfrac{q}{2}) \\
&= \tfrac{q}{2} - \tfrac{p}{2} = \tfrac{1}{2} \times (q - p) \\
&= \tfrac{1}{2} \times \mu(p \leqslant x \leqslant q),
\end{aligned}
$$

and similarly for the other cases $\mu(S_{i+}.(0 \leqslant x \leqslant 1) \wedge S_{j-}.(p \leqslant x \leqslant q))$.

Thus, each past-and-future step $S_{i+}.A \wedge S_{j-}.B$ decreases the size of $A \wedge B$ at least by $\frac{1}{2}$. Each limit component $J_{\sigma,\tau}(\sigma, \tau \in \Sigma^{\infty})$ contains at most one state: J is atomic. Since J is also full, each $J_{\sigma,\tau}(\sigma, \tau \in \Sigma^{\infty})$ contains exactly one state. The dyadic map is topologically transitive and sensitive to initial conditions. This well-known result has been re-derived here to illustrate the use of the proposed criteria for fullness and for atomicity.

2.7.3 Intuitive Analysis. The above analysis can be understood as follows. The domain of the dyadic map consists of the two half-segments $0 \leqslant x \leqslant \frac{1}{2}$ and $\frac{1}{2} \leqslant x \leqslant 1$, and the co-domain is the whole segment $0 \leqslant x \leqslant 1$. The inverse of the dyadic map contracts the whole segment by two, whereas the direct dyadic map expands the half-segments by two. Thus the combined effect of the direct and of the inverse map contracts the whole segment by two: the intersection of large segments and small segments yields small segments. Two applications of this combined (past-and-future) map yields the four segments $0 \leqslant x \leqslant \frac{1}{4}$, $\frac{1}{4} \leqslant x \leqslant \frac{2}{4}$, $\frac{2}{4} \leqslant x \leqslant \frac{3}{4}$, and $\frac{3}{4} \leqslant x \leqslant 1$. For n such applications, we obtain 2^n segments of length 2^{-n}. These smaller segments cover the whole segment $0 \leqslant x \leqslant 1$ and converge towards single points.

This reasoning can also be used to analyze the inverse of the dyadic map. In general, the invariant and the properties of fullness and atomicity are the same for a system and for its inverse.

3 Attraction

Let us analyze how a system can go from some given states to some other ones, viz how it can "progress" in the space of states. The results of this section also use [10].

3.1 Concepts

Firstly, let us present the intuitive ideas behind attraction. Classically, attraction happens when iterating a system like the one represented by the first graphics (on the left): asymptotic attraction to a (fixed) state, $\frac{1}{2}$.

$$S_1 = [\, 0 \leqslant x \leqslant 1 \rightarrow x := \tfrac{1}{3} + \tfrac{x}{3} \rightarrow \tfrac{1}{3} \leqslant x \leqslant \tfrac{2}{3} \,]$$

Another interesting kind of attraction happens when iterating a system like the second one (graphics represented on the right): attraction to a set of states, here an interval, $[\frac{1}{2}, 1]$.

$$S_2 = [\; 0 \leqslant x \leqslant \tfrac{2}{3} \rightarrow x := x + \tfrac{1}{3} \rightarrow \tfrac{1}{3} \leqslant x \leqslant 1$$
$$\square \;\; \tfrac{1}{2} \leqslant x \leqslant 1 \rightarrow x := x \qquad \rightarrow \tfrac{1}{2} \leqslant x \leqslant 1 \,]$$

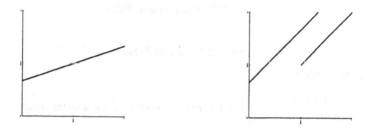

Fig. 1. The graph of S_1 (left) contains a fixed state $\frac{1}{2}$, while the graph of S_2 (right) has a fixed set of states $[\frac{1}{2}, 1]$

Attraction is to termination what infinite iterations are to finite ones. A system S beginning in a pre-assertion P terminates on a post-assertion Q if, for each initial state in P, S necessarily terminates after a finite number of iterations and must then reach a state in Q. A predicate Q **attracts** a predicate P by a system S if, after each realizable infinite iteration beginning in P, the resulting state belongs to Q:

$$\forall \sigma \in \Sigma^\infty : S_{\sigma+}.P \Rightarrow Q$$

We call this concept **weak attraction**. We can compare it with partial correctness of programs: if the program terminates, then it is correct. Here, if some infinite future σ exists, then P is attracted by Q when executing S.

What is the problem with this definition? Nothing prevent the case where there is no realizable infinite future, viz $S_{\sigma+}.P$ is always **false**. Why did we present it first? Simply because it gives the essence of the phenomenon of attraction. We will now give three other possible definitions of attraction, each of them adding a condition avoiding the future of P to be empty.

We speak of **simple attraction** when *at least one* state of P has a potential infinite future:

$$(\forall \sigma \in \Sigma^{\infty} : S_{\sigma+}.P \Rightarrow Q) \wedge (P \wedge J_+ \neq \textbf{false})$$

So we are sure that $S_{\sigma+}.P \neq \textbf{false}$ for some σ, viz $S_+^{\infty}.P \neq \textbf{false}$.

We can ask for more. We speak of **strict attraction** when P is not empty and *every* state of P has a potential infinite future:

$$(\forall \sigma \in \Sigma^{\infty} : S_{\sigma+}.P \Rightarrow Q) \wedge (P \Rightarrow J_+) \wedge (P \neq \textbf{false})$$

Finally, we speak of **full attraction** when, *for each possible trace*, there exists *at least one* state of P with that trace as a potential infinite future:

$$\forall \sigma \in \Sigma^{\infty} : (S_{\sigma+}.P \Rightarrow Q) \wedge (S_{\sigma+}.P \neq \textbf{false})$$

We call it full attraction because the second part of the conjunction looks like the definition of fullness of the negative invariant J_-:

$$\forall \sigma \in \Sigma^{\infty} : S_{\sigma+}.\textbf{true} \neq \textbf{false}$$

viz

$$\forall \sigma \in \Sigma^{\infty} : J_{\sigma,} \neq \textbf{false}$$

In summary, we have:

$$\left. \begin{array}{l} \text{full attraction} \\ \text{strict attraction} \end{array} \right\} \Rightarrow \text{simple attraction} \Rightarrow \text{weak attraction}$$

It is clear that full attraction and strict attraction are based on complementary conditions and are thus unrelated. It could be different if we had defined the full attraction as:

$$\forall \sigma \in \Sigma^{\infty} : (S_{\sigma+}.P \Rightarrow Q) \wedge (S_{\sigma+}.P \neq \textbf{false}) \wedge (P \Rightarrow J_+) \wedge (P \neq \textbf{false})$$

In this case, full attraction implies strict attraction.

What is the most classical definition, in the context of dynamical systems? In [2], for instance, we find the following definition: the closed invariant set $A \subset \mathbf{R}^m$ is an **attracting set** for the function $g : \mathbf{R}^m \to \mathbf{R}^m$ iff there is some neighbourhood U of A such that

$$\forall x \in U : (\forall n \geqslant 0 : g^n(x) \in U) \wedge (\lim_{n \to \infty} g^n(x) \in A)$$

It means that every point starting from a neighbourhood U of A, stays in U ad *infinitum* and converges to A through g. Every point in U has thus an infinite future. There is no mention of "potential future": the functions considered classically are deterministic and, at each step, there is only one way of computing the next state. So if at a given instant, a state belongs to U, its present is the only state accessible from its past and its future is completely determined by its present. Considering this, we can now return to the question beginning this paragraph: it seems that the most classical definition is the strict attraction. In what follows, we will only work with this concept (which we will call **attraction**), unless stated otherwise.

For instance, $(x = 0)$ attracts $(0 \leqslant x \leqslant 1)$ by the following system:

$$S = [\quad 0 \leqslant x \leqslant 1 \to x := \frac{x}{2} \to 0 \leqslant x \leqslant \frac{1}{2}$$
$$\quad [] \ 0 \leqslant x \leqslant 1 \to x := \frac{x}{3} \to 0 \leqslant x \leqslant \frac{1}{3} \ ]$$

The **basin of attraction** of Q by S is the largest predicate that is attracted by Q when iterating S. It is clear that termination on Q can be subsumed under attraction, by adding the silent transition $Q \to x := x$ to the considered program. It has the same function as the "idling" transition in elementary transition systems: there is no effect [11].

3.2 Criteria of Attraction

We now present sufficient conditions to prove a predicate Q attracts a predicate P by a system S, as we did for fullness and atomicity.

Proposition 4. *To prove strict attraction, it suffices to find a family Ψ of nonempty predicates and a function μ from states to nonnegative reals such that:*

1. $P \in \Psi$
2. $\forall i \in \Sigma, A \in \Psi : S_{i+}.A \in \Psi$
3. $\forall A \in \Psi : (\mu(A) = 0) \Rightarrow (A \Rightarrow Q)$
4. $\exists k : 0 \leqslant k < 1 : \forall i \in \Sigma : \forall A \in \Psi : \mu(S_{i+}.A) \leqslant k \times \mu(A)$
5. $P \neq \mathbf{false}$ and $P \Rightarrow J_+$

There is thus an infinite series of predicates in Ψ which begins with P and which converges into Q. Let us outline the proof of this proposition.

Proof. It is easy to prove by induction:

$$\forall n \geqslant 0 : \forall \sigma \in \Sigma^n : S_{\sigma+}.P \in \Psi$$

The basic case is given by (1) and the induction is based on (2). Using (4), we can then prove:

$$\forall n \geqslant 0 : \forall \sigma \in \Sigma^n : \mu(S_{\sigma+}.P) \leqslant k^n \times \mu(P)$$

and thus

$$\forall \sigma \in \Sigma^\infty : \mu(S_{\sigma+}.P) = 0.$$

Finally, using (3) gives

$$\forall \sigma \in \Sigma^\infty : S_{\sigma+}.P \Rightarrow Q.$$

□

Remark. It is possible to give more refined criteria to prove attraction [10]. They are all based on the two following central keys:

1. approximation of the predicate P and its successive iterations,
2. definition of a decreasing function μ.

3.3 Attraction by Invariants

Consider the Cantor map

$$S = [\quad 0 \leqslant x \leqslant 1 \to x := \tfrac{x}{3} \quad \to 0 \leqslant x \leqslant \tfrac{1}{3}$$
$$\square \quad 0 \leqslant x \leqslant 1 \to x := \tfrac{x}{3} + \tfrac{2}{3} \to \tfrac{2}{3} \leqslant x \leqslant 1]$$

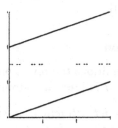

Fig. 2. The Cantor map and its invariant set (suggested in the middle)

Its invariant J is the "middle-third" Cantor set in [0,1] [12]; it is full and atomic.

Let us prove J attracts the domain $P = (0 \leqslant x \leqslant 1)$ of S. In this system, we have $J = J_-$ and $P = J_+$. We use the following proposition:

Proposition 5. If $J = J_- \wedge J_+ = J_-$, viz $J_- \Rightarrow J_+$, then the invariant J attracts J_+: the limit source J_+ contains the limit sink J_- and is attracted by it.

Proof. The definition of the negative invariant, $J_- = \forall n, S_+^n.\textbf{true}$, can be translated into J_- attracts **true**: $\forall \sigma \in \Sigma^\infty, S_{\sigma+}.\textbf{true} \Rightarrow J_-$. Thus every predicate P' is also attracted by J_-, because $P' \Rightarrow \textbf{true}$. Hence J_+ is attracted by J because $J = J_-$. We have also strict attraction if $J_+ \neq \textbf{false}$. □

Here, we also have strict attraction of P to $J_- = J$ because $P \Rightarrow J_+$ (in this case, we have $P = J_+$) and the positive invariant is not empty.

The inverse of the Cantor map has the same Cantor-set as invariant. Yet, this invariant does not anymore attract the domain $0 \leqslant x \leqslant 1$: the inverse Cantor-map is repulsing.

In fact, the invariant of the direct Cantor-map equals its negative invariant, whereas the invariant of the inverse Cantor-map equals its positive invariant. Recall a negative invariant results from infinite traces, and a positive invariant begins infinite traces (§2.3).

As an example, consider the system

$$S = [\quad 0 \leqslant x \leqslant \tfrac{1}{2} \wedge 0 \leqslant y \leqslant \tfrac{1}{2} \to (x,y) := (2x, \tfrac{y}{4})$$
$$\to 0 \leqslant x \leqslant 1 \wedge 0 \leqslant y \leqslant \tfrac{1}{8}$$
$$[] \quad \tfrac{1}{2} \leqslant x \leqslant 1 \wedge 0 \leqslant y \leqslant \tfrac{1}{2} \to (x,y) := (2x - 1, \tfrac{y}{4} + \tfrac{4}{8})$$
$$\to 0 \leqslant x \leqslant 1 \wedge \tfrac{4}{8} \leqslant y \leqslant \tfrac{5}{8}$$
$$[] \quad 0 \leqslant x \leqslant \tfrac{1}{2} \wedge \tfrac{1}{2} \leqslant y \leqslant 1 \to (x,y) := (2x, \tfrac{y}{4} + \tfrac{2}{8})$$
$$\to 0 \leqslant x \leqslant 1 \wedge \tfrac{3}{8} \leqslant y \leqslant \tfrac{4}{8}$$
$$[] \quad \tfrac{1}{2} \leqslant x \leqslant 1 \wedge \tfrac{1}{2} \leqslant y \leqslant 1 \to (x,y) := (2x - 1, \tfrac{y}{4} + \tfrac{6}{8})$$
$$\to 0 \leqslant x \leqslant 1 \wedge \tfrac{7}{8} \leqslant y \leqslant 1$$
$$]$$

The positive invariant J_+ is by definition stable under S_-, viz preserved in the future, and it includes the full square $0 \leqslant x \leqslant 1 \wedge 0 \leqslant y \leqslant 1$. Thus it includes J_-, generated by S_+ and preserved in the past; J_- consists of horizontal segments of length one. Since $J_- \Rightarrow J_+$, we have $J = J_-$. The invariant J attracts the square $0 \leqslant x \leqslant 1 \wedge 0 \leqslant y \leqslant 1$ since J_- attracts it.

If we modify the above system so that S_- is contracting along x by 4 instead of 2, the positive invariant J_+ may become a strict subset of the full square, viz a set of scattered vertical segments. Then, J_- is not included in J_+ anymore, and we cannot prove, using the present approach, that J attracts $0 \leqslant x \leqslant 1 \wedge 0 \leqslant y \leqslant 1$. We should then use the criteria in §3.2.

Besides proving a full and atomic invariant is attracting, we may wish to prove that a given attracting predicate generates interesting dynamics, i.e. is sensitively dependent on initial conditions and topologically transitive. To verify this, one could use variants of the criteria (§§2.5, 2.6) for detecting fullness and atomicity. The problem again is to discover adequate families of approximating predicates and an adequate convergence function.

Some people dislike the use of bi-infinite traces: they feel parallel systems may well run indefinitely in the future, but must have a well-defined beginning. Here, we require complete duality between past and future: if termination is removed, then initialization is removed too. However, termination is now generalized into attraction, viz infinite termination. A similar, dual generalization could be defined: initialization could be generalized to repulsion, viz infinite initialization (§4.1).

4 Composition

In this section, we investigate basic operations on systems: inversion, sequential composition, sum, and product. The sum and the product are first presented as free compositions (i.e. without interaction between the components). Connected forms of composition will be discussed in a following section. As we said in the introduction, our main objective is to develop composition laws on systems such that it could be possible to predict the properties of a composed system from the analysis and the properties of the different components. We will also establish a link between our sum (resp. product) and a model of asynchronous (resp. synchronous) systems.

Before presenting the different forms of composition, let us remark an important aspect of the following definitions. We express each system as an infinite iteration of a finite number of guarded commands. The operational semantics of such systems is the nondeterministic choice of one guarded command at each iteration. The compositions we propose act on the guarded commands involved. Each composed system still has only one global iteration loop which contains the modified guarded commands. One could say that the composition takes its effect inside the loop. There is no composition of several loops, either sequentially or concurrently.

4.1 Inversion

We will first define the inversion of systems (e.g. [13]). It is clear, and useful as shown later. Let S be the system to be inversed:

$$S = [\; [] \; i \in \Sigma_S : \; b_i \to x := e_i \to c_i \;]$$

where the post-guards $c_i = S_{i+}.b_i$. Since every function e_i is assumed to be invertible, we may write S^{-1}, the inverse system of S:

$$S^{-1} = [\; [] \; i \in \Sigma_{S^{-1}} : \; c_i \to x := e_i^{-1} \to b_i \;]$$

To sum up,

Proposition 6.

$$([] i \in \Sigma_S : S_i)^{-1} = [] i \in \Sigma_{S^{-1}} : (S_i^{-1})$$
$$(S_i : b_i \to x := e_i \to c_i)^{-1} = (S_i^{-1} : c_i \to x := e_i^{-1} \to b_i)$$

The reader can easily verify that $\Sigma_S = \Sigma_{S^{-1}}$, $S_+^{-1} = S_-$ and $S_-^{-1} = S_+$.

In fact, before entering into more details, let us propose a general **duality principle**:

S	S^{-1}
S_+	S_-^{-1}
J_+	J_-
attraction	repulsion
contraction	expansion
future	past

It is clear that the state space of S^{-1} is the same as that of S. It is also clear that the positive (resp. negative) invariant of S is equivalent to the negative (resp. positive) invariant of S^{-1}. The invariants of S and S^{-1} are the same. The criteria developed for fullness and atomicity (§§2.5, 2.6) stay valid when considering inverse systems. The attraction is somewhat modified because its definition uses the predicate transformer S_+. Attraction for the inverse system

S^{-1} of S uses S_+^{-1} which is equivalent to S_-. So attraction for S^{-1} (or for S to the past) means inverse attraction or repulsion for S (to the future). Thus, the fact Q strictly attracts P by S^{-1} may be expressed as follows:

$$(\forall \sigma \in \Sigma^\infty : S_{\sigma-}.P \Rightarrow Q) \wedge (P \Rightarrow S_+^\infty.\textbf{true}) \wedge (P \neq \textbf{false})$$

It means that, starting from a predicate included in Q, and iterating forward a infinite number of times, the system S reaches states of the predicate P. Moreover, the predicate P is not empty and its states have a potential infinite past. There is a repulsion from Q to P by S iff the system S^{-1} attracts P to Q. We will see below some algebraic rules composing inversion with other operations.

4.2 Sequential Composition

Let us now define another simple composition, namely the sequential one [14]. It is the equivalent to the mathematical composition of functions, for instance $x := f(g(x))$. There are two notations for expressing the sequential composition of two systems S and T (i.e. application of S followed by an application of T): $S;T$ and $T \circ S$. The former insists on the sequential aspect of the composition, the latter looks like the mathematical composition. Let us first define two systems S and T:

$$S = [\,[]\ i \in \Sigma_S\ :\ a_i \rightarrow x := e_i \rightarrow b_i\,]$$

$$T = [\,[]\ j \in \Sigma_T\ :\ c_j \rightarrow y := f_j \rightarrow d_j\,]$$

The system S works on the state space X. A state $x \in X$ is a vector of state variables $x_1, \ldots, x_n$. Similarly, T works on Y. A state $y \in Y$ is a vector of state variables $y_1, \ldots, y_m$. Let us call Z the state space obtained by set-union of the axes of X and Y. Thus we take the set-union (i.e. without repetition) of the variables of x and y and we call it z. These variables $z_1, \ldots, z_p$ define the state space Z. We can now give the form of the sequential composition of S and T:

$$S;T = [\,[]\ k \in \Sigma_{S;T}\ :\ (a \wedge c.e)_k \rightarrow z := (f.e)_k \rightarrow (d)_k\,]$$

where $\Sigma_{S;T} = \Sigma_S \times \Sigma_T$,
    k is a way of renaming the couples $(i,j) \in \Sigma_S \times \Sigma_T$,
    $(a \wedge c.e)_k$ means $a_i.z \wedge c_j.e_i.z$,
    $(f.e)_k$ means $f_j.e_i$, and $(d)_k$ means d_j.
    It is easy to prove that $(S;T)_-.P = S_-.T_-.P$ and $(S;T)_+.P = T_+.S_+.P$ for any predicate P on Z. From this, it is also straightforward to deduce the following proposition:

Proposition 7. *The invariant of* $S;T$ *can be expressed as*

$$J^{S;T} = \forall n \geqslant 1 : (S_-.T_-)^n.\textbf{true} \wedge (T_+.S_+)^n.\textbf{true}$$

We leave the details to the interested reader. It is also easy to see that

Proposition 8.

$$(S;T)^{-1} = (T^{-1};S^{-1})$$

The state spaces of S and T can be equal, partially disjoint, or completely disjoint. Following the overlap is partial or complete (resp. null), the dynamics and the invariant of the sequential composition can be very difficult (resp. easy) to determine. This is analyzed below, in the case of sum.

4.3 Sums of Systems

Obtaining the sum of two systems is simple. Let us make some cooking: take the guarded-commands of each system and put them in a new one, all together; the new system is the sum of our two initial systems.

The behaviour of each system is, for each step, the choice of a valid guard and execution of the corresponding command. If no guard is valid, the system stops its execution. If several guards are valid, there is a nondeterministic choice between the different guards. Any system containing more than one guarded-command can be considered as a sum of simpler ones. So the sum of systems can be seen as a nondeterministic, quasi-parallel model of asynchronous parallelism: one transition is taken at a time, whatever the number of processes involved.

We first present the ideas by a few examples, then we define the sum more formally and we analyze its dynamics.

4.3.1 Examples of Sums.

4.3.1.1 Dyadic Map Plus Valley Map. Consider the dyadic map S and the valley map T:

$$S = [\quad 0 \leqslant x \leqslant \tfrac{1}{2} \to x := 2x \qquad \to 0 \leqslant x \leqslant 1$$
$$[]\ \tfrac{1}{2} \leqslant x \leqslant 1 \to x := 2x - 1 \to 0 \leqslant x \leqslant 1\ ]$$

$$T = [\quad 0 \leqslant x \leqslant \tfrac{1}{2} \to x := 1 - 2x \to 1 \geqslant x \geqslant 0$$
$$[]\ \tfrac{1}{2} \leqslant x \leqslant 1 \to x := 2x - 1 \to 0 \leqslant x \leqslant 1\ ]$$

The components $J_{\sigma,\tau}$ for S (§2.3) and T are the same, and thus S and T have the same invariant $J = 0 \leqslant x \leqslant 1$. Adding S and T, we obtain

$$S + T = [\quad 0 \leqslant x \leqslant \tfrac{1}{2} \to x := 2x \qquad \to 0 \leqslant x \leqslant 1$$
$$[]\ 0 \leqslant x \leqslant \tfrac{1}{2} \to x := 1 - 2x \to 1 \geqslant x \geqslant 0$$
$$[]\ \tfrac{1}{2} \leqslant x \leqslant 1 \to x := 2x - 1 \to 0 \leqslant x \leqslant 1\ ]$$

The invariant of the sum $S+T$ is the same as those of the component systems S and T; moreover, both S and T are expansive in the future, i.e. contracting in the past. As a consequence, the nondeterministic sum $S + T$ has essentially the same dynamics as S and as T: the invariants of $S + T$, S, and T all are full and atomic.

4.3.1.2 Dyadic Map Plus its Inverse. The inverse of the dyadic map is

$$S^{-1} = [\quad 0 \leqslant x \leqslant 1 \to x := \tfrac{x}{2} \qquad \to 0 \leqslant x \leqslant \tfrac{1}{2}$$
$$\square\ 0 \leqslant x \leqslant 1 \to x := \tfrac{x}{2} + \tfrac{1}{2} \to \tfrac{1}{2} \leqslant x \leqslant 1\,]$$

The invariants of the dyadic map and of its inverse are identical. We may combine them in the sum $S + S^{-1}$ having the same invariant (we will see below that it is not always the case):

$$S + S^{-1} = [\quad 0 \leqslant x \leqslant \tfrac{1}{2} \to x := 2x \qquad \to 0 \leqslant x \leqslant 1$$
$$\square\ \tfrac{1}{2} \leqslant x \leqslant 1 \to x := 2x - 1 \to 0 \leqslant x \leqslant 1$$
$$\square\ 0 \leqslant x \leqslant 1 \to x := \tfrac{x}{2} \qquad \to 0 \leqslant x \leqslant \tfrac{1}{2}$$
$$\square\ 0 \leqslant x \leqslant 1 \to x := \tfrac{x}{2} + \tfrac{1}{2} \to \tfrac{1}{2} \leqslant x \leqslant 1\,]$$

Here, S and S^{-1} are expansive in the future and in the past, respectively. Thus $S + S^{-1}$ permits a 2-periodic trace generating the bi-infinite sequence of predicates

$$\cdots \to 0 \leqslant x \leqslant \frac{1}{2} \to 0 \leqslant x \leqslant 1 \to 0 \leqslant x \leqslant \frac{1}{2} \to \cdots$$

In this sequence, the predicate $0 \leqslant x \leqslant \tfrac{1}{2}$ is invariant, in the eternal past as well as in the eternal future. Hence, the components of the invariant do not always converge towards single states, and the invariant of $S + S^{-1}$ is *not* atomic. This is essentially due to the fact $S + S^{-1}$ contains the following subsystem S_{sub}, the invariant of which is not atomic:

$$S_{sub} = [\quad 0 \leqslant x \leqslant \tfrac{1}{2} \to x := 2x \to 0 \leqslant x \leqslant 1$$
$$\square\ 0 \leqslant x \leqslant 1 \to x := \tfrac{x}{2} \to 0 \leqslant x \leqslant \tfrac{1}{2}\,]$$

Similar sums can be composed on the basis of the logistic map, which is a transformation of the tent map [1], and on the basis of the dyadic map and the valley map. For instance, we obtain a "star map" by adding the dyadic-, tent-, and valley-maps, and their inverses: its invariant is the same as for each of these basic systems, but not its dynamics. Dynamics is preserved under sums only if the component systems do contract predicates in the same direction (past or future).

4.3.1.3 Inverse Dyadic Map Plus Inverse Valley Map. Consider

$$S^{-1} + T^{-1} = [\quad 0 \leqslant x \leqslant 1 \to x := \tfrac{x}{2} \qquad \to 0 \leqslant x \leqslant \tfrac{1}{2}$$
$$\square\ 1 \geqslant x \geqslant 0 \to x := \tfrac{1-x}{2} \to 0 \leqslant x \leqslant \tfrac{1}{2}$$
$$\square\ 0 \leqslant x \leqslant 1 \to x := \tfrac{x+1}{2} \to \tfrac{1}{2} \leqslant x \leqslant 1\,]$$

Since S^{-1} and T^{-1} are contracting in the same direction, their sum has a full and atomic invariant $J = (0 \leqslant x \leqslant 1)$. This observation corresponds to the fact that the inverse of a sum is the sum of the inverses:

Proposition 9.

$$(S + T)^{-1} = S^{-1} + T^{-1}$$

4.3.2 Analysis of Sums.

In general, consider two systems S and T:

$$S = [\,\Box\ i \in \Sigma_S\ :\ a_i \rightarrow x := e_i \rightarrow b_i\,]$$

$$T = [\,\Box\ j \in \Sigma_T\ :\ c_j \rightarrow y := f_j \rightarrow d_j\,]$$

The sum $S + T$ of these systems is easy to define:

$$S + T = [\,\Box\ k \in \Sigma_{S+T}\ :\ g_k \rightarrow z := h_k \rightarrow m_k\,]$$

where $\Sigma_{S+T} = \Sigma_S \bigcup \Sigma_T$, $\Sigma_S \bigcap \Sigma_T = \varnothing$,

$$k \in \Sigma_S \Rightarrow g_k = a_k, h_k = e_k, m_k = b_k$$

and

$$k \in \Sigma_T \Rightarrow g_k = c_k, h_k = f_k, m_k = d_k$$

As for the sequential composition, the state variable z is obtained by fusion of the variables x and y. It means that the state spaces of S and T can be equal, partially disjoint, or completely disjoint, as it was the case for sequential composition. Following the overlap is complete (resp. null), the dynamics and the invariant of the sum can be very difficult (resp. easy) to determine.

The invariant of the sum is a predicate containing the invariants of S and T and many other terms mixing S and T in the future and in the past. We may just write something like:

Proposition 10. *If the components of a sum are not disjoint, then*

$$J^S \vee J^T \Rightarrow J^{S+T}$$

Proof. Let us try to get an expression of the invariant of the sum of two systems S and T as defined above. We know (§2.3) that:

$$J^S = \forall n \geqslant 1 : S_-^n.\mathbf{true} \wedge S_+^n.\mathbf{true}$$

$$J^T = \forall n \geqslant 1 : T_-^n.\mathbf{true} \wedge T_+^n.\mathbf{true}$$

Of course we can write

$$J^{S+T} = \forall n \geqslant 1 : (S+T)_-^n.\mathbf{true} \wedge (S+T)_+^n.\mathbf{true}$$

It is easy to prove that $(S+T)_-.P = S_-.P \vee S_+.P$. It follows that $(S+T)_-^n.P = (S_- \vee T_-)^n.P$. We can thus rewrite the invariant as

$$J^{S+T} = \forall n \geqslant 1 : (S_- \vee T_-)^n.\mathbf{true} \wedge (S_+ \vee T_+)^n.\mathbf{true}$$

and this expression contains the disjunction of J^S and J^T. $\qquad\Box$

As we mentioned above, we cannot say anything better about the invariant of the sum if there is an overlap between the state variables of the components systems. Even in the simple case where the components are a system S and its own inverse, we would like that $J^{S+S^{-1}} = J^S = J^{S^{-1}}$ but the first equality does not always hold, while the second is trivially true. For instance, take the system:

$$S = [\quad 0 \leqslant x \leqslant \tfrac{1}{2} \to x := x \qquad \to 0 \leqslant x \leqslant \tfrac{1}{2}$$
$$[]\; \tfrac{1}{2} \leqslant x \leqslant 1 \to x := x - \tfrac{1}{2} \to 0 \leqslant x \leqslant \tfrac{1}{2}\,]$$

We easily see that the invariant of S is $(0 \leqslant x \leqslant \tfrac{1}{2})$. The invariant of its inverse is of course the same. Nevertheless, the invariant of $S + S^{-1}$ is $(0 \leqslant x \leqslant 1)$.

We may thus not conclude that in general the invariant of a sum of systems having the same invariant is the latter. Moreover, as shown above, the dynamics of this sum does not necessarily correspond to the dynamics of the component systems.

Indeed, the composition of systems by sums amounts to introduce new possibilities in the dynamics, whereas the concepts of fullness and atomicity (§§2.5, 2.6) are based on necessities: *all* bi-infinite traces must determine singleton components of invariants. We can thus introduce complementary definitions of *potential* fullness and atomicity: we would require that *some* bi-infinite traces determine singleton predicates and that these predicates cover the whole invariant. With such weaker definitions, the dynamics of the sum would correspond to that of the components. This complementary, straightforward approach is left for further work.

Let us examine the case of systems without any common variable; the results become easier to establish. The sum of two systems S and T acts like the scheduler in a multi-tasking environment of a one-processor computer. The sum gives the hand to S or T alternatively. It results in a nondeterministic asynchronous execution of S and T together. Suppose the scheduler is fair, i.e. it does not ignore any system or its guarded-commands. Intuitively, it seems normal that the invariant of the sum be the invariant of S and the invariant of T. More formally, we have the following proposition [3]:

Proposition 11. *When there is no interaction between the components of the sum, then*

$$J^{S+T} = J^S \wedge J^T$$

Proof. It is easy to see that:

$$(S + T)^n_-.(\mathbf{true_x} \wedge \mathbf{true_y}) = \vee^n_{i=0}(S^i_-.\mathbf{true_x} \wedge T^{n-i}_-.\mathbf{true_y})$$

This disjunction simply explores all the possibilities of interleaved executions of S and T: all iterations choice T; one iteration for S, $n-1$ for T; ...; n iterations for S, zero for T. When $n \to \infty$, the disjunction explores many cases where one of the two systems iterates finitely many times, while the other one iterates infinitely

---

[3] To express the cartesian product of two predicates A and B, you simply have to write $A \wedge B$. For instance, $(x, y) \in ([0, 1] \times \mathbb{N}) \equiv (0 \leqslant x \leqslant 1) \wedge (y \in \mathbb{N})$.

many times. Then, if the scheduler of S and T (i.e. of the transitions of the sum) is fair, the only possible case is the one in which S *and* T iterate infinitely many times. In this case, the invariant *is* the conjunction of the invariant of S and T. □

In the last example introducing the sum, we have observed the fact that the inverse of a sum is the sum of the inverses:

Proposition 12.

$$(S+T)^{-1} = S^{-1} + T^{-1}$$

4.4 Product of Systems

After the sum of systems, we now present the product of systems [15]. Given two systems, their product results in their parallel execution. There are two major differences with respect to the sum:

1. the parallel execution of the systems in a product is purely synchronous,
2. the state spaces of the systems are disjoint [4]

The behaviour of the product is thus easy to describe. At each step, one valid guard is chosen in each system and the corresponding commands are executed simultaneously. The state spaces being disjoint, we will see that the invariant of the product and its properties are easy to guess. The product can be seen as a nondeterministic model of synchronous parallelism.

4.4.1 Analysis of Products. Consider two systems on distinct state-spaces:

$$S = [\, [] \ i \in \Sigma_S : \ b_i \to x := e_i \,]$$

$$T = [\, [] \ j \in \Sigma_T : \ b'_j \to y := e'_j \,]$$

Their **product** combines them in parallel, so to speak; it is defined by

$$S \times T = [\, [] \ (i,j) \in \Sigma_{S \times T} : \ (b \wedge b')_{(i,j)} \to (x,y) := (e,e')_{(i,j)} \,]$$

where $\Sigma_{S \times T} = \Sigma_S \times \Sigma_T$, $(b \wedge b')_{(i,j)} = (b_i \wedge b'_j)$ and $(e,e')_{(i,j)} = (e_i, e'_j)$.

Proposition 13. *The invariant $J^{S \times T}$ of the product is the conjunction of the invariants of the systems S and T ;*

$$J^{S \times T} = J^S \wedge J^T$$

---

[4] We will see later that the state spaces can be linked through connected forms of composition.

Proof.

$$(S \times T)_-.(P(x) \wedge Q(y))$$
$$= \forall i \in \Sigma_1, j \in \Sigma_2 : b_i(x) \wedge b'_j(y) \wedge P(e_i(x)) \wedge Q(e'_j(y))$$
$$= [\forall i \in \Sigma_1 : b_i(x) \wedge P(e_i(x))] \wedge [\forall j \in \Sigma_2 : b'_j(y) \wedge Q(e'_j(y))]$$
$$= S_-.P(x) \wedge T_-.Q(y)$$

The case of $(S \times T)_+.(P(x) \wedge Q(y))$ is similar. $\square$

The same decomposition holds for the components of the invariant of a product of systems:

Proposition 14.

$$J^{S \times T}_{\sigma \times \tau, \sigma' \times \tau'} = J^S_{\sigma, \sigma'} \wedge J^T_{\tau, \tau'} = (J^S_{\sigma,} \wedge J^S_{,\sigma'}) \wedge (J^T_{\tau,} \wedge J^T_{,\tau'})$$

Proof.

$$(S \times T)_{\sigma \times \tau, \sigma' \times \tau'}.(P(x) \wedge Q(y))$$
$$= S_{\sigma,\sigma'}.P(x) \wedge T_{\tau,\tau'}.Q(y)$$
$$= (S_{\sigma_+}.P \wedge S_{\sigma'_-}.P) \wedge (T_{\tau_+}.Q \wedge T_{\tau'_-}.Q)$$

where

$$(\sigma_1 i_1 \times \sigma_2 i_2) = (\sigma_1 \times \sigma_2)(i_1 \times i_2)$$

$\square$

Moreover, the following proposition holds:

Proposition 15. *If the invariants J^S and J^T are full and atomic, then the invariant $J^{S \times T}$ of the product is also full and atomic. Sensitive dependence on initial conditions and topological transitivity are preserved under product.*

Proof. For fullness, we use §2.5 with

$$\Phi^{S \times T} = \Phi^S \wedge \Phi^T, \ \Psi^{S \times T} = \Psi^S \wedge \Psi^T$$

For atomicity, we use §2.6 with

$$\mu^{S \times T}((A^S_1 \wedge A^T_2) \wedge (B^S_1 \wedge B^T_2)) = \mu^S(A^S_1 \wedge B^S_1) + \mu^T(A^T_2 \wedge B^T_2)$$
$$k^{(S \times T)} = \max(k^{(S)}, k^{(T)})$$

$\square$

Unlike sums, products of systems do preserve fullness and atomicity: products have a conjunctive effect. This is essentially due to the *conjunctive* character of the definitions of fullness and atomicity; see the discussion in §4.3.2.

Attraction is also straightforward for product. The conjunctive effect still plays an important role.

Proposition 16. *If Q attracts P by S and Q' attracts P' by T, then $Q \wedge Q'$ attracts $P \wedge P'$ by $S \times T$.*

Proof. We know that $\Sigma_{S \times T} = \Sigma_S \times \Sigma_T$. We also have $\Sigma_{S \times T}^{\infty} = \Sigma_S^{\infty} \times \Sigma_T^{\infty}$. Thus, every trace σ of $\Sigma_{S \times T}^{\infty}$ can be decomposed in a couple of traces (σ_1, σ_2) of Σ_S^{∞} and Σ_T^{∞}. We note it with "$=$": $\sigma = (\sigma_1, \sigma_2)$. We may write $(S \times T)_{\sigma+}.(P \wedge P') = (S_{\sigma_1+}.P) \wedge (T_{\sigma_2+}.P')$. The hypotheses give $S_{\sigma_1+}.P \Rightarrow Q$ and $T_{\sigma_2+}.P' \Rightarrow Q'$. We also have the conjunction: $S_{\sigma_1+}.P \wedge T_{\sigma_2+}.P' \Rightarrow Q \wedge Q'$. Of course, we have $P \wedge P' \neq \textbf{false}$ and $P \wedge P' \Rightarrow J_+^S \wedge J_+^T = J_+^{S \times T}$. Strict attraction is thus obtained easily. $\square$

It is possible to compose inversion with product. The following proposition is easily proved:

Proposition 17.

$$(S \times T)^{-1} = S^{-1} \times T^{-1}$$

4.4.2 Examples of Products of Systems.

4.4.2.1 Dyadic Map Times Itself. Given the dyadic map S in §4.3.1.1, its square is

$$
\begin{aligned}
S(x) \times S(y) = [\quad &0 \leqslant x \leqslant \tfrac{1}{2} \wedge 0 \leqslant y \leqslant \tfrac{1}{2} \rightarrow (x,y) := (2x, 2y) \\
&\rightarrow 0 \leqslant x \leqslant 1 \wedge 0 \leqslant y \leqslant 1 \\
[]\quad &0 \leqslant x \leqslant \tfrac{1}{2} \wedge \tfrac{1}{2} \leqslant y \leqslant 1 \rightarrow (x,y) := (2x, 2y - 1) \\
&\rightarrow 0 \leqslant x \leqslant 1 \wedge 0 \leqslant y \leqslant 1 \\
[]\quad &\tfrac{1}{2} \leqslant x \leqslant 1 \wedge 0 \leqslant y \leqslant \tfrac{1}{2} \rightarrow (x,y) := (2x - 1, 2y) \\
&\rightarrow 0 \leqslant x \leqslant 1 \wedge 0 \leqslant y \leqslant 1 \\
[]\quad &\tfrac{1}{2} \leqslant x \leqslant 1 \wedge \tfrac{1}{2} \leqslant y \leqslant 1 \rightarrow (x,y) := (2x - 1, 2y - 1) \\
&\rightarrow 0 \leqslant x \leqslant 1 \wedge 0 \leqslant y \leqslant 1]
\end{aligned}
$$

Henceforth, we often write $S \times S$ for $S(x) \times S(y)$: the obvious separation of state spaces is left understood.

The domain of $S \times S$ consists of four squares. Each of these initial squares is mapped on the whole domain, which is a square four times as big; conversely, each initial square has four smaller squares as pre-images. The limits in the past of the initial squares are points, and their union covers all the domain $0 \leqslant x \leqslant 1 \wedge 0 \leqslant y \leqslant 1$. Thus, the invariant of $S \times S$ is full and atomic, as is the invariant of S.

4.4.2.2 Dyadic Map Times its Inverse Consider the system

$$
\begin{aligned}
S \times S^{-1} = [\quad &0 \leqslant x \leqslant \tfrac{1}{2} \wedge 0 \leqslant y \leqslant 1 \rightarrow (x,y) := (2x, \tfrac{y}{2}) \\
&\rightarrow 0 \leqslant x \leqslant 1 \wedge 0 \leqslant y \leqslant \tfrac{1}{2} \\
[]\quad &0 \leqslant x \leqslant \tfrac{1}{2} \wedge 0 \leqslant y \leqslant 1 \rightarrow (x,y) := (2x, \tfrac{y}{2} + \tfrac{1}{2}) \\
&\rightarrow 0 \leqslant x \leqslant 1 \wedge \tfrac{1}{2} \leqslant y \leqslant 1 \\
[]\quad &\tfrac{1}{2} \leqslant x \leqslant 1 \wedge 0 \leqslant y \leqslant 1 \rightarrow (x,y) := (2x - 1, \tfrac{y}{2}) \\
&\rightarrow 0 \leqslant x \leqslant 1 \wedge 0 \leqslant y \leqslant \tfrac{1}{2} \\
[]\quad &\tfrac{1}{2} \leqslant x \leqslant 1 \wedge 0 \leqslant y \leqslant 1 \rightarrow (x,y) := (2x - 1, \tfrac{y}{2} + \tfrac{1}{2}) \\
&\rightarrow 0 \leqslant x \leqslant 1 \wedge \tfrac{1}{2} \leqslant y \leqslant 1]
\end{aligned}
$$

The domain of $S \times S^{-1}$ consists of two vertical rectangles, and can thus be decomposed into four identical squares. Each of these initial squares is mapped into two copies of a flattened, horizontal rectangle of size $1 \times \frac{1}{4}$ w.r.t. axes (x, y); the co-domain is thus decomposed into four horizontal rectangles, each of which is the image of two of the initial squares.

Similarly w.r.t. the past, the inverse of the given map transforms each of the initial squares into two vertical rectangles of size $\frac{1}{4} \times 1$, and the four squares are thus transformed into four such vertical rectangles; each rectangle is the image of two squares.

The combined, forwards-and-backwards transition is obtained by superposing the direct map and its inverse: the combined transition maps the four initial squares into the intersection of the four horizontal rectangles and the four vertical ones, i.e. into sixteen squares.

At each step, the number of squares composing some iterate $J^{(n)}$ of the invariant is multiplied by four. The resulting invariant has thus the same structure as in §4.4.2.1: the dynamics of the dyadic map times its inverse equals the dynamics of the dyadic map times itself.

The above product $S \times S^{-1}$ includes the classical baker-transformation [1], which is a typical example of hyperbolic system. The present decomposition of such a map in terms of a product allows a clear and simple analysis of the seemingly complex dynamics. In our approach, the analysis of $S \times S^{-1}$ is essentially the same as that of $S \times S$ in the previous paragraph.

4.5 Smale's Horseshoe

In this section, we analyze a more complex, well-known example: the Smale Horseshoe (see [16] and [2], §4.1, pp. 420–437). It will permit us to see how to combine the different types of composition described up to now and their properties. We will also see that a top-down analysis of dynamical systems may be possible with an adequate decomposition of the system into simpler ones. This example is also summarized in [17].

The map we will work with is a simplified version of the original one studied in [16]. It constitutes a prototypical map possessing a chaotic invariant set. Let us describe the expression of this horseshoe-like map f, which is so called due to the shape of its codomain. Let $0 < \lambda < \frac{1}{2}$ and $\mu > 2$.

$$
\begin{aligned}
A_x &= 0 \leq x \leq 1, & B_y &= 0 \leq y \leq \tfrac{1}{\mu}, \\
A_y &= 0 \leq y \leq 1, & C_y &= 1 - \tfrac{1}{\mu} \leq y \leq 1, \\
H_0 &= A_x \times B_y, & D &= A_x \times A_y, \\
H_1 &= A_x \times C_y.
\end{aligned}
$$

$$
f(x, y) = \begin{cases} (\lambda x, \mu y) & \text{on} & H_0 \\ (-\lambda x + 1, -\mu y + \mu) & \text{on} & H_1 \end{cases}
$$

The graphical representation of f is in Fig. 3.
We rewrite f as a guarded-command program F:

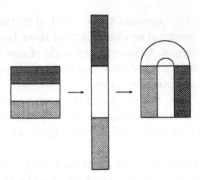

Fig. 3. Graph of f, from the leftmost picture to the rightmost one

$$F = [\quad (x,y) \in H_0 \rightarrow (x,y) := (\lambda x, \mu y)$$
$$\square \; (x,y) \in H_1 \rightarrow (x,y) := (-\lambda x + 1, -\mu y + \mu) \,]$$

We remark immediately that the state spaces are disjoint: x and y act independently. So, we may easily answer the question: How can we decompose F into simpler subsystems? A way of decomposing it is the following:

$$F = R \times S + V \times W$$

where

$$
\begin{array}{lll}
R(x) = \lambda x & \text{on} & A_x \\
S(y) = \mu y & \text{on} & B_y \\
V(x) = -\lambda x + 1 & \text{on} & A_x \\
W(y) = -\mu y + \mu & \text{on} & C_y
\end{array}
$$

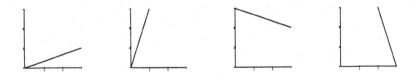

Fig. 4. From left to right, graphs of R, S, V, and W

R and V are contracting in the future, and S and W are contracting in the past, i.e. dilating in the future. The products $R \times S$ and $V \times W$ are two examples of **hyperbolic systems**, i.e. systems which are contracting in different temporal directions (past and future). The invariants of these systems are easily computed: $J^R = (x = 0)$, $J^S = (y = 0)$, $J^V = (x = 1)$ and $J^W = (y = 1)$. All of them are trivially full and atomic. By propositions 13 and 15, $J^{R \times S} = (0,0)$ and $J^{V \times W} = (1,1)$ are also full and atomic. To characterize the invariant $J^{R \times S + V \times W}$ of the global system F, and its structure, we can use the following general proposition:

Proposition 18. *The sum of two systems having compatible dynamics (in terms of expansion, contraction or any combination for multi-dimensional systems) and different full and atomic invariants, gives a system with a full and atomic invariant, which is a closed, totally disconnected, perfect set (Cantor set).*

By "compatible dynamics", we mean: g and h have the same dimension; in each dimension, g and h are both contracting or dilating; the sum is globally contracting in the future $(lip(g)+lip(h) < 1)$ or in the past $(lip(g^{-1})+lip(h^{-1}) < 1)$. The Lipschitz constant $lip(g)$ is defined as follows: $lip(g) = \sup_{x \neq y} \frac{d(g(x),g(y))}{d(x,y)}$.

Here, the global invariant $J^{R \times S + V \times W}$ of the sum is full and atomic, and is a two-dimensional Cantor set contained in D.

The question remains open to know in general in which cases the sum preserves the properties of fullness and atomicity. This is left for further work, as the same questions about product and sequential composition.

To obtain the inverse of the system F, we may use the definition of inversion presented above (§4.1). From the two interesting composition laws using inversion:

$$(S + T)^{-1} = S^{-1} + T^{-1} \qquad \text{and} \qquad (S \times T)^{-1} = S^{-1} \times T^{-1}$$

we deduce $F^{-1} = X^{-1} \times Y^{-1} + V^{-1} \times W^{-1}$. The invariant and dynamical properties of F^{-1} are the same as those of F.

5 Discussion

We present here some extensions and discuss interesting issues that could be tackled in the proposed framework.

5.1 Fairness and Periodicity

First of all, we would like to point out a relationship between our model of systems, and properties such as fairness and periodicity. We said above that fullness implies fairness but not conversely, in general. This is simply because all behaviours (evolutions, iterations) are possible and the set of all behaviours contains, among others, fair ones. The reader can also easily notice that if a system is deterministic and fair, it is not automatically periodic. If the system is deterministic, it has only one choice, i.e. one valid guard, at each iteration step. If it is fair, an infinitely often valid guard will be considered infinitely often and its corresponding command will be executed. So it is interesting to distinguish between state-fairness and transition-fairness. The first one coupled with determinism implies a periodicity because the system comes back to a given state and from this, the behaviour repeats itself. If the system is only transition-fair, nothing can be deduced about periodicity of states. If the system is nondeterministic, there is also a possible separation between these concepts of state-fairness and transition-fairness.

5.2 Duality Past-Future

Secondly, one may question the usefulness of the systematic duality past-future we present here. There are several aspects to consider:

1. Purely theoretically, the duality entails symmetric results.
2. Some definitions do not need duality; for instance, attraction is defined as a one-way property. But repulsion is a dual concept.
3. Certain systems can be studied without considering their behaviours in the past (as an example, the one-way analysis of the logistic map in [1]).
4. On the other hand, some systems do need the duality to reveal their complexity. Why? An informal argument is the following. When the system iterates to the future, its past grows. If one wants to evaluate an invariant, it is thus necessary to take its past into account, as well as its future, to get all its dynamics. This is essential in hyperbolic systems such as Smale's horseshoe.
5. Finally, the consideration of both infinite past and future can be seen as the analysis of a differential system in which we ignore the transient solutions.

5.3 Finite vs Infinite Traces

Thirdly, let us discuss the definitions of invariants and attraction presented above. Both were presented in an infinitary context: the invariant is a predicate defining the states which have potentially infinite future and past; the definition of attraction says that after an infinite future starting from a given predicate, the system reaches some other predicate and stays in it. In applications, one could be tempted to relax these definitions and introduce two variations:

1. A finite future and/or a finite past.
2. More constrained invariants: replace $J_\mp = S_\pm^\infty.\mathbf{true}$ by $J_\mp = S_\pm^\infty.P$ for some P.

Concerning the invariant of a system S, we could introduce:

$$J(P, n, Q, m) = S_+^n.P \wedge S_-^m.Q$$

It simply express the set of states having a potential finite past (of length n) starting in P and a potential finite future (of length m) arriving in Q. This generalizes the idea of pre- and post-conditions in systems.

As to attraction, another definition could be, for a fixed n (we only reformulate the central part of the definition, namely the weak-attraction):

$$\forall \sigma \in \Sigma^n : S_{\sigma+}.P \Rightarrow Q$$

Here we precise the number of steps to reach the states of Q. One could also try to solve the two following problems: to find the predicates P or Q such that $\forall \sigma \in \Sigma^\infty : Q = S_{\sigma+}.P$, and to find the traces $\sigma \in \Sigma^\infty$ such that $Q = S_{\sigma+}.P$.

5.4 Connected Compositions

Finally, let us discuss the composition of systems. We have mentioned implicit and explicit interactions between systems. Let us first recall some interesting properties of implicit interaction. If the state spaces of two subsystems are disjoint, the execution of $S; T$ alternatively gives the hand to S and T. In this case, we observe a conjunctive effect also present in null-overlap sums and in products: $J^{S;T} = J^S \wedge J^T$, and fullness, atomicity, and attraction are preserved. We gave some details of these points in the analysis of sums and products. Thus, if the overlap is null, we may write $S + T \equiv S \times T \equiv S; T \equiv T; S$.

Explicit interaction is what we already mentioned in section 4. We call it connected forms of composition, as opposed to the free forms of composition we defined there. Indeed, the sum and product we presented only consist in the juxtaposition (i.e. without explicit interaction) of given systems. We do not mean by this that the result is always simple; we have seen that is it not the case (§4.5) ! We mean that there is a priori no explicit interaction between the components juxtaposed in the new system. In the case of the product, an intersection between the state spaces would be irrelevant because of the synchronous aspect of the composition. In the case of the sum, an interaction may exist if the state spaces of the summands intersect each other; the interaction is here implicit, hidden in one of the summands. We would like to introduce a more explicit interaction between the components of a system, under the form of a predicate linking the variables of these components, or under the form of an invariant of the composed system. Many ideas are possible but we do not yet see which is the best to modelize complex interconnected systems like neural networks or cellular automata (see [18] for a deep insight into neural networks and cellular automata, [10] for a first idea of connected compositions).

6 Related Work

Three aspects can be examined: mathematics, computing science, neurosciences. Our objective is only to present here different approaches related to our framework, not to claim any exhaustiveness.

6.1 Nonlinear Dynamical Systems

The reader has certainly noticed many references like [1] or [2]. These are good sources for an introduction to dynamical systems. Devaney studies discrete deterministic dynamical systems iterating to the future; he uses analytic tools and symbolic dynamics (coarse-grained observation of dynamical systems). He also proposes a definition of chaos we adapted to identify the properties of fullness and atomicity. Wiggins studies continuous and discrete dynamical systems in a common framework. He also uses symbolic dynamics as a tool for understanding the behaviour of complex dynamical systems. He presents a definition of chaos very close to that of Devaney. He concentrates his attention on the phenomena of bifurcation and on hyperbolic systems, and he considers both future and past, as we justified in §5.2.

Many other mathematicians now study nonlinear dynamical systems from a number of viewpoints such as symbolic dynamics [19], ergodic theory [20], cellular automata (see §6.4 below). We try to unify these concepts in a general framework for studying high-dimensional dynamical systems.

The analysis of fractals is also close to our approach. Hutchinson [5] proposed the notion of invariant of union functions for studying the self-similarity of fractals. A union function is the union of a finite number of contracting maps, and is equivalent to our predicate transformer S_+ of a system S. His invariant is thus related to our notion of invariant. The difference between our approach and Hutchinson's one is that he only considers contracting maps iterating forward. We work with both future and past and we consider a finite number of one-to-one functions. Barnsley [21] also studies the space of fractals and the dynamics of IFS (Iterated Function System) on fractal sets (Cantor-like sets, totally disconnected ones, etc.). Barnsley's IFS are identical to Hutchinson's union functions.

6.2 Formal languages

Interesting developments are to be considered in symbolic dynamics: it appears worthwhile to study dynamical systems as generators of particular languages and characterize the complexity of these languages by using grammars and automata [22].

Generalizations of the shift dynamical system are proposed in [23] and undecidability results are deduced. Formal language theory happens to be an interesting theory to characterize dynamical systems. Results appear in [24], where you will see a definition of invariance which is quite close to ours and a notion coming from ergodic theory also very close to our notion of atomicity. The author studies the transition to chaos and the complexity of dynamical systems through formal languages and the Chomsky hierarchy. A similar but complementary approach is the following one. One-dimensional dynamical systems and finite automata are studied together in [25]. The author analyzes unimodal systems and makes use of Turing machines. He proposes conditions on systems to be regular or recursive.

This work yields interesting results concerning topological conjugacies and several properties of dynamical systems (periodicity, bifurcations, etc.). The study of dynamical systems with predicate transformers permits to introduce symbolic dynamics and thus to use results from formal language theory.

6.3 Program Semantics

There are many results that can be related to the proposed approach.

First of all, we may not forget the logic-based theory of programming of Dijkstra [3, 26]: the reader has certainly seen a correspondence between the wp-calculus and the predicate transformers we use. Using the dual $\tilde{S}_-$ of S_- such that $\tilde{S}_-.P = \neg S_-.\neg P$, we can define wp as $wp.S.P = S_-.P \wedge \tilde{S}_-.P$. The use of the theory developed in [26] would help in strengthening and deepening the results presented here. Also based on the logic-based theory of programming, some authors study inversion of programs, and this leads to a general construction method of algorithms (see e.g. [13]). This way of using inversion is closely related to our approach: we make use of inversion as an operator on systems, which we compose with other operators to compose and analyze more complex systems.

An interesting study of system properties has been made, ten years ago, by Sifakis [6] after [8, 14]. He analyzed simple forms of systems, namely transition systems. He defined predicate transformers on these systems and concepts like invariance, terminating trajectory, infinite trajectory (a little weaker than our definition of positive invariant), etc. He mainly used fixed-point theorems to analyze the properties of these concepts, exactly as we did to calculate the invariants by successive iterations of our predicate transformers.

The theories of both Hoare and Milner (see [27] and [28]) also influenced our approach in the sense they first define a solid mathematical framework of processes in which they elaborate composition laws. They propose algebras of synchronous imperative processes [27] or asynchronous functional processes [28]. They use traces as a major tool for observing the behaviour of concurrent processes. All aspects of our framework are already in their theories: fixed-points, recursion as infinite iteration, compositions, nondeterminism, traces, etc. These concepts seem to be invariants (!) for studying evolutive systems such as communicating sequential processes, dynamical systems or transition systems.

We consider that a trace corresponds to the observation of the behaviour of a system. Mazurkiewicz [29] follows the same idea when he proposes a theory of traces as sequential observations of concurrent systems. A difference is that we have explicitly defined the systems producing traces while he takes the traces as given. His theory mainly explores asynchronous systems.

Arnold [15] studies transition systems in which he introduces different forms of composition, all synchronous: the free product (identical to our product without interaction) and the synchronized product. The latter is a connected form of composition, based on a mechanism of synchronization. It is seen as a subcase of the free product because it adds communication and synchronization constraints. The author attempts to formalize the definition of the constraints and the resulting product but he confesses, as we do, that the modelization is not yet perfectly developed and much of the work must still be made by hand.

6.4 Neurosciences

Finally, we devote a specific section to this subject lying at the crossroads of many disciplines (mathematics, computing science, biology, physics, etc.): neurosciences. Many people study neuromimetic systems. The challenge is twofold:

1. Intrinsic: a good modelization of the neurons, the brain, and the nervous system, could help us to understand them better.

2. Extrinsic: some people have discovered that the models used to represent neurons can also be used to solve algorithmic problems such as optimization, pattern recognition, classification and function approximation [30, 31], and also to serve as models for many evolutive complex systems (ecology [32], spin glasses [33], etc.).

It is no use to present here all the related work in the fields of neural networks and cellular automata. There are many ways of studying such systems: theoretically and experimentally. We will just mention two recent and interesting approaches, based on theoretical considerations. Blum and Wang [34] study neural networks and their dynamics by considering small systems which are more understandable and by composing these. They focus their attention on bifurcation properties, and on oscillations produced by small systems and composed ones. Marcus, Waugh and Westervelt [35] adopt a global viewpoint for extracting analytic results about fixed-points, stability, and convergence of different neural networks. Why do we present those works? Because we also consider small systems composed into more complex ones and we try to study the behaviour of these complex systems with various analytic tools. The way the neurons are interconnected is a kind of connected product, and the concepts of invariance, stability and convergence are also very important in our approach. Moreover, the problems of bifurcation can be studied from a synthetic viewpoint: how to construct systems with predefined behaviours ?

7 Conclusions

The dynamical structure of seemingly intricate systems can be clarified by recomposing these systems using sums and products. Often, the invariants can be composed in a similar way, and properties such as fullness, atomicity, and attraction can be derived by combining the same properties for the components. In the case of sums, the compositionality of dynamics holds for a weakened version of dynamics, to be defined on the basis of possibilities instead of necessities.

Attraction depends on contraction by forward transitions. In the simple case, all forward transitions contract strictly, along all axes. In the general, hyperbolic case, forward transitions may be contracting along some axes but not all; it is then more difficult to prove that an invariant attracts the domain of a system, or that an attracting predicate is sensitively dependent on initial conditions and topologically transitive.

The paper presents a number of simple, basic techniques for detecting attraction and for composing systems. This already allowed to clarify the understanding of simple examples, and of a number of more substantial case-studies. The applicability of these techniques must be investigated further so as to strengthen them where appropriate. An algebra of system composition should be constructed; it could integrate among others inversion, sequential composition, and generalized forms of product. Attraction could then be achieved by design, not just verified a posteriori.

We plan to extend our framework and include new kinds of functions in systems. Of course, we will have to guarantee properties like invariance, and make use of fixed point theorems. We are also working on the formalization of composition operators. First results appear in [17], together with properties concerning invariants of sums of systems.

On the whole, the analysis of the dynamics of a composed system by composing the analyses of the dynamics of the components proves a promising approach.

8 Acknowledgements

The authors are grateful to P. Kúrka, M. Simons, G. Troll, and to members of IFIP WG 2.3 on Programming Methodology for useful comments and suggestions.

References

1. Devaney, R.L. *An Introduction to Chaotic Dynamical Systems*. Addison-Wesley, 2nd ed., 1989.
2. Wiggins, S. *Introduction to Applied Nonlinear Dynamical Systems and Chaos*, *TAM* 2. Springer-Verlag, 1990.
3. Dijkstra, E.W. *A Discipline of Programming*. Prentice Hall, 1976.
4. Sintzoff, M. Invariance and contraction by infinite iteration of relations. In Banâtre, J.P. and Le Metayer, D., (eds), *Research Directions in High-Level Parallel Programming Languages*, LNCS 574, pp. 349–373. Springer-Verlag, 1992.
5. Hutchinson, J.E. Fractals and self similarity. *Indiana University Mathematics Journal*, 30(5):713–747, 1981.
6. Sifakis, J. A unified approach for studying the properties of transition systems. *Theoretical Computer Science*, 18:227–258, 1982.
7. Tarski, A. A lattice-theoretical fixpoint theorem and its applications. *Pacific Journal of Mathematics*, 5:285–309, 1955.
8. van Lamsweerde, A. and Sintzoff, M. Formal derivation of strongly correct concurrent programs. *Acta Informatica*, 12:1–31, 1979.
9. Francez, N. *Fairness*. Texts and Monographs in Computer Science. Springer-Verlag, 1986.
10. Geurts, F. and Lombart, V. Etude des systèmes de transitions discrets. Unité d'Informatique, U.C.Louvain, June 1992. Travail de fin d'études.
11. Manna, Z. and Pnueli, A. *The Temporal Logic of Reacative and Concurrent Systems : Specification*. Springer-Verlag, 1992.
12. Dugundji, J. *Topology*. Wm.C. Brown Publishers, 2nd ed., 1989.
13. Chen, W. and Udding, J.T. Program inversion: More than fun! *Science of Computer Programming*, 15:1–13, 1990.
14. Sintzoff, M. Ensuring correctness by arbitrary postfixed-points. In *Proc. 7th Symp. Math. Found. Comput. Sci.*, LNCS 64, pp. 484–492. Springer-Verlag, 1978.
15. Arnold, A. *Systèmes de Transitions Finis et Sémantique des Processus Communicants*. Masson, 1992.
16. Smale, S. Diffeomorphisms with many periodic points. In Cairns, S.S., (ed.), *Differential and Combinatorial Topology*, pp. 63–80. Princeton University Press, 1965.

17. Sintzoff, M. and Geurts, F. Compositional analysis of dynamical systems using predicate transformers (summary). In *Proc. of 1993 International Symposium on Nonlinear Theory and its Applications, Hawaii*, 1993.

18. Goles, E. and Martinez, S. *Neural and Automata Networks, Dynamical Behavior and Applications*. Mathematics and Its Applications. Kluwer Academic Publishers, 1990.

19. Hao, B.L. *Elementary Symbolic Dynamics and Chaos in Dissipative Systems*. World Scientific, 1989.

20. Bedford, T., Keane, M., and Series, C., (eds). *Ergodic Theory, Symbolic Dynamics and Hyperbolic Spaces*. Oxford Science Publications, 1991.

21. Barnsley, M.F. *Fractals Everywhere*. Academic Press, 1988.

22. Ginsburg, S. *Algebraic and Automata-Theoretic Properties of Formal Languages, Fundamental Studies In Computer Science* 2. North-Holland/American Elsevier, 1975.

23. Moore, C. Generalized one-sided shifts and maps of the interval. *Nonlinearity*, 4:727–745, 1991.

24. Troll, G. Formal languages in dynamical systems. Tech. Rep. 47, SFB 288, T.U.Berlin, 1993.

25. Kůrka, P. One-dimensional dynamics and factors of finite automata. Tech. rep., Department of Mathematical Logic and Philosophy of Mathematics, Charles U., Prague, 1993.

26. Dijkstra, E.W. and Scholten, C.S. *Predicate Calculus and Program Semantics*. Texts and Monographs In Computer Science. Springer-Verlag, 1990.

27. Hoare, C.A.R. *Communicating Sequential Processes*. International Series in Computer Science. Prentice Hall, 1985.

28. Milner, R. *Communication and Concurrency*. International Series in Computer Science. Prentice Hall, 1989.

29. Mazurkiewicz, A. Basic notions of trace theory. In de Bakker, J.W., de Roever, W.P., and Rozenberg, G., (eds), *Linear Time, Branching Time and Partial Order in Logics and Models for Concurrency*, LNCS 354, pp. 285–363. Springer-Verlag, 1988.

30. Hopfield, J.J. Neural networks and physical systems with emergent collective computational abilities. *Proc. of the National Academy of Sciences*, 79:2554–2558, 1982.

31. Hopfield, J.J. Neurons with graded response have collective computational properties like those of two-state neurons. *Proc. of the National Academy of Sciences*, 81:3088–3092, 1984.

32. Phipps, M. From local to global: the lesson of cellular automata. In DeAngelis, D. and Gross, L., (eds), *Individual-Based Approaches in Ecology: Concepts and Models*. Chapman and Hall, 1992.

33. Weisbuch, G. *Dynamique des systèmes complexes, Une introduction aux réseaux d'automates*. InterEditions, 1989.

34. Blum, E.K. and Wang, X. Stability of fixed points and periodic orbits and bifurcations in analog neural networks. *Neural Networks*, 5:577–587, 1992.

35. Marcus, C.M., Waugh, F.R., and Westervelt, R.M. Nonlinear dynamics and stability of analog neural networks. *Physica D*, 51:234–247, 1992.

Springer-Verlag
and the Environment

We at Springer-Verlag firmly believe that an international science publisher has a special obligation to the environment, and our corporate policies consistently reflect this conviction.

We also expect our business partners – paper mills, printers, packaging manufacturers, etc. – to commit themselves to using environmentally friendly materials and production processes.

The paper in this book is made from low- or no-chlorine pulp and is acid free, in conformance with international standards for paper permanency.

Lecture Notes in Computer Science

For information about Vols. 1–812
please contact your bookseller or Springer-Verlag

Vol. 849: R. W. Hartenstein, M. Z. Servít (Eds.), Field-Programmable Logic. Proceedings, 1994. XI, 434 pages. 1994.

Vol. 850: G. Levi, M. Rodríguez-Artalejo (Eds.), Algebraic and Logic Programming. Proceedings, 1994. VIII, 304 pages. 1994.

Vol. 851: H.-J. Kugler, A. Mullery, N. Niebert (Eds.), Towards a Pan-European Telecommunication Service Infrastructure. Proceedings, 1994. XIII, 582 pages. 1994.

Vol. 852: K. Echtle, D. Hammer, D. Powell (Eds.), Dependable Computing – EDCC-1. Proceedings, 1994. XVII, 618 pages. 1994.

Vol. 853: K. Bolding, L. Snyder (Eds.), Parallel Computer Routing and Communication. Proceedings, 1994. IX, 317 pages. 1994.

Vol. 854: B. Buchberger, J. Volkert (Eds.), Parallel Processing: CONPAR 94 – VAPP VI. Proceedings, 1994. XVI, 893 pages. 1994.

Vol. 855: J. van Leeuwen (Ed.), Algorithms – ESA '94. Proceedings, 1994. X, 510 pages.1994.

Vol. 856: D. Karagiannis (Ed.), Database and Expert Systems Applications. Proceedings, 1994. XVII, 807 pages. 1994.

Vol. 857: G. Tel, P. Vitányi (Eds.), Distributed Algorithms. Proceedings, 1994. X, 370 pages. 1994.

Vol. 858: E. Bertino, S. Urban (Eds.), Object-Oriented Methodologies and Systems. Proceedings, 1994. X, 386 pages. 1994.

Vol. 859: T. F. Melham, J. Camilleri (Eds.), Higher Order Logic Theorem Proving and Its Applications. Proceedings, 1994. IX, 470 pages. 1994.

Vol. 860: W. L. Zagler, G. Busby, R. R. Wagner (Eds.), Computers for Handicapped Persons. Proceedings, 1994. XX, 625 pages. 1994.

Vol. 861: B. Nebel, L. Dreschler-Fischer (Eds.), KI-94: Advances in Artificial Intelligence. Proceedings, 1994. IX, 401 pages. 1994. (Subseries LNAI).

Vol. 862: R. C. Carrasco, J. Oncina (Eds.), Grammatical Inference and Applications. Proceedings, 1994. VIII, 290 pages. 1994. (Subseries LNAI).

Vol. 863: H. Langmaack, W.-P. de Roever, J. Vytopil (Eds.), Formal Techniques in Real-Time and Fault-Tolerant Systems. Proceedings, 1994. XIV, 787 pages. 1994.

Vol. 864: B. Le Charlier (Ed.), Static Analysis. Proceedings, 1994. XII, 465 pages. 1994.

Vol. 865: T. C. Fogarty (Ed.), Evolutionary Computing. Proceedings, 1994. XII, 332 pages. 1994.

Vol. 866: Y. Davidor, H.-P. Schwefel, R. Männer (Eds.), Parallel Problem Solving from Nature - PPSN III. Proceedings, 1994. XV, 642 pages. 1994.

Vol 867: L. Steels, G. Schreiber, W. Van de Velde (Eds.), A Future for Knowledge Acquisition. Proceedings, 1994. XII, 414 pages. 1994. (Subseries LNAI).

Vol. 868: R. Steinmetz (Ed.), Multimedia: Advanced Teleservices and High-Speed Communication Architectures. Proceedings, 1994. IX, 451 pages. 1994.

Vol. 869: Z. W. Raś, Zemankova (Eds.), Methodologies for Intelligent Systems. Proceedings, 1994. X, 613 pages. 1994. (Subseries LNAI).

Vol. 870: J. S. Greenfield, Distributed Programming Paradigms with Cryptography Applications. XI, 182 pages. 1994.

Vol. 871: J. P. Lee, G. G. Grinstein (Eds.), Database Issues for Data Visualization. Proceedings, 1993. XIV, 229 pages. 1994.

Vol. 872: S Arikawa, K. P. Jantke (Eds.), Algorithmic Learning Theory. Proceedings, 1994. XIV, 575 pages. 1994.

Vol. 873: M. Naftalin, T. Denvir, M. Bertran (Eds.), FME '94: Industrial Benefit of Formal Methods. Proceedings, 1994. XI, 723 pages. 1994.

Vol. 874: A. Borning (Ed.), Principles and Practice of Constraint Programming. Proceedings, 1994. IX, 361 pages. 1994.

Vol. 875: D. Gollmann (Ed.), Computer Security – ESORICS 94. Proceedings, 1994. XI, 469 pages. 1994.

Vol. 876: B. Blumenthal, J. Gornostaev, C. Unger (Eds.), Human-Computer Interaction. Proceedings, 1994. IX, 239 pages. 1994.

Vol. 877: L. M. Adleman, M.-D. Huang (Eds.), Algorithmic Number Theory. Proceedings, 1994. IX, 323 pages. 1994.

Vol. 878: T. Ishida; Parallel, Distributed and Multiagent Production Systems. XVII, 166 pages. 1994. (Subseries LNAI).

Vol. 879: J. Dongarra, J. Waśniewski (Eds.), Parallel Scientific Computing. Proceedings, 1994. XI, 566 pages. 1994.

Vol. 880: P. S. Thiagarajan (Ed.), Foundations of Software Technology and Theoretical Computer Science. Proceedings, 1994. XI, 451 pages. 1994.

Vol. 881: P. Loucopoulos (Ed.), Entity-Relationship Approach – ER'94. Proceedings, 1994. XIII, 579 pages. 1994.

Vol. 882: D. Hutchison, A. Danthine, H. Leopold, G. Coulson (Eds.), Multimedia Transport and Teleservices. Proceedings, 1994. XI, 380 pages. 1994.

Vol. 883: L. Fribourg, F. Turini (Eds.), Logic Program Synthesis and Transformation – Meta-Programming in Logic. Proceedings, 1994. IX, 451 pages. 1994.

Vol. 884: J. Nievergelt, T. Roos, H.-J. Schek, P. Widmayer (Eds.), IGIS '94: Geographic Information Systems. Proceedings, 1994. VIII, 292 pages. 19944.

Vol. 885: R. C. Veltkamp, Closed Objects Boundaries from Scattered Points. VIII, 144 pages. 1994.

Vol. 886: M. M. Veloso, Planning and Learning by Analogical Reasoning. XIII, 181 pages. 1994. (Subseries LNAI).

Vol. 887: M. Toussaint (Ed.), Ada in Europe. Proceedings, 1994. XII, 521 pages. 1994.

Vol. 888: S. A. Andersson (Ed.), Analysis of Dynamical and Cognitive Systems. Proceedings, 1993. VII, 260 pages. 1995.

Vol. 889: H. P. Lubich, Towards a CSCW Framework for Scientific Cooperation in Europe. X, 268 pages. 1995.

Vol. 890: M. J. Wooldridge, N. R. Jennings (Eds.), Intelligent Agents. Proceedings, 1994. VIII, 407 pages. 1995 (Subseries LNAI).

Vol. 891: C. Lewerentz, T. Lindner (Eds.), Formal Development of Reactive Systems. XI, 394 pages. 1995.